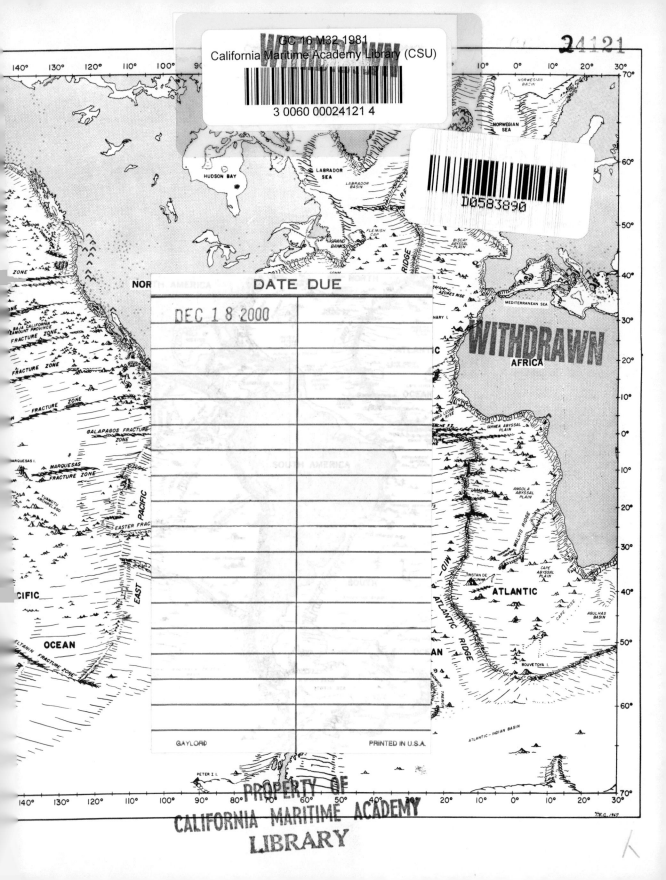

ELEMENTS OF OCEANOGRAPHY

SECOND EDITION

J. MICHAEL McCORMICK

Montclair State College

JOHN V. THIRUVATHUKAL

Montclair State College

 SAUNDERS COLLEGE PUBLISHING

Philadelphia   New York   Chicago
San Francisco   Montreal   Toronto
London   Sydney   Tokyo   Mexico City
Rio de Janeiro   Madrid

Address orders to:
383 Madison Avenue
New York, NY 10017
Address editorial correspondence to:
West Washington Square
Philadelphia, PA 19105

This book was set in Palatino by Hampton Graphics, Inc.
The editors were Don Jackson, Janis Moore and Mike Boyette.
The art director and text and cover designer was Nancy E. J. Grossman.
The production manager was Tom O'Connor.
New artwork was drawn by Mario Neves.
The printer was Von Hoffmann Press.

Cover photo: Surf breaking at Cape Kiwanda, Oregon Coast, by James Randklev.

LIBRARY OF CONGRESS

CATALOG CARD NO.: 80–53920

McCormick, J. Michael, & John V. Thiruvathukal
  Elements of oceanography.

Philadelphia, Pa.: Saunders College
416 p.
8101   801010

ELEMENTS OF OCEANOGRAPHY   ISBN 0-03-057806-X

1234 114 987654321

CBS COLLEGE PUBLISHING
Saunders College Publishing
Holt, Rinehart and Winston
The Dryden Press

Preface to the Second Edition

Much progress has been made in oceanography since the first edition of *Elements of Oceanography* was published. For example, startling new discoveries of life in areas around deep sea hot springs have been made. In addition, deep sea mining of minerals is close to becoming an economic reality since the U.S. Congress passed the Deep Seabed Hard Mineral Resources Act in 1980.

This text has been extensively revised and contains a new chapter on sediments that includes material on oil and mineral deposits. Many illustrations have been improved, and several line drawings have been replaced by photos. This text includes study questions and additional recent suggested readings. The theme and mission of the book, however, remain the same, and it is dedicated to the non-scientist.

In addition to all the people who reviewed the first edition, we would like to extend our deep gratitude to the many individuals who contributed suggestions for improving the text, especially to the following scientists and teachers who reviewed the entire new manuscript: Leo Berner, Texas A & M University; Benno Breninkmeyer, S. J., Boston College; Rodney Feldman, Kent State University; Herbert Frolander, Oregon State University; Nickolas Holland, Scripps Institution of Oceanography; Bruce Molnia, U.S. Geological Survey; Gray Multer, Fairleigh Dickinson University; Ray Smith, Eastern Connecticut State College; Ray Staley, Florida State University; Jack Travis, University of Wisconsin; and Charles Walters, Kansas State University.

We would also like to thank the many professors who made specific comments and suggestions for improving the book, especially Robert Able, New Jersey Marine Sciences Consortium; Ernest Angino, University of Kansas; Alfred Carsola, Grossmont and Mesa Colleges; Joseph Chase, Bridgewater State College; Richard Cole, Brookdale Community College; David Darby, University of Minnesota; Bruce Dod, Wayland Baptist College; W. W. Drexler, Shippensburg State College; Robert Feller, University of Washington; P. A. Haefner, Jr., Virginia Institute of Marine Science; Roy Hundley, St. Petersburg Jr. College; Ernest Jerome, Malaspina College; T. W. Lins, Mississippi State University; Alan Lipkin, Sonoma State College; David Love, Washington State University; J. P. Manker, Georgia Southwestern College; A. Melvin, Trenton State College; Chris Mungall, Texas A & M University;

William Schroeder, University of Alabama; James Tanner, Florida Institute of Technology; and Alan M. Young, University of South Carolina.

The staff of Saunders College Publishing has continued to be very encouraging, helpful, and patient throughout the process of revision. Special thanks go to Joan Garbut and Donald Jackson.

Preface to the First Edition

The oceans, although they hold vast resources of food, minerals, petroleum, and energy, are under ever-increasing environmental stress. The prospects for gaining further benefit from the seas depend on human activities and political decisions. These are processes in which we all take part. Therefore, it is of tremendous importance for all of us to be aware of the nature and potentials of the oceans and of their frailty. In that sense, this book is aimed at the college student as citizen, voter, and future doctor, lawyer, teacher, artist, politician, and so on.

This text is designed for a beginning oceanography course and does not require any previous scientific background. The metric system of measurements is emphasized throughout the text, and appropriate conversion tables are given at the end of the book. Technical terms are defined where appropriate as well as in the Glossary. We have endeavored to describe the essentials of the nature of the oceans and their potential for wise exploitation as well as the dangers that lie in their misuse. We hope that this has been done in a way that gives relevance to the various principles of oceanography.

Since no text is written in a vacuum, we are indebted to many of our colleagues in the scientific community. Many of the photographs were graciously provided by researchers in diverse fields of oceanography and marine biology. We are especially indebted to Allan Bé, Lamont-Doherty Geological Observatory; Harold "Wes" Pratt, National Marine Fisheries Service; David Stein, Oregon State University; Douglas P. Wilson, The Laboratory, Plymouth, England; M. N. A. Peterson, Scripps Institution of Oceanography (Deep Sea Drilling Project); Paul Hargraves, James Kennett, and John Sieburth, University of Rhode Island; and Susumu Honjo, Woods Hole Oceanographic Institution.

In addition, Vicky Briscoe of the Woods Hole Oceanographic Institution and Anne Nixon of the Lamont-Doherty Geological Observatory were especially helpful in providing us with institutional material for many of the illustrations. We would also like to acknowledge the help of Dean Bumpus and Gilbert Rowe of the Woods Hole Oceanographic Institution and Oswald Roels of Columbia University, who provided important information.

Portions or all of the manuscript were read and sometimes reread and criticized by Larry Leyman of Fullerton College, James McCauley of Oregon State University, Thomas Worsley of the University of Washington, James Marlowe of Miami-Dade Community College, Richard Mariscal of Florida State University, and R. Gordon Pirie of the University of Wisconsin—Milwaukee. Their help is gratefully appreciated. We, of course, must take full responsibility for any errors that remain. Our students are also appreciated for their willingness to serve as critics of some of the early manuscript. In addition, we would appreciate receiving comments and suggestions from the readers of this text.

All the way through this project the staff of W. B. Saunders Co. has been encouraging and helpful. Many of the fine drawings were done by Grant Lashbrook and John Hackmaster. Special thanks are due Richard Lampert, editor, and Amy Shapiro, manuscript editor, who have shown incredible patience, endurance, and good sense. We also extend a special note of gratitude to our wives, Beverly and Mary, whose encouragement and help kept us going.

Contents

ELEMENTS OF OCEANOGRAPHY

Introduction

AN OVERVIEW OF OCEANOGRAPHY

Oceanography, the study of the oceans, is a multidisciplinary science in which geology, chemistry, physics, and biology interact. The advent of modern oceanography had to await scientific developments in these fields. Many oceanographic problems are interdisciplinary and require cooperation among scientists of varied backgrounds. Even engineering, economics, and political science are used increasingly to solve problems related to the use and misuse of the oceans (Fig. 1.1).

Geological oceanography deals with the study of the ocean floor, including the sediments and the underlying rocks. Modern geological investigations have revealed much about the dynamics of the ocean floor and the origin of the ocean basins. This research has led to a revolutionary view of the earth itself. Until the 1960's relatively few scientists

The *Atlantis*, the first ship of the Woods Hole Oceanographic Institution. (Courtesy of WHOI.)

FIGURE 1.1 Artist's conception of a proposed offshore nuclear power plant. It is essential to thoroughly understand oceanographic processes and possible environmental impacts before choosing a site, building, and operating such a plant. (Courtesy of Public Service Electric and Gas Company.)

believed that whole ocean basins and continents could migrate. This new view of the earth, called *plate tectonics,* can explain the existence and locations of most of the features of the ocean floor, such as oceanic trenches and undersea mountain ranges, as well as many of the mountain ranges, volcanic regions, and earthquake zones of the world.

The oceans cover 71 per cent of the earth's surface. Because most of the major nonrenewable natural resources—petroleum and minerals—are found in only limited quantities on land, our quest for them on and beneath the ocean floor becomes increasingly important. We are just beginning to turn our attention to the ocean bottom in search of new resources. Already, about a quarter of the world's petroleum production comes to us from beneath the ocean floor, and this proportion is expected to increase considerably in the future. Offshore mining of minerals currently is minor in scope; however, its potential is great. The ocean floor is known to contain vast quantities of mineral deposits, such as *manganese nodules,* which are rounded manganese-rich deposits, generally smaller than a softball, containing not only manganese but also copper, nickel, and cobalt (Fig. 1.2).

Chemical oceanography is concerned with the distribution in time and space of chemical constituents of sea water as well as with chemical reactions in living and non-living systems, especially in relation to the composition of sea water.

The chemical character of sea water has been strongly influenced by billions of years of river input from land. As fresh water flows over the rocks and soil, minute amounts of minerals are dissolved and carried to the sea. Most of these mineral salts are left behind when sea water evaporates to form clouds. The total salt content and volume of the oceans

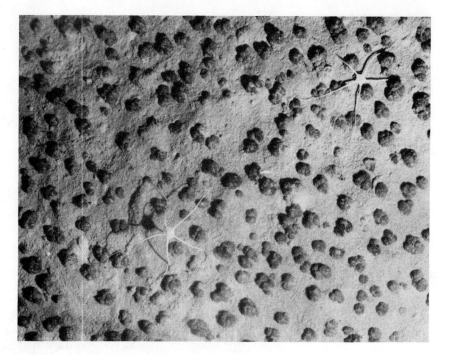

FIGURE 1.2 Manganese nodules on the sea floor (about 4000 meters). Note the two brittle stars. (Courtesy Smithsonian Institution Oceanographic Sorting Center.)

have increased over billions of years, but the process has been a very slow one. It is generally believed that the salt concentration (salinity) of the oceans has not changed greatly over the last several hundred million years.

Living plants and animals consume chemical nutrients as they grow and give them back when they die and decompose. The skeleta and shells of organisms are also derived from chemicals supplied by the sea. In addition, other chemical reactions take place between the sea and the minerals on the sea floor. Manganese nodules are probably a result of such processes. Today, minerals such as magnesium are being extracted from sea water and manganese nodules may soon be mined from the sea floor. Still, the chemical resources of the sea have barely been touched.

Human efforts to exploit the land, as well as the dumping of industrial and domestic waste, have altered the composition of the river water entering the sea. Chemical pollution thus threatens the commercial fisheries in many areas as well as the quality of the recreational beaches and coastal water. In fact, no part of the oceans is totally immune from the harmful effects of human activity.

The field of *physical oceanography* encompasses the study of waves, tides, and currents, as well as all other processes that depend on the physical nature of sea water. It is because of the unique nature of water that the sea transmits and absorbs light energy, that ice floats, that waves erode the shore, and that marine life flourishes. An understanding of the properties of sea water is essential to a study of the processes that take

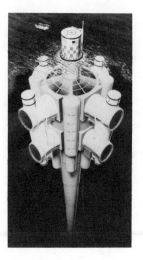

FIGURE 1.3 Artist's conception of an Ocean Thermal Energy Conversion (OTEC) power plant. Energy would be derived from the heat stored in the upper layers of the ocean. (Courtesy of the Department of Energy and Lockheed Missle and Space Co., Inc.)

place in the sea. The interaction of air and sea has a major impact on global weather and climate. In fact, most of the rain and snow that fall on land entered the air through evaporation at sea.

Energy from the sun is the primary driving force for most of the physical processes in the sea. Most of the light that is absorbed by the sea is converted into heat. This, of course, takes place only near the sea surface. Heat also may be lost at the surface, primarily by evaporation and radiation from surface water back toward space. These heat-related processes modify the temperature and salinity of surface water and provide the basis for deep ocean circulation generated by the sinking of denser water. The movement of ocean water also is caused by the force of wind, which is ultimately the result of unequal heating of the air over the earth.

Our lives are greatly influenced by physical processes in the sea. Surface currents influence shipping routes. Destructive waves erode the shore and damage coastal buildings. Life along some shores is influenced by the tides. In a few locations tides even generate power. In the future, energy in the sea water, such as heat and energy in waves, may be tapped for human benefit (Fig. 1.3).

Biological oceanography is the study of life in the sea and of its distribution, production, and ecology. The production of plants, their consumption by herbivores, and the herbivores' subsequent consumption by carnivores provides a wealth of food for humans. Efficient mariculture (sea farming) and fishing depends on extensive knowledge of the ecology of marine organisms (Fig. 1.4).

Each species of marine life has specific environmental requirements which must be met for it to survive. These prerequisites for life include proper ranges of temperature, salinity, and oxygen, as well as inorganic and organic nutrients. Furthermore, ocean circulation has a profound

FIGURE 1.4 *A*, The port of Callao, Peru, showing fishing boats at the dock, awaiting a run of anchovy. *B*, One of the big tanks in which anchovies are unloaded upon arrival at the plant. From there, they enter the processing channels, to come out only in the form of fish meal. (FAO photos. *A* by S. Larrain; *B* by R. Coral.)

influence on the distribution of drifting organisms, including the larvae of some bottom dwellers. Organisms on the sea floor also require a proper geological environment.

The oceans provide a great variety of habitats, with one type blending into others. Thus, marine life is unevenly distributed and the successful production of a species varies widely from one region to another.

A proper understanding of biological oceanography requires not only a knowledge of biology, but also a familiarity with geological, chemical, and physical processes that determine the character of the marine environment.

The food resources of the oceans are also greatly influenced by human activities. Although many species of marine life are under-utilized, others are over-fished or may be soon. In addition, domestic and industrial pollution threatens fisheries and mariculture near the shore. Deriving food from the sea demands a knowledge of oceanographic processes and wise management of resources.

HISTORICAL DEVELOPMENTS

Although people have used oceans for thousands of years, it was not until the eighteenth century that humanity began systematically studying the oceans. Progress in oceanography has accelerated since then, but a great deal remains to be learned. Much of the earliest interest in the oceans was oriented toward navigation. In the nineteenth century, interest diversified to include economic problems, such as the transatlantic cable crossings and fisheries, as well as purely scientific curiosity.

Oceanography was still a descriptive field. The twentieth century saw a new era of international cooperation through efforts such as the International Council for the Exploration of the Sea, the International Geophysical Year, and the International Indian Ocean Expedition.

Early Exploration

The first explorations of the oceans were primarily to search for new trade routes and riches and to describe the shapes of the seas. By 1000 B.C., the Mediterranean Sea was fairly well known, at least near the shore and in the vicinity of islands and shallows. Many of the early sailors had a fear of leaving sight of land. Homer (about 850 B.C.) envisaged the world comprising the lands around the Mediterranean, with an all-encircling unexplored ocean beyond (Fig. 1.5). The earth was believed to be flat and lying under the "dome of the heavens," which rose from the outer edge of the ocean.

The Phoenicians were the best navigators in the Mediterranean about 1000 B.C. They knew how to navigate by the stars and are believed to have sailed to Spain (founding the city of Cádiz) and possibly even as far north as the coast of Cornwall in England, the location of ancient tin mines. About 600 B.C., King Necho of Egypt reportedly sent a crew of Phoenicians down the Red Sea and around Africa. They were said to have returned by way of the Strait of Gibraltar about three years later. Accounts of this journey were not taken seriously or were not known to early scholars.

Later scholars, from about 400 B.C. to 450 A.D., thought the earth was a sphere. The evidence, which could be observed by travelers, included the following:

1. At a given hour on a given date, shadows get longer as one travels northward.
2. The visible constellations change as one travels north or south.
3. During an eclipse of the moon, the shadow on the moon's face is that of a spherical earth.
4. The mast is the last part of a ship to be seen as it sails over the horizon.

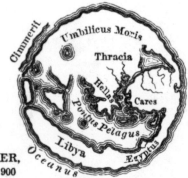

FIGURE 1.5 Homer's conception of the world (about 850 B.C. (Courtesy of Bettmann Archive.)

Pytheas was the first Greek to explore the British Isles around 325 B.C. He accurately estimated the circumference of Britain and the distance from Marseilles to northern Britain. He also sailed north past the Arctic Circle and reached land at either Iceland or Norway. He had developed a useful method of estimating latitude, but lacking an accurate timepiece (chronometer), he was unable to determine longitude. This proved to be a major stumbling block to ocean exploration. Pytheas was also one of the few Greek scholars to note the effect of the moon on the tides. The writings of Pytheas have been lost. His exploits are known only from secondary sources, limiting our understanding of his accomplishments.

Eratosthenes (about 200 B.C.) developed a method for measuring the circumference of the Earth by using the angles at which the sun's rays struck the earth simultaneously at Syene (Aswan) and Alexandria, in Egypt. Although the concept was good, his estimate was in error because the distance between the two cities could not be determined with accuracy at that time. There is also some uncertainty as to the length of his unit of measurement, the *stade*.

Ptolemy (about 150 A.D.) is credited with measuring longitude eastward of a reference longitude 2° west of the Canary Islands. Accurate measurement, especially of great distances had to await the development of an accurate chronometer. Ptolemy and other scholars of the time assumed that unexplored land linked Southeast Asia and Africa (Fig. 1.6), across what is known today to be the southern part of the Indian Ocean. This concept persisted through the Dark Ages until 1497, when the Portuguese explorer Vasco da Gama reached India by sailing around Africa and through the Indian Ocean.

FIGURE 1.6 Ptolemy's conception of the world (about 150 A.D. (Courtesy of Bettmann Archive.)

After the fall of the Roman Empire and through the Dark Ages, the flat-earth concept regained its prominence. Hence, when Columbus proposed to sail west to India there were those who still believed that he would fall off the edge of the earth.

Other important early events in charting the seas are summarized in Table 1.1. These events provided the backbone for the scientific study of the seas.

The Beginning of Scientific Study of the Seas

In 1770, Postmaster General Benjamin Franklin, in order to improve mail service between England and the colonies, authorized the production of a chart of the Gulf Stream, in the Atlantic Ocean. This work enabled navigators to choose the fastest route by taking advantage of favorable currents. Franklin's chart of the Gulf Stream was all that was available to navigators until the nineteenth century (Fig. 1.7). He believed that ocean currents were caused mainly by wind and that some currents were deflected by obstacles such as continents.

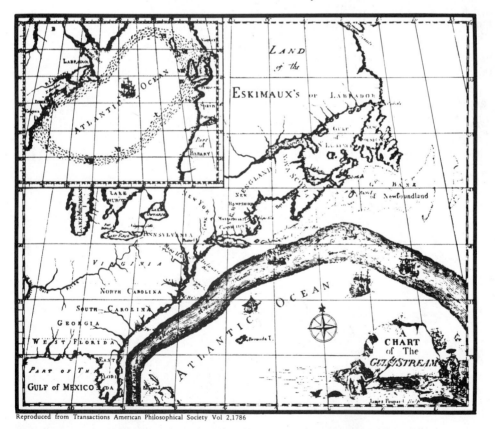

Reproduced from Transactions American Philosophical Society Vol 2,1786

FIGURE 1.7 Ben Franklin's chart of the Gulf Stream. (From *Transactions of the American Philosophical Society*, 1786.)

Table 1.1 MAJOR ADDITIONAL EVENTS IN SEA EXPLORATION

1000	Leif Ericson is believed to have crossed the Atlantic Ocean to Canada
1492	Christopher Columbus crosses the Atlantic to West Indies
1500	Pedro Alvares Cabral discovers Brazil
1513	Juan Ponce de Leon describes the Florida Current
1513–18	Vasco Nuñez de Balboa sights the Pacific Ocean, and, a few years later, sails and explores the Pacific coast south of the Isthmus of Panama
1515	Peter Martyr discusses the origins of the Gulf Stream
1519–22	Ferdinand Magellan circumnavigates the earth and attempts deep-sea soundings in the Pacific but does not reach bottom
1569	Gerardus Mercator develops a projection of the world on which north-south and east-west directions are represented by straight lines. This type of projection is still used for navigational charts (See Appendix A)

Captain James Cook made three extensive voyages through the Atlantic, Indian, and the Pacific oceans (Fig. 1.8) between 1768 and 1779. He charted vast areas, especially in the South Pacific. His journeys in the Pacific took him as far south as 71° 10′S latitude and north to 70° 44′N latitude. He would have gone further but was stopped by ice pack both

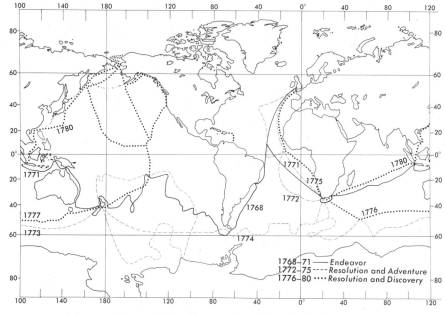

FIGURE 1.8 The voyages of Captain Cook.

times. During his second and third voyage he successfully determined longitude using chronometers* and celestial navigation. This was more than 2000 years after Pytheas developed the concept of latitude. Cook's charting established the overall shape of the world ocean (except for the Antarctic fringes) for the first time, and thus brought to a close what many call the "age of exploration."

Cook also demonstrated the practicality and desirability of including scientists in voyages of ocean exploration. The work of the botanists, Joseph Banks and Daniel Solander, who were with Cook aboard his first ship, the *Endeavour*, was followed by voyages of scientists with other captains. Charles Darwin's travels aboard the *Beagle* (1831–6) and J. D. Hooker's voyage with James Ross to the Antarctic (1839–43) were probably made possible by Cook's success.

Lieutenant Matthew Fontaine Maury, a navigator on board the USS *Falmouth,* had been frustrated in his attempt to find detailed wind and current charts. When in 1842 he was made Superintendent of the Depot of Charts and Instruments (the Depot was eventually divided into the Naval Hydrographic Office and the Naval Observatory), he decided to compile the types of charts he had sought earlier. Maury collected data on winds and currents from the logs of naval ships and plotted the information on a chart of the Atlantic Ocean. What he saw was a generally clockwise flow of water around a relatively calm Sargasso Sea. In 1855, he published *Physical Geography of the Sea,* considered to be the first major oceanography text.

One of the first expeditions to investigate the biology of the oceans in addition to collecting navigational data was that of the *Beagle,* with Charles Darwin aboard. The *Beagle* sailed around South America to the Galapagos Islands off Ecuador. During the voyage, Darwin collected enough data to last him a lifetime. His conclusions about natural selection and evolution gained him a place in the history of biology, and his theory of the origin of coral reefs is still widely accepted, with few modifications.

In 1840 Edward Forbes, a professor at the University of Edinburgh, reported that his dredging investigations in deep water indicated that the abundance of life decreased with depth, until an "azoic zone" was reached at about 600 meters. This view is now known to be incorrect. In fact, Sir John Ross (in 1818) had already collected animals from a depth of nearly 2000 meters. However, Forbes was a well-known marine biologist, and his views received widespread attention and encouraged others to attempt even deeper dredging.

---

*Christian Huygens, in 1660, built his first chronometer for use at sea. However, it was not accurate enough for determination of longitude. The chronometer was greatly refined in the 1770's by John Harrison of England and by others. Truly accurate measurement of longitude, however, could not be made until the chronometer could be checked periodically against land-based time via wireless transmission in 1904.

From 1839 to 1843, James Ross (a nephew of John Ross), on the ships *Erebus* and *Terror,* sailed the Antarctic Ocean and charted its southern limits. This was the first expedition to do more than a few deep-sea soundings. He also collected animals from the sea floor and found life at all depths.

It might appear that such deep dredging was of little practical importance to the governments that sponsored these expeditions. However, toward the end of the nineteenth century there was a sudden interest in the laying of transatlantic telegraph cables. More information was needed regarding depths, deep currents, bottom composition, and especially animals that might bore into and destroy the protective covering of the cable. In the 1850's deep-sea telegraph cables brought to the surface were found to be encrusted with marine organisms, thus finally refuting Forbes' concept of an "azoic zone" below 600 meters.

In 1847, Louis Agassiz, a zoologist, accompanied the U.S. Coast Survey on its work in Massachusetts Bay. This was the first of a series of expeditions in which he was to participate with the Coast Survey. He collected sediments and animals from the sea floor and later studied the nature of coral reefs. His work gained prestige for himself and the Coast Survey and probably did much to encourage government spending on oceanographic research.

Professor Wyville Thompson of the University of Belfast, Ireland, led expeditions to the sea near Britain in 1868 and 1869. These expeditions produced evidence of a great variety of marine life off the coast, and temperatures in the depths as cold as 0°Celsius (°C) were recorded. The reports that followed, combined with a British fear of being outdone by the Americans, led to the greatest deep-sea expedition of all time, the *Challenger* expedition (1872–6). It was the first major scientific journey organized solely for the study of the oceans. Thompson traveled nearly 130,000 kilometers in the HMS *Challenger* (Fig. 1.9), along a winding course around the world. The results, bearing on physical, chemical, geological, and biological oceanography, took 23 years to publish and filled 50 heavy volumes. In spite of equipment that was primitive by today's standards, the crew was able to bring back samples of mud and animals from depths of more than 8000 meters. Some of the more striking accomplishments of the expedition included confirmation of the now-recognized fact that the major chemical constituents of sea water are found in constant proportions from one place to another in the open sea. In addition, knowledge of the drifting and bottom-dwelling life was expanded tremendously, and almost 5000 new species were identified.

In the late nineteenth century, it was known that driftwood and the remains of an American ship crushed by ice had drifted from the Siberian Sea through the Arctic Ocean to the east coast of Greenland. This led the Norwegian explorer Fridtjof Nansen to plan an expedition to study the circulation of Arctic waters. Nansen believed that a ship could be constructed that would withstand the crushing forces of the ice. In spite of

FIGURE 1.9 The HMS *Challenger*, which conducted oceanographic work around the world in the nineteenth century. (Courtesy of Bettmann Archive.)

the considerable skepticism of other experts, he persuaded the Norwegian government to help finance the trip. His ship, the *Fram* (Fig. 1.10), was a 38-meter, double-ended, three-masted schooner with a reinforced hull four feet thick.

In 1893, Nansen sailed northeast until the *Fram* was frozen into the ice north of Siberia. Eventually the *Fram* drifted northwest to 84°N latitude, about 600 kilometers (km) from the North Pole. At this point, Nansen left the ship with Frederick Johansen and tried to reach the North Pole by dog sled. Meanwhile the *Fram,* with the rest of the crew, drifted more or less southwest toward Spitzbergen (north of Norway). Nansen and Johansen had to turn back before reaching the Pole and luckily were picked up by an English expedition near Franz Josef Land. The *Fram* had drifted less than two km a day over depths greater than 3000 meters. The entire voyage lasted about three years. Very careful records were kept of the drift, wind, and water depth. Nansen observed that the ship tended to drift not quite with the wind but somewhat to the right of the wind. In 1902, V. W. Ekman reported that this phenomenon was due to the effect of the earth's rotation (See Appendix C, Coriolis force) on all objects moving over its surface. This knowledge has been of great value in the study of ocean currents.

FIGURE 1.10 Nansen's ship, the *Fram*, in the Arctic ice. (Courtesy of Norwegian Information Service.)

Between 1911 and 1918, Alexander Behn in Germany, Reginald Fessenden in the United States, and Pierre Langvin in France independently developed and tested methods of determining the ocean depth using the echo (reflection) of sound waves off the bottom. In 1914, Fessenden was awarded the first patent for an *echo-sounding* device.

The first practical electronic echo sounder was developed by the U.S. Navy in 1922. This development must be considered one of the most important milestones in charting the relief (bathymetry) of the ocean floor. This method quickly replaced the old lead-line method used by the early explorers of the sea.

Toward the end of World War I, the Allies started to develop a device that could be aimed in any direction and could locate submarines and other objects below the sea surface. This work was continued after the war first by Britain and then by the United States. The instrument eventually became known as *SONAR* (*SO*und *NA*vigation and *R*anging). This system is similar in principle to but distinct in function from the echo sounder.

In Table 1.2 are listed some additional milestones in the history of oceanography.

The Development of Fisheries Research

Although people have obtained food from the oceans for thousands of years, it was not until the 1860's that governments began to consider

Table 1.2 ADDITIONAL MILESTONES IN OCEANOGRAPHY

1802	Nathaniel Bowditch publishes his *New American Practical Navigator*. This book has been revised many times and remains a standard work on navigation to this day.
1818	John Ross dredges animals from nearly 2000 meters while exploring Baffin Bay for a passage to the Pacific
1835	Gaspard Gustave de Coriolis publishes a classic paper on the effects of the earth's rotation on fluids in motion
1843	Alexander Dallas Bache becomes superintendent of the U.S. Coast Survey* and expands work on physical oceanography and the nature of sediments and bottom-dwelling life
1865	Johann George Forchhammer analyzes samples of sea water from a variety of locations and finds that the ratios between the major salt constituents are nearly constant from sample to sample

*Later called the U.S. Coast and Geodetic Survey and now incorporated into the National Oceanic and Atmospheric Administration (NOAA) of the Department of Commerce.

financing of expeditions to study the erratic catches of fish in European waters. Some experts believed that the problem was due to "overfishing." Others thought that varying currents or other characteristics of the ocean waters were responsible for the poor catches observed in certain years. Research was aimed at determining natural causes for the appearance or disappearance of the fish.

A Norwegian zoologist, G. O. Sars, believed that the fish, especially the young ones, were carried by currents. He studied the biology and migrations of cod and herring. Sars discovered that cod eggs float and drift with the currents and that the young fish feed on plankton (drifting organisms which for the most part are very small). Plankton also are the food for other young and many adult fish, such as herring. Sars believed that the presence of water that was good for the plankton at a time when plankton were needed by the newly hatching fish might result in an especially successful brood of fish. These fish would mature and grow during a period of a few years and could eventually result in large commercial catches.

In the late nineteenth century Johannes Peterson, a Danish zoologist, successfully tagged flatfish by attaching two buttons to the fish with a silver wire. He applied the method of tagging and recapture to observe the migration of fish. Similar techniques are used today.

The Norwegian biologist Johan Hjort brought statistical techniques to the study of fish production in the early twentieth century. He determined the changing age distribution of fish populations over a period of years. He recognized certain broods of fish from one year to the next as they grew and matured. These data indicated that a particularly good year for fishing might be due to conditions present in the breeding grounds years before. Also, early detection of a large new brood could be used to predict future fishing success. Since young fish eat plankton, he reasoned, conditions favorable to plankton growth should be required for a sizable brood of fish.

Scientists working on the fisheries problem realized that fish migrate through the waters of many nations and that a solution to fisheries problems would require the cooperation of all the northern European fishing nations. In 1899, King Oscar II of Sweden called an international conference to discuss the situation. Representatives of Great Britain, Germany, Denmark, Norway, and Sweden attended. By 1902, Finland, Holland, and Russia had joined the others to form the International Council for the Exploration of the Sea. The Council coordinated efforts in physical and biological oceanography as they applied to the fishery problem. The Council is still active today but on a much reduced scale.

Although the work of the Council was ambitious, the fisheries problem was never completely solved. During this period of intensified research, however, much basic information was acquired about currents, water characteristics, and the ecology of plankton and fish.

Modern Oceanography

The first major oceanographic expedition to be conducted in a truly systematic way was that of the *Meteor,* a German research vessel. The *Meteor* surveyed the physical, geological, chemical, and biological aspects of the Atlantic Ocean from 1925 to 1927 in a series of transects, sampling the water at predetermined depths. The most up-to-date technology available at the time, including echo sounders, was used. The vertical profiles of temperature, salinity, and oxygen that were obtained from this expedition for the Atlantic Ocean are still widely quoted.

Various governments have continued to engage in oceanographic expeditions and research. In the United States various federal agencies have intensified their oceanographic activities in recent years. These include the U.S. Coast Guard, U.S. Navy, U.S. Geological Survey, and the National Oceanic and Atmospheric Administration. Private and public research institutions and universities are also engaged in major oceanographic work, frequently with considerable governmental financial support. Progress in oceanography in the United States has been greatly enhanced by the work at three institutions formed specifically for that purpose. Scripps Institution of Oceanography of the University of California, in La Jolla, California, was established in 1905. The Woods Hole Oceanographic Institution was founded in 1930 in Cape Cod, Massachusetts (Fig. 1.11). The Lamont Doherty Geological Observatory was formed in 1948 and is now part of Columbia University in New York. These institutions have engaged in worldwide voyages of scientific discovery and have become first-rate academic institutions as well, educat-

FIGURE 1.11 The Woods Hole Oceanographic Institution. Note the modern research vessels in the harbor. Compare these to the *Atlantis*, shown in the chapter opening photo. (Courtesy of WHOI.)

ing oceanographers for the future. Today, more than a dozen universities also conduct large-scale research and education in oceanography.

In 1966 the United States Congress passed the Sea Grant College and Programs Act, which outlined the government's commitment to long-term non-military funding of marine science education and research. Colleges and private institutions are eligible for support from the Sea Grant Program. This program is similar in many ways to the Land Grant College Program initiated in the last century. Both are committed to education and public service regarding natural resources.

The National Oceanic and Atmospheric Administration (NOAA) was formed in 1970 in an effort to bring together and better coordinate the oceanographic and meteorologic activities of various departments of the U.S. government. NOAA is part of the Department of Commerce and consists of nine agencies, including the National Ocean Survey, National Marine Fisheries Service, National Weather Service, and the Office of Sea Grant.

Oceanography today is truly international. Many nations have again realized that international cooperation results in a greater economy of finances, avoids duplication of effort, and may speed the solution of problems. The International Council for the Exploration of the Sea, discussed earlier, was one of the first such efforts. More recently, the

FIGURE 1.12 The deep sea drilling ship *Glomar Challenger*. (Courtesy of Deep Sea Drilling Project, National Science Foundation.)

International Geophysical Year (1957–8) and the International Indian Ocean Expedition (organized under UNESCO, 1959–65) involved many nations in the pursuit of knowledge of the seas.

The Deep Sea Drilling Project, supported by the National Science Foundation and managed by the Scripps Institution of Oceanography, started operation in 1968. This project has enabled scientists to study rocks down to depths of 1000 meters or more below the sea floor. The results of this study tend to confirm the idea that the ocean floor is indeed very young, less than 200 million years in age, as compared with the oldest continental rocks, which are about 4 billion years old. This work is conducted from a specially constructed drilling ship, the *Glomar Challenger* (Fig. 1.12).

The Deep Sea Drilling Project became the International Program of Ocean Drilling in 1975, and has involved the support and participation of the Soviet Union, France, Japan, the United Kingdom, the Federal Republic of Germany, as well as several institutions in the United States.

In 1968, President Lyndon B. Johnson proposed, and the United Nations General Assembly endorsed, the concept of an International Decade of Ocean Exploration (IDOE) for the 1970's. Its goal was to improve the use of the oceans and their resources for the benefit of all. In 1969, the National Science Foundation was given the task of planning, managing, and funding the U.S. role in IDOE. Numerous research projects flourished, involving the participation and cooperation of 36 countries. The research role of IDOE is being continued into the 1980's as the

Coordinated Ocean Research and Exploration Section (CORES), funded by the National Science Foundation.

As with many areas of human knowledge oceanographic information has also been used for selfish purposes by various nations. As we learn more about fisheries biology, mining of the sea floor, use of military submarines, and even the possibility of using the sea floor as a site for weapons, setting ground rules for the use of the oceans has become more critical and at the same time more difficult.

THE POLITICAL OCEAN

The oceans and shores are increasingly influenced by the uses to which they are put. Present technology leaves no part of the ocean immune from human interference. Conflicts range from the right of a private citizen to own an ocean beach to the right of a nation to mine a portion of the sea floor.

Conflict Along the Coast

BEACHES: PUBLIC OR PRIVATE?

Local problems develop when groups of people have conflicting views of the use of the coastal ocean. For example, conflicts arise between advocates of private and public ownership of beaches. Sunbathing, swimming, surfing, fishing, and ocean-watching are all activities that make use of the beach and are certainly important to the coastal tourist industry.

Laws regarding the ownership of beaches vary from state to state and among nations. Beaches on the coasts of California, Oregon, Washington, Mexico, and Puerto Rico are open to the general public, and free access to virtually all of these beaches is maintained by law. On the other hand, many beaches along the Atlantic coast of the United States are controlled by individuals or municipalities. There is a trend toward making these beaches public, caused by the expense of maintaining the beaches. Natural processes such as storms frequently erode beaches, and states may provide funds for beach restoration. The state of New Jersey, however, will not fund beach erosion control projects unless the municipality has a plan for public access to the beach. Texas allows "private" ownership down to the line of mean high tide; however, this private property is open to the public as far as 200 feet (about 60 meters) inland from that line, or to the line of permanent vegetation. Conversely, in many resorts in Florida, beaches are virtually inaccessible to anyone except paying guests at beachfront hotels. In such cases, the expense of beach maintenance is borne by the hotels themselves.

COASTAL ZONE MANAGEMENT

Some coastal areas, although unsuitable for homes or industrial sites, are developed unwisely. Developers frequently fill marshes, build canals which can become stagnant, or even build on unstable barrier islands or land above cliffs. These areas frequently are subject to coastal landslides or to inundation by storm waves (Fig. 1.13). The necessary rebuilding may require millions of dollars of public funds. What is needed is better land-use planning to avoid unnecessary property damage and loss of life in the future.

Coastal regions are subject to a wide variety of uses that may be in conflict. Some bays have been partially filled to create more land for port facilities and industrial development, destroying in the process productive breeding grounds for clams, oysters, and fish. The filling of San Francisco Bay is an example: In 1835, the bay occupied roughly 1760 square km (660 square miles). By 1975, the area had decreased by some 40 per cent, eliminating about 520 square km of mostly shallow-water environments. Today, through local and national coastal management efforts as well as pollution abatement, some of these problems are being alleviated and future problems avoided.

POLLUTION

The welfare of the coastal regions depends in part on activities that take place far inland. For example, pollution of streams, even in land-

FIGURE 1.13 Apartment building damaged by storm waves along the New Jersey coast in 1962. (Courtesy of the Army Corps of Engineers.)

locked states, may pollute the coastal ocean. Pollution may affect the breeding success of migratory salmon and hence the number of adult salmon in the oceans. Pollution may be a cause of toxic red tides along the coast.

Pollution of the coast may result from such activities as ocean dumping of sewage and industrial waste, including PCB's (polychlorinated biphenyls) and mercury, oil spills from ships, and offshore oil drilling (Fig. 1.14). The ships that carry oil are much larger than in years past, and the magnitude of potential oil spills is consequently much greater than before. The transfer of oil to and from these supertankers frequently requires new offshore oil ports (Fig. 1.15) in deeper water than is found

FIGURE 1.14 On April 4, 1975, the 170-meter (557 ft) oil tanker *Spartan Lady* broke in two and sank in heavy seas about 200 km southeast of New York. The dark streaks in the photos are oil spilling onto the sea. (Courtesy of U.S. Coast Guard.)

FIGURE 1.15 *A,* An Arabian American Oil Company offshore oil port at Ras Tanura, Saudi Arabia, with several tankers at various stages of loading. *B,* The largest tanker ever built in the United States, which is more than 300 meters long and draws more than 20 meters of water when loaded, is too deep to enter any United States port. After its maiden voyage it transferred its oil to smaller tankers bound for the United States at Rotterdam, Netherlands. (*A* courtesy of ARAMCO; *B* courtesy of the American Petroleum Institute.)

in most coastal harbors. Offshore drilling for oil probably will increase in the years ahead, and pollution associated with it may occur unless sufficient safeguards are maintained.

In the United States, both coastal states and the federal government have been under pressure from environmentalists to devise and enforce very strict regulations. This concern was greatly heightened, if not originated, by two particularly severe accidents involving petroleum—the wreck of the tanker *Torrey Canyon* in 1967 and the spill from a Union Oil well in Santa Barbara Channel off the California coast in early 1969.

Just how much damage is really done by human-caused pollution is questioned by some scientists. It is not clear that the damage from a disastrous oil spill is irreversible. It has been reported that by 1972 the marine ecosystem in the Santa Barbara Channel had returned to normal, and similar observations have been made regarding the beaches fouled by the *Torrey Canyon*. This is chiefly because the oil is ultimately metabolized by marine bacteria. At least some of the oil on many beaches results from natural seepage through faults in the sea floor. In the late 18th century, Captain George Vancouver reported that he sailed for several days through a substance that looked and smelled like tar off the California coast near the Santa Barbara Channel. There are a minimum of 60 known natural seeps in the Santa Barbara Channel. The situation in the Gulf of Mexico is even more dramatic. Some 12,000 natural seeps exist in an area of about 2300 square km (900 square miles) in the northwestern part of the gulf. Almost 10 per cent of the oil entering the sea comes from natural seeps.

Still, some regulation is needed, and local governments must cooperate with the national regulating agencies. National anti-pollution legislation is aimed at protecting the environment for marine food resources. Fish pay no attention to political boundaries and may migrate thousands of kilometers and through coastal waters of several states or even nations. Their survival depends on clean water and controlled fishing in all of these waters. In other words, their survival depends on cooperation among states and among nations.

Territorial Seas

Historically, the leading naval powers have stressed the freedom of the high seas with limited jurisdiction over coastal waters. The seventeenth-century Dutch legal scholar Hugo Grotius was one of the earliest proponents of this concept. In 1609, he published an essay, *Mare Liberum,* which presented the arguments in favor of freedom of the seas. He is frequently referred to as the "father of international law."

The narrow territorial sea claimed by the early naval powers was only 3 nautical miles (5.55 km), the distance that could be controlled by cannonball fire. Frequently these were nations that wanted free access to widely distributed colonies. Nations with restricted access to the sea or with weaker naval forces have strived to protect their interests with broad territorial waters over which they would retain sovereignty. The concept of a three-mile territorial sea was widely accepted until fairly recently, when some nations extended their territorial control to as much as 200 nautical miles (370 km). Some of the commonly claimed limits have been 3, 6, 10, 12, 18, and 200 nautical miles.

> ". . . under no circumstances, we believe, must we ever allow the prospects of rich harvests and mineral wealth to create a new form of colonial competition among the maritime nations. We must be careful to avoid a race to grab and to hold the lands under the high seas. We must ensure that the deep seas and the ocean bottoms are, and remain, the legacy of all human beings."
>
> Lyndon B. Johnson, July 13, 1966

The high seas have been regarded as belonging to no nation but free to be used by all. In 1967, Ambassador Arvid Padro of Malta proposed to the United Nations an extension of this concept, allowing that the sea bed beyond the limits of national jurisdiction be internationalized and used for peaceful purposes and for the benefit of all mankind. He further proposed that the United Nations be given jurisdiction over the international sea bed. These proposals were eventually incorporated into the 1975 Draft Treaty of the Law of the Sea Conference in Geneva.* The Draft Treaty proposed a 12-nautical-mile (22.2 km) territorial sea for all nations that have a coastline. The treaty would give the right to unimpeded passage through the many international straits whose widths are less than 24 nautical miles.

Under provisions of the proposed Law of the Sea Treaty, coastal nations would be given authority to protect their offshore marine environment from pollution by enforcing regulations regarding safety of ships. They could detain and inspect vessels and impose penalties against violators.

Economic Zones

Some countries, in addition to their territorial seas, have claimed economic zones that extend national jurisdiction, mainly concerning fish stocks and pollution, out to 200 miles from land. This control, which has been exercised for some years, is essentially as proposed in the Draft Treaty of the Law of the Sea Conference. This Treaty considers all of the natural resources (living and non-living) within 200 miles of shore, regardless of the width of the continental shelf, as belonging to the coastal nations. The economic zones cover about 35 per cent of the ocean

---

*Since the 1975 conference in Geneva, further modifications, proposals, and compromise have been introduced at subsequent meetings of the Law of the Sea Conference, but no formal agreement has been reached.

area. If a nation has a continental shelf wider than 200 miles, it would also own and control the resources of the sea bed out to the edge of the shelf.

Uses of the Sea Bed

The sea bed is of great economic importance to us because of the wealth of natural resources that it contains. However, its ownership has been in doubt. In 1945, President Truman proclaimed that the United States owned the resources of the continental shelf adjacent to the United States. In 1953 the Submerged Lands Act granted to the individual states jurisdiction out to three miles, except for Texas and Florida, which were given control to nine miles into the Gulf of Mexico. In the same year the Outer Continental Shelf Act was passed, granting federal control over the rest of the shelf.

The Geneva Convention of the Continental Shelf, declared in 1958 that the resources of the continental shelf belong to the adjacent nation, in agreement with the Truman Proclamation. When two countries share a continental shelf, they are to divide the shelf between them. Some shelves are as much as 1500 km wide, providing resources far out to sea. Other nations have no shelf at all. Therefore, the realm is further declared to extend "to a depth of 200 meters or, beyond that limit, to where the depth of the superjacent waters admits of the exploration of the natural resources of the said areas . . .." Today, this statement has led to problems because no part of the ocean is technologically beyond the reach of exploration. Resources such as petroleum and manganese nodules may be found far from shore and are attractive lures for exploitation by advanced technology. Thus, conflicts among nations may arise, and further modification of this convention is certainly warranted.

As stated earlier, the Draft Treaty proposed that coastal nations would own the resources of the sea bed out to 200 miles from shore (the Economic Zone) or out to the edge of the continental shelf, whichever is farther. The Draft Treaty also proposed the creation of an International Seabed Authority to deal with resources of the sea floor located beyond the areas of ownership and control of the coastal nations, to conduct its own research and mining in the open sea, and to issue permits for exploration and mining of the international sea bed. The more than 60 per cent of the ocean area that is represented by the international sea bed is to be used for the benefit of all.

In 1980 the U.S. enacted the Deep Seabed Hard Mineral Resources Act establishing procedures for the development of mineral resources in the deep sea bed. This legislation gives legal backing to the traditional concept that the resources of the international sea bed belong to whomever retrieves them first. Other nations may take a similar approach, resulting in inevitable conflicts. Perhaps the only solution is an international law of the sea.

PETROLEUM IN THE ANTARCTIC

Natural gas and oil are believed to exist under the Ross Sea off Antarctica. It is estimated that there may be up to 115 trillion cubic ft (32.6 trillion cubic meters) of gas and 45 billion barrels (7.15 billion cubic meters) of oil in the Antarctic. This is more than four times the estimated gas and petroleum reserves of Alaska's North Slope.

The Antarctic is immune from commercial exploitation under a 1959 treaty made at a time when it was believed there was nothing of commercial value on the continent. The treaty is scheduled to expire in 1989. However, it may be tempting to break the treaty earlier since there would be no leases or royalties to pay. The future of the continent remains unclear.

Fisheries Conflicts

Local fishing disputes sometimes arise between commercial and sports fishermen. The sports fishermen frequently claim that the commercial fishermen deplete the game-fish stocks by their efficient techniques or that they catch too many of the fish that serve as food for the game fish (Fig. 1.16). International fishing disputes may also arise, as when foreign fishing vessels disturbed lobster traps along the East Coast of the United States in the early 1970's. Fishing fleets (Fig. 1.17) from several countries, including the Soviet Union, Poland, Japan, and Italy, fished off the coast of the United States, and ships were occasionally seized for catching lobsters and escorted to shore. Some captains were even fined. In addition, American lobstermen have reportedly been "harassed" by foreign fishing ships trawling through their lobster pots in quest of fish. This problem has been partially resolved through a bilateral agreement between the Soviet Union and the United States, in

FIGURE 1.16 A very efficient purse seine surrounding a large school of menhaden in Long Island Sound. Sports fishermen have protested that this technique removes too many of the fish that serve as food for game fish. Most game fish, however, have varied diets, and the menhaden population appears to have been rather stable in recent years. (Photo by Charles R. Meyer.)

FIGURE 1.17 The Polish base ship *Pomorze*, with a side-trawler alongside, off the coast of Virginia. (Official U.S. Coast Guard Photo.)

which the Soviets have agreed to avoid disturbing or collecting creatures of the continental shelf, including lobsters, and to permit boarding of their vessels by United States enforcement officials. In addition, a U.S.–U.S.S.R. Fishery Claims Board was established to deal with conflicts that arise and to award damages where appropriate.

United States fishermen catch tuna in the waters off Peru and Ecuador. Both countries claim a 200-mile fishing limit. Ecuador also owns the Galapagos Islands and so claims fishing rights as far as 800 miles out to sea. Many boats have been seized and the owners and captains fined $400,000 or more, depending on the capacity of the boat. Some United States tuna fishermen stay outside the 200-mile limit of other countries, which causes them to take more time to fill their boats with fish. Others consider the risk to be worthwhile and continue to fish in the rich waters off South America. Some even pay high fees for Ecuadorian licenses that permit them to fish as close as 40 miles from shore.

In 1966, the United States extended its exclusive fishing rights from three to 12 miles, and a further extension of this limit to 200 miles was approved by the U.S. Congress in January 1976. This law, the Fisheries Conservation and Management Act, went into effect March 1, 1977. It is a unilateral action by the United States intended to protect the fish stocks off U.S. shores and to ensure an ample catch for American fishermen. The law implies acceptance of the 200-mile limit claimed by such countries as Ecuador and opposed by U.S. tuna fishermen. The law requires foreign fishing vessels workings between 200 and three miles off the U.S. shore to have a special permit and to abide by regulations set by the U.S. Government. (The waters within three miles of shore are under the control of the adjacent state.) Foreign fishermen must keep a log of the trawls made and the number and kinds of fish taken. Observers may be placed aboard the ships to monitor their fishing activities, and the Coast Guard makes periodic inspections.

Governmental control of fisheries activities, including net design and catch limits, is necessary to maintain the living stock of fish and invertebrates, especially over the continental shelf. This control requires international cooperation, not only through observance of economic zones but also through such international commissions as the International Pacific Halibut Commission, the Inter-American Tropical Tuna Commission, and the International Commission for North Atlantic Fisheries. International cooperation is especially essential in controlling the fisheries for migratory fish, such as tuna.

Freedom for Ocean Research

There has been great concern regarding the freedom of ocean research in an ocean restricted by 200-mile economic zones (more than 35 per cent of the ocean area). Some coastal nations claiming jurisdiction over a broad economic zone have refused permission to the United States and other developed nations wishing to do ocean research in the shelf waters of those countries. There is some suspicion that oceanographic ships have been used for covert surveillance. The proposed Law of the Sea Treaty provides freedom of research in the economic zone beyond the territorial sea with the following restrictions:

1. The research must be fundamental in nature.
2. Scientists and observers from the coastal nation must be provided the opportunity to participate in the research.
3. The data and findings from the research must be made available to the coastal nation.

If the research is directly related to the resources of the economic zone, the coastal nation may either refuse permission or reach an agreement with the visiting nation to their mutual advantage. These restrictions, of course, may result in considerable red tape and could delay the proposed research. On the other hand, international cooperation, especially with smaller, developing nations, might be further promoted. Such cooperation has proved successful in the past. For example, in 1974, scientists and technicians from 72 countries participated in the Atlantic Tropical Experiment of the Global Atmospheric Research Program. The work required coordination among workers of varied backgrounds, speaking many languages, and working from 40 ships, 12 aircraft, and numerous land stations. The host country was Senegal, a small nation in westernmost Africa,which has the best port facilities in that part of the Atlantic, the port of Dakar. Participating nations included Brazil, Canada, Finland, France, East and West Germany, Mexico, Netherlands, Portugal, the Soviet Union, the United Kingdom, and the United States. A great deal of information was gathered regarding the origin of tropical rain clouds, surface ocean characteristics, upwelling, and the equatorial

FIGURE 1.18 The *R. V. Chain* from the Woods Hole Oceanographic Institution was one of the many research vessels participating in programs of the International Decade of Ocean Exploration. (Courtesy of Woods Hole Oceanographic Institution.)

undercurrent. The success of this venture is an encouraging note for further international research.

Many other international projects have been launched in the 1970's, especially through the International Decade of Ocean Exploration (IDOE). These projects have involved scientists and ships (Fig. 1.18) from many nations and have included research on marine resources (biological and geological), chemistry, pollution, ocean circulation, and weather. Several developing coastal nations, such as Senegal, Gabon, Brazil, Peru, and Ecuador, and some landlocked nations such as Switzerland and Bolivia have participated in these projects. Some of these projects have obvious economic overtones and have involved research in the territorial waters claimed by such coastal states as Peru.

Research and exploitation, even in the open ocean, will be restricted in the future. The proposed Law of the Sea Treaty suggests that the International Seabed Authority be notified of any proposed research in the open ocean. However, permits would be required for much of the research involving the international sea bed. If, indeed, there is a net benefit for all people, the inconveniences experienced in the years to come will be worthwhile.

What About the Future?

One message of this book is that the oceans are not only interesting but also valuable and worth protecting from unnecessary damage. They have a role to play in supplying energy (petroleum, thermal energy, and tidal energy), food (conventional fisheries and mariculture), avenues of navigation and trade, minerals, and even recreation. It is a sad but well-established fact that people constantly squabble and occasionally go to war over virtually anything that has value, and throughout history this has certainly been true of the oceans. The threat we all face today is that the oceans are becoming more valuable—petroleum is more scarce and more expensive; we need and can obtain more food from the sea; and our ability to extract minerals from the oceans and sea beds continues to improve. Thus, the threat of conflict or simply of selfish misuse of marine resources is constantly increasing.

There is an increasing awareness of these dangers, particularly through the United Nations, and signs of international cooperation during the mid-1970's have been heartening. Similarly, the governments of most coastal nations have begun to take steps to assure the continued health of the marine environment under their jurisdiction.

The question now is whether these hopeful trends will continue. The answer is beyond the scope of this or any other textbook. It really lies with informed individuals and their governments.

SUMMARY

The foundations of oceanography rest on the early explorations of the Phoenicians and Greeks and their contributions to geography and navigation. Virtually no progress was made in oceanography during the Dark Ages (476–1450 A.D.) The voyages of Columbus and Vasco da Gama led to a new age of exploration, primarily in search of new trade routes. Thus the geography of the world ocean was fairly well known when the *Challenger* set sail in 1872 on the great oceanographic voyage that covered 70,000 nautical miles. The numerous smaller expeditions that followed resulted in a basic descriptive understanding of the oceans.

The oceanographic work done in the early twentieth century was directed toward the dynamic aspects of oceanography and the development of sophisticated sampling and analytic tools, such as the echo sounder. Some marine biologists focused their efforts on solving fisheries problems and in the process developed a mechanism of international cooperation, the International Council for the Exploration of the Seas.

Modern oceanography continues to focus on investigation of the dynamic ocean processes and the distribution of life and mineral resources. The worldwide need for food, energy, and raw materials adds economic incentive to these pursuits.

The technical competence of many nations and the awareness of most nations of the value of the seas has necessitated even greater international cooperation and has led to the International Law of the Sea Conferences. The oceans provide a common area of concern to all nations, coastal and landlocked alike.

Human activities on land and sea contribute undesirable wastes such as sewage, industrial chemicals, and oil to the oceans. The activities of one nation may adversely affect the economy of another, as when pollutants are carried by currents into the territorial waters of another nation. This and other aspects of maritime conflicts are being addressed by international conferences in an effort to produce a workable Law of the Sea Treaty.

IMPORTANT TERMS

Azoic zone	Manganese nodule
Beagle	Mariculture
Challenger	*Meteor*
Chronometer	NOAA
CORES	Plankton
Echo sounder	Salinity
Economic zone	Sea Grant Program
Endeavour	SONAR
IDOE	Territorial sea
Law of the Sea Draft Treaty	

STUDY QUESTIONS

1. How have the motives for oceanographic voyages evolved from earliest times to the present?
2. What early visible evidence indicated that the earth was not flat?
3. What is an echo sounder? What special benefit does SONAR provide to current oceanographic research?
4. What was the importance of Darwin's observations aboard the *Beagle*?
5. List four areas where conflicts between people or nations occur in the use of the oceans and coastal waters.
6. What are the goals of the Deep Sea Drilling Project?
7. What are the three fundamental provisions of the Law of the Sea Draft Treaty regarding oceanographic research in territorial waters?
8. Discuss the scientific contributions of the *Challenger* expedition.
9. What problems can be expected to surface in the future regarding control of the economic resources of the open sea?
10. Discuss the significance of the proposed universal 200-mile economic zones.
11. Discuss the role of the International Council for the Exploration of the Sea in the study of fisheries problems in the early twentieth century.

Anderson, A. 1973 (November). The Rape of the Seabed. *Saturday Review/World.*

Bailey, Herbert S., Jr. 1953 (May). The Voyage of the *Challenger. Scientific American.*

Cadwalader, George. 1973. Freedom for Science in the Oceans. *Science.182*:15–20.

Cohen, Robert. 1979. Energy from Ocean Thermal Gradients. *Oceanus. 22*:12–22.

Deacon, Margaret. 1971. *Scientists and the Sea, 1650–1900, a Study of Marine Science.* London, Academic Press.

Eckert, Ross D. 1979. *The Enclosure of Ocean Resources: Economics and the Law of the Sea.* Stanford, Hoover Institution Press, Stanford University.

Eiseley, Loren C. 1956 (February). Charles Darwin. *Scientific American.*

Gullion, Edmund A. (ed.). 1968. *Uses of the Seas.* Englewood Cliffs, N.J., Prentice-Hall.

Hammond, Allen L. 1975. Probing the Tropical Firebox: International Atmospheric Science. *Science. 188*:1195–8.

––––––––––––. 1976. Deep Sea Drillers: Entering a New Phase. *Science, 191*:168–9.

Hedberg, Hollis D. 1979. Ocean Floor Boundaries. *Science 204*:135–44.

Idyll, C. P. (ed.). 1972. *The Science of the Sea, a History of Oceanography.* New York, E. P. Dutton.

Meyer, Charles. 1975 (May). The Menhaden War. *National Fisherman.*

Ross, Carol. 1976 (February). Passage of 200-Mile Limit Seen Adding New Problems. *National Fisherman.*

Schiller, Ronald. 1975 (November, December). The Grab for the Oceans, Parts I and II. *Reader's Digest.*

Schlee, Susan. 1973. *The Edge of an Unfamiliar World, a History of Oceanography.* New York, E. P. Dutton.

Shapley, Deborah. 1975. Now a Draft Sea Law Treaty—But What Comes After? Science, *188*:918.

Wenk, Edward, Jr. 1972. *The Politics of the Ocean.* Seattle. University of Washington Press.

Wooster, Warren S. 1980. The Endless Quest. *Oceanus 23(1)*:68–71.

Geology of the Ocean Basins

2

As the need for exploration and exploitation of natural resources grows, the knowledge of what lies beneath the oceans (71 per cent of the earth's surface) becomes increasingly important. Unfortunately, the major nonrenewable natural resources—petroleum and minerals—are found in only limited quantities near the earth's surface. We are rapidly exhausting many of the land resources and are just beginning to turn our attention to the ocean bottom in search of new resources.

The drifting apart of Africa (left) and Arabia (right) resulted in the creation of the Gulf of Aden (bottom) and the Red Sea (top). (Photo courtesy of NASA.)

Many problems in oceanography, such as the origin of sea water and the origin of ocean basins, cannot be studied without an understanding of what the interior of the earth is like, which in turn requires some knowledge of the origin of the earth itself. Except for the rocks brought up from a few deep wells on land and in shallow-water areas, there is no direct information about the earth's interior. Some of these wells have penetrated to depths of about 10 km (about 30,000 ft), but this is an insignificant value compared with the average radius of the earth (6370 km). The deepest holes drilled in deep water have penetrated only about one km below the sea floor. Even volcanoes cannot provide information about the earth's deep interior, because they are believed to originate within about 100 km of the earth's surface, and because the material that they extrude has been contaminated by the upper layers.

Present ideas about the origin of the ocean basins are the result of a revolutionary theory in the earth sciences, the *plate tectonics theory,* which was formulated in the 1960's, based primarily on data collected about the ocean basins themselves.

ORIGIN OF THE EARTH

There are many theories about the origin of the earth. The *dust-cloud hypothesis* is currently popular among astronomers. This theory states that about 4.5 billion years ago, the earth (as well as other members of the solar system) developed from a diffuse cloud of dust and gas. This so-called dust cloud was made up mostly of hydrogen but also contained at least all of the elements found on earth and perhaps others no longer present. Individual members of the solar system, including the planets and their satellites, were formed by contractions taking place within local mass concentrations in the dust cloud. Most of the materials condensed at the main center of the cloud and eventually became the sun. The sun's temperature and density increased owing to gravitational contraction and eventually became high enough to initiate the nuclear reactions by which the sun now radiates energy into space. Similarly, the interior of the primitive earth became very hot, owing to gravitational contraction and radioactive heat generation, although not sufficiently hot or dense to start nuclear reactions, and it melted. The heavier metallic elements, with iron probably dominating the mixture, settled toward the center of the earth, leaving the lighter materials above it. The result is a layered earth. Solar radiation increased the surface temperatures of planets in inverse proportion to their distance from the sun. This heating, coupled with the radiation pressure from *solar wind* (a continuous stream of charged particles emanating from the sun), caused the earth to lose large quantities of the lighter elements and the remaining dust-cloud cover.

The earth, then, was completely molten at some time during its early history and has been losing heat ever since. However, only a portion of the heat lost by the earth today is the original heat; the other portion is believed to be derived from the disintegration of radioactive materials concentrated near the outer part of the earth. Although the earth is believed to be about 4.5 billion years old, the oldest known rock is only about 4 billion years old. The absence of older rocks may be due to the molten early history of the earth or to their subsequent destruction or alteration by geological processes.

As the interior of the earth cooled, various minerals crystallized and aggregated to form the different rocks of the crust. Cooling at the surface helped to retain the atmosphere and eventually caused the condensation of the water now filling the ocean basins. The origin of the earth's atmosphere and of sea water are discussed in Chapter 4.

INTERNAL STRUCTURE OF THE EARTH

Physical Nature of the Earth's Interior

Figure 2.1 illustrates a cross section of the earth, showing its four major layers: the *crust,* the *mantle,* the *outer core,* and the *inner core.* The crust varies in thickness from about six km to more than 70 km. Below the crust and extending to a depth of 2900 km is the mantle. The outer core

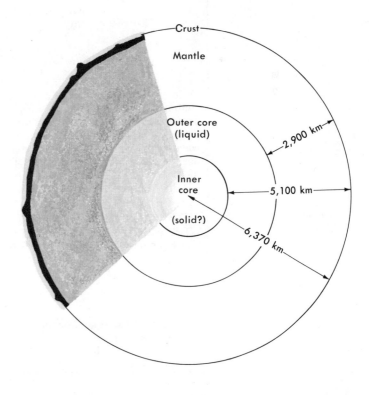

FIGURE 2.1 Cross-section of the earth, showing its major regions as interpreted from the study of earthquake waves. Note that the crust is not drawn to scale. In proportion to the rest of the figure, the crust is about as thick as the line forming the outer circle.

extends from a depth of 2900 to 5100 km and resembles a liquid, whereas the inner core, extending from 5100 to 6370 km (the center of the earth), most resembles a solid.

Theories about the physical nature of the earth's interior are based on studies of earthquakes and nuclear explosion, during which numerous shock waves or *seismic waves* are produced at the source. The three main types of seismic waves are *primary* or P waves, *secondary* or S waves, and *surface* waves. The P and S waves travel within the earth, whereas the surface waves, as their name implies, travel along its surface. We will discuss the behavior of P and S waves to illustrate the nature of the earth's interior.

About one million earthquakes are produced within the earth each year, and hundreds of seismographs located all over the world record the seismic waves produced by these earthquakes. P waves travel faster than S waves and hence are recorded first by seismographs. In addition, P waves can travel through solids and liquids, whereas S waves can be transmitted only through solid media. The physical difference between the two types of waves is that in the case of P waves, the particles of the

---

A paradoxical relationship exists between the velocity of seismic waves and the density of rocks. From the following equations for the velocity of P and S waves, as the density increases the wave velocity should decrease.

$$\text{P-wave velocity} = \sqrt{\frac{\text{incompressibility} + 4/3 \text{ rigidity}}{\text{density}}}$$

$$\text{S-wave velocity} = \sqrt{\frac{\text{rigidity}}{\text{density}}}$$

From experiments on rocks near the surface of the earth, however, quite the contrary is true; velocity of seismic waves increases with increasing rock density. This, of course, must imply that the effects of the elastic properties (especially rigidity) of the rocks have much greater effects than density on the wave velocity. Furthermore, the density must increase as the elastic properties increase, but at a slower rate. The table below shows some examples of rocks, their densities and P wave velocities.

TYPICAL DENSITIES AND P-WAVE VELOCITIES FOR ROCKS NEAR THE EARTH'S SURFACE, AS DETERMINED IN THE LABORATORY (1 km/sec = 3,600 km/hr = 2,237 miles/hr)

Material	Density (gm/cc)	P-wave velocity (km/sec)
Marine mud	1.8	1.8
Consolidated sedimentary rock	2.5	5.0
Granite	2.7	5.7
Oceanic basalt	3.0	6.8

medium vibrate back and forth in the same direction that the wave travels, while in the case of the S wave, the particles vibrate at right angles to the direction of the wave. The P-wave velocity within any medium is dependent on the density and the elastic properties (rigidity and incompressibility) of the medium, whereas the S-wave velocity depends only on density and rigidity. It is because liquids have no rigidity that they are unable to transmit S waves.

Figure 2.2 shows the paths of the shortest travel times of some typical P waves within the earth. If the earth's interior were homogeneous and the P-wave speed remained constant within it, all seismic waves would travel in straight lines rather than in the curved paths shown in the figure. The curved paths are produced by *refraction* or bending of the seismic waves. This refraction is due to a continuous increase in the wave speed within each zone (see Appendix B for a discussion of wave refraction). The sharp downward deflection of the wave at a depth of 2900 km (point a on Fig. 2.2) apparently is caused by an abrupt decrease in the wave speed there. The wave is sharply deflected upward at a depth of 5100 km (point b on Fig. 2.2), apparently by an abrupt increase in the wave speed at this depth. The paths of S waves are similar to those of the P waves, except that no S waves have been observed to travel below a depth of 2900 km. This would lead to the conclusion that the region below this depth (the core) behaves as a liquid. However, seismologists interpret the abrupt increase in the P-wave speed at the depth of 5100 km as

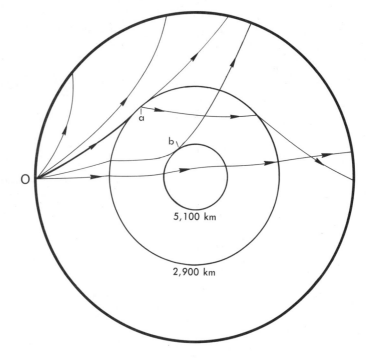

FIGURE 2.2 Paths of some P waves within the earth. O is the source (earthquake) from which these waves radiate. The P wave at a depth of 2900 km (point a) is sharply deflected downward by an abrupt decrease in the wave speed. The sharp deflection upward at a depth of 5100 km (point b) is caused by an abrupt increase in the wave speed at this depth. In each zone, continuous increase in wave speed with depth causes the waves to be refracted in curved paths.

5,100 km

2,900 km

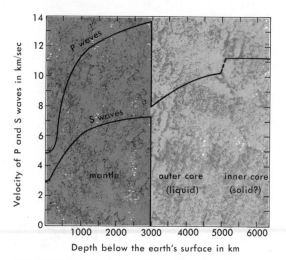

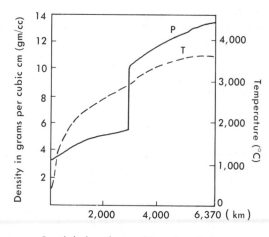

FIGURE 2.3   *A,* Variation of P- and S-wave velocities within the earth. P-wave velocity decreases abruptly at a depth of 2900 km and increases abruptly at a depth of 5100 km. No S waves are detected in the core (below 2900 km). (Modified after Garland, 1979.)   *B,* Density (solid-line) of the earth as interpreted from seismic data. Water has a density of 1 gram per cubic cm (gm/cc). The dashed line gives a possible temperature curve for the earth's interior. (Modified after Garland, 1979.)

indicating that the inner core is probably solid. They postulate that the enormous pressure near the earth's center forces it into the solid state, even though it is probably the hottest zone. Figure 2.3A shows the variations of P- and S-wave speeds within the earth obtained from earthquake studies. Figure 2.3B is a density model of the earth as interpreted from seismic data.

Chemical Nature of the Earth's Interior

Seismic studies of the earth are indirect and provide information only about the physical nature of the earth's interior. We may never be certain about its exact chemical makeup. However, clues to its chemical composition come from an entirely different field, the study of meteorites. Meteorites are considered by many astronomers to be formed from the shattered remains of a planet that was once part of our solar system. All of these planets are thought to have originated from the same dust cloud. Hence, the composition of meteorites may be similar to that of the interior of the earth. The two common types of meteorites falling on the earth are stony and metallic meteorites. Stony meteorites are composed of iron and magnesium silicates (minerals containing silicon and oxygen) and resemble certain igneous rocks (that is, rocks formed from the cooling of molten materials). The density and seismic properties of these meteorites are comparable to those of the mantle under appropriate conditions, and geologists believe that it is likely that they are in fact very similar. The metallic meteorites, mixtures of approximately 90 per cent iron and 10 per cent nickel, resemble the core in their physical properties.

The origin of meteorites is still a controversial problem in astronomy. But if the meteorites give us an accurate picture of the earth's interior, we can say that the mantle is composed of iron-magnesium silicates and that the core is a mixture of nickel and iron. Other geophysical data also support the meteoritelike composition of the earth's interior. For instance, some geophysicists believe that an electrically conducting metallic core would help to explain the origin of the earth's magnetic field. If one assumes an initially molten earth, it may be easier to visualize the settling of heavier metallic materials toward the center of the earth. Of course, all of these ideas, no matter how elegant or scientifically compelling, are speculative. Many other substances could also account for the observed seismic speeds and the resulting densities of the mantle and the core.

Crust of the Earth

The crust of the earth is the incredibly thin skin covering the mantle. Proportionately, it is far thinner than the skin of an apple. It is, however, the most complex and variable part of the earth. The existence of the crust as a separate entity was first recognized in 1909 by Andrija Mohorovičić, a Yugoslavian seismologist. He observed that the speeds of P and S waves increase abruptly when going from the crust to the mantle. For example, P waves travel at a speed of about 7.5 km/sec in the lowest part of the crust, but just below it, in the upper mantle, the speed is about 8 km/sec. The "boundary" between the crust and the upper mantle where the seismic speed changes is called the *Mohorovičić discontinuity,* commonly known as the *Moho.*

Crustal thickness varies considerably. In oceanic areas, it is usually between 6 and 15 km, averaging about 10 km. This includes the ocean depth, which averages about 3.8 km. The crust in continental areas averages about 35 km in low-lying areas but may be greater than 70 km in mountainous regions. However, the crust is unusually thin in some parts of the world, such as in portions of the northwestern United States, where it is less than 20 km thick. Figure 2.4 shows a typical cross section of the crust across continents and oceans. Seismic studies indicate two distinct layers within the crust under continents, whereas oceanic crust consists of a single major layer, excluding the sediment cover. The upper continental crust (about half the thickness of the crust) has an average density of about 2.7 gm per cubic cm (gm/cc), or 2.7 times that of water. The lower continental crust, which is believed to be similar to the oceanic crust, has an average density of about 3.0 gm/cc.

The chemical makeup of the crust is poorly known. No drill hole has penetrated the whole crust, either in continental or oceanic areas. Deep-sea drillings indicate the presence of basalt (a dark-colored igneous rock)

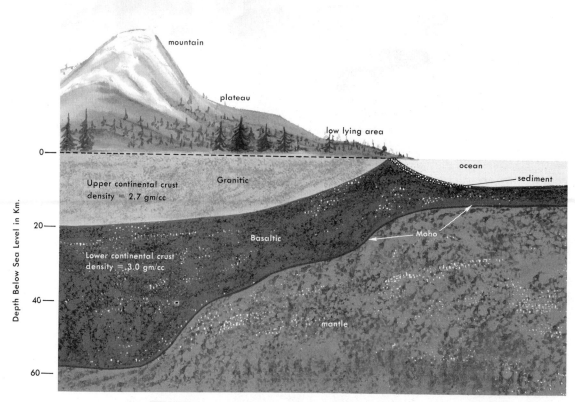

FIGURE 2.4   Typical cross-section of the crust of the earth. The continental crust has two layers, whereas the oceanic crust consists of one major layer. In continental areas, the crustal thickness increases with increasing elevation. The Moho is almost a reflection of the topography. The upper continental crust is granitelike, and the lower continental crust as well as the oceanic crust is basaltlike.

below the oceanic sediments. These drill holes, however, have penetrated only about a kilometer of rock. It is generally believed that the upper continental crust is composed of granitelike rocks and that the lower continental crust as well as the oceanic crust are composed of basaltlike rocks. Granite and basalt are common near the surface of the earth and have the densities and seismic velocities observed in the crust.

The Mohole Project

In 1959, the United States government undertook an expensive study, known as the *Mohole Project,* to drill a hole through the crust of the earth, so that rock samples from the crust and perhaps the upper mantle could be obtained. As the crust is much thinner under the oceans, it was decided to drill the Mohole in an oceanic area. After extensive research, a drilling site was selected near the Hawaiian Islands, where the crust was believed to be only 6 km thick. Considerable effort was made to perfect the technology needed to drill such a hole in the ocean. The problem included not only drilling at the high temperatures and pressures that exist deep within the earth but also the stabilization of the drilling platform. Although some progress was made, and two small test holes were drilled, large cost overruns caused the project to be abandoned in 1965. The Soviet Union has announced that they have started drilling several Mohole-like holes. At the present writing their degree of success is unknown.

Ocean Margin Drilling

The United States government has embarked on a new deep sea drilling program for the 1980's, for drilling deep holes along the coastal areas of the Pacific and Atlantic Oceans. Seismic studies have revealed the existence of very thick sedimentary deposits along the continental slopes off the Atlantic coast of the United States. It has been suggested that these sediments may contain significant reserves of petroleum.

The 10 year drilling project is expected to cost around $700 million. The expenditure is to be shared equally by the government and by major U.S. industries in petroleum related fields. It is also expected that there may be some contributions from the non-U.S. institutions. Present plans call for the drilling of about 15 holes. The deepest and most expensive one is to be 7 km long, off the coast of New Jersey, at an estimated cost of 30 million dollars.

The *Glomar Explorer* (Fig. 2.5), the ship that was originally built for the use of the U.S. Central Intelligence Agency, is to be converted for drilling in the deep sea. The remodeling will include many safeguards such as blowout prevention, riser (casing), and well head control equipment. When converted, the *Explorer* will be the world's largest drilling ship, capable of operating at water depths of two to four km. Most

FIGURE 2.5 An aerial view of the Hughes *Glomar Explorer* during sea trials. (Photo courtesy of U.S. Navy.)

importantly, the drill will be able to penetrate several kilometers into the sea floor and extract cores. The actual drilling, however, is not scheduled to begin until the mid-1980's. By this time, the operations of the present drilling ship *Glomar Challenger* will be phased out.

The Lithosphere and the Low-Velocity Layer

In addition to the major zones within the earth, another region, the *low-velocity layer,* or *asthenosphere,* exists in the upper mantle, generally between about 70 and 300 km below the earth's surface. However, the low-velocity layer approaches the sea floor near the mid-ocean ridges. The layer is best developed under the oceans and poorly developed under the continents. The low-velocity layer is characterized by slower seismic speeds than the layers above and below it, implying that the materials of the low-velocity layer may be plastic or even partially molten.

That part of the earth that is above the low-velocity layer is called the *lithosphere.* It may be as much as 150 km or more in thickness beneath older continental areas, and about 100 km in thickness in other continental areas. The lithosphere should not be confused with the crust of the earth; it includes the crust and the uppermost mantle. The low-velocity layer is extremely important in explaining both the origin of volcanoes and the drifting of continents, as will be discussed later in this chapter.

Isostasy

Land areas of the earth are constantly undergoing erosion. At the present rate of erosion, most of the continental areas of today would be completely eroded down to sea level in a few tens of millions of years. How, then, could have continents survived so long? Geophysical studies such as measurement of the gravitational field of the earth indicate that mountainous areas do not generally represent greater accumulations of mass than the mass under lowland areas. Similarly, ocean basins do not represent mass deficits. Mountains and oceans represent only "apparent" excesses and deficits of mass, respectively, near the surface of the earth; the major portions of the earth are in mass equilibrium. This means that the mass under a square meter of the earth's surface is the same everywhere. This concept of mass balance within the earth is known as *isostasy.*

Seismic studies indicate that the lithosphere in mountainous areas, in addition to having a thicker crust, has a deeper "root" extending into the low-velocity layer. The lithosphere may be thought of as "floating" on the "plastic" low-velocity layer, as icebergs float in the ocean. Just as a large iceberg extends deeper into the water and rises higher into the air than a small iceberg, a mountainous lithosphere extends deeper into the

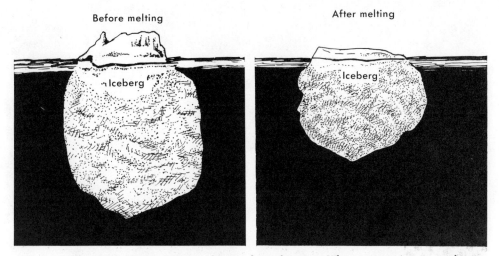

FIGURE 2.6   The melting iceberg analogy of isostasy. When a mountain range undergoes erosion, the bottom of the lithosphere rises.

low-velocity layer than does the lithosphere in lowland or oceanic areas. When a mountain range undergoes erosion, the lithosphere becomes thinner. Not only does the mountaintop become lower but also the bottom of the lithosphere (the "root") rises. This is analogous to what happens to a melting iceberg (Fig. 2.6). A similar mechanism of mass balance maintains isostasy in an area which is receiving large quantities of sediments.

CONTINENTAL DRIFT, SEA-FLOOR SPREADING, AND PLATE TECTONICS

Continental Drift

Until the early 1960's, most geological concepts were explained on the basis of the permanency of continents and ocean basins. That is, it was believed that continents and oceans remained in place throughout geological time, except for some continental areas that were repeatedly submerged under shallow seas and for the growth of certain continents along their edges. Some scientists, however, held the view that all the continents were once joined together as a supercontinent called *Pangaea* (Fig. 2.7), which broke apart, and that these continental fragments have been drifting over the earth ever since. *Laurasia* and *Gondwanaland* were names given, respectively, to the northern and southern hemispheric portions of Pangaea.

The greatest proponent of continental drift was a German scientist, Alfred Wegener, who in 1912 revived the concept and provided consid-

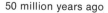
more than 200 million years ago

50 million years ago

1.5 million years ago

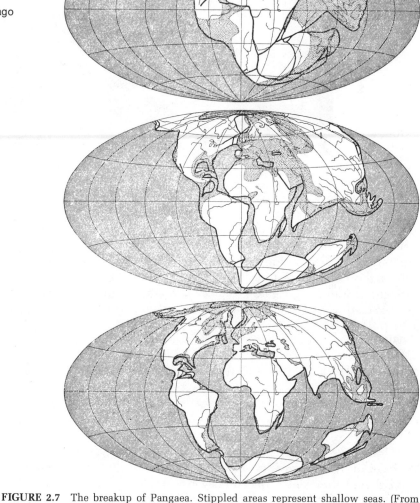

FIGURE 2.7 The breakup of Pangaea. Stippled areas represent shallow seas. (From Wegener, A.: The Origin of Continents and Oceans. New York, Dover Publications, Inc., 1969.)

erable evidence in support of it. Two years earlier, an American scientist, Frederick B. Taylor, had advanced a similar idea but was not as successful as Wegener in publicizing his theory. However, ideas remotely resembling "continental drift" had been proposed as far back as 1620 by the English philosopher Francis Bacon. According to Wegener, the breakup of the supercontinent started about 200 million years ago. His most notable piece of evidence was the fact that continents seemed to fit

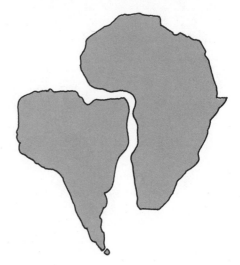

FIGURE 2.8 Fitting of South America and Africa. Most continents may be fitted in this way, providing the most notable evidence for continental drift.

together like pieces of a jigsaw puzzle (Fig. 2.8). Additional supporting arguments offered by Wegener and others for continental drift included the following: (1) geological similarities between continents (similar rock assemblages and mountain structures of the same age); (2) similar fossil assemblages of older than 200 million years in many continents; (3) the presence of coal in Antarctica and ancient coral reefs in high latitudes (both impossible to explain unless these regions were once located in low latitudes or warm climatic regions); and (4) evidence of glaciation more than 200 million years ago (Fig. 2.9) in areas now located near the equator. In addition, continental drift was able to explain the formation of folded mountain ranges such as the Himalayas by the collision of moving land masses. Although much of this evidence seemed reasonable, many geologists were unwilling to accept continental drift, mainly because of the lack of a known mechanism capable not only of breaking up the supercontinent but also of making the continents move large distances. Thus many of the arguments in favor of continental drift were considered coincidental relationships, or were explained in other ways.

Sea-Floor Spreading

Since the time of Wegener, much has been learned about the sea floor, including such large-scale features as undersea mountain ranges and deep, narrow trenches. Furthermore, evidence has been uncovered of the distribution and ages of the sediment and rocks of the sea floor and the distribution of volcanic activity and submarine earthquakes. Much of this evidence was mystifying until the 1960's, when a revolutionary hypothesis called *sea-floor spreading* was formulated. This hypothesis states that the ocean floor is moving, or spreading, away from the mid-ocean mountain ranges. The discussion that follows will explain this

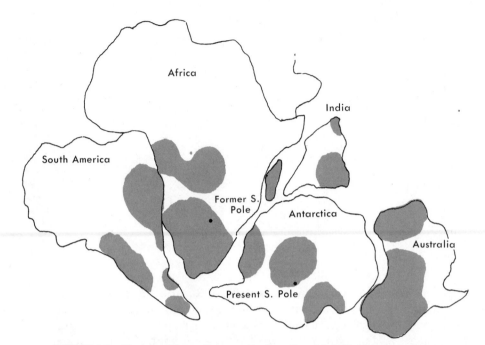

FIGURE 2.9 Glacial evidence in support of continental drift. The shaded areas were glaciated more than 200 million years ago.

concept and its importance to our understanding of the oceans as well as the continents.

The largest topographic feature on the sea floor is the *mid-ocean ridge* system (Fig. 2.10). Mid-ocean ridges form a nearly continuous volcanic mountain range, running through all the oceans and having a total length of about 25,000 km. Only the Mid-Atlantic Ridge, however, is actually located near the center of an ocean. The axes of mid-ocean ridges are sometimes characterized by longitudinal *rift valleys,* whose depths may exceed that of the nearby ocean floor (Fig. 2.11A). Mid-ocean ridges may have widths of as much as 3000 km and may rise to about three km above the surrounding sea floor. The ridges may rise above sea level to form islands, for instance, the Azores, Tristan da Cunha, and Iceland. The rocks near the axes of the mid-ocean ridges are the youngest, and are increasingly older the farther they lie from the ridges. The crests of the mid-ocean ridges are not only the sites of numerous volcanoes and shallow earthquakes but also the locations of unusually high heat flow from the earth's interior. For example, recent investigations from the submersible, *ALVIN,* on the East Pacific Rise, near the Galapagos Islands, have revealed submarine hot springs that release water at temperatures of about 380°C. This water is rich in mineral nutrients and, through the production of bacteria, supports a rich community of bottom dwelling animals.

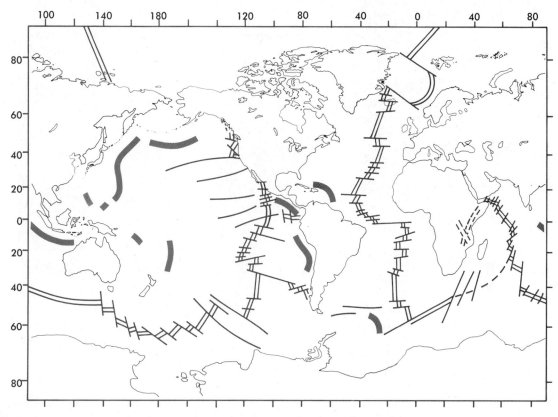

FIGURE 2.10   Locations of mid-ocean ridges and oceanic trenches. Heavy lines indicate trenches and double lines indicate ridges. Note the fracture zones offsetting the ridges. The double dashed line in Africa represents the East African Rift. (After Isacks et al., 1968, and Le Pichon, 1968.)

The *oceanic trenches* (Fig. 2.11*B*) represent the deepest portions of the oceans. The greatest depth, the Challenger Deep in the Marianas Trench, is about 11,500 meters, deeper than Mount Everest is tall. Trenches are never found in the middle of the oceans, being located instead near and parallel to continents and island chains, especially in the Pacific Ocean. Like the mid-ocean ridges, the trench regions are characterized by volcanoes and earthquakes. However, unlike those of the mid-ocean ridges, trench earthquakes include both shallow ones and the deepest known ones. Some originate at depths as great as 700 km.

The formation of ridges, trenches and many other features of the ocean floor remained a mystery until the 1960's. In 1960 and 1962, the American geologist Harry H. Hess proposed the new and revolutionary hypothesis called *sea-floor spreading,* a term coined by another American geologist, Robert S. Dietz, in 1961. This hypothesis accounts for many of the features of the ocean floor as well as the evolution of the

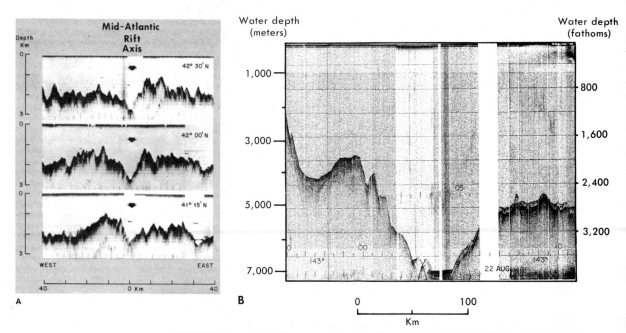

FIGURE 2.11   *A*, Profiles across the rift valley of the Mid-Atlantic Ridge obtained using seismic reflection techniques. *B*, A similar profile across the Aleutian Trench. (*A* from National Academy of Science, 1972; *B* courtesy of Lamont-Doherty Geological Observatory R/V, Conrad Cruise # 12, 1969.)

ocean basins themselves. According to this concept, pairs of convection currents in the mantle rise from beneath mid-ocean ridges and spread laterally away from them. Oceanic trenches are produced where the convection currents descend into the earth. Figure 2.12 illustrates sea-floor spreading in a hypothetical ocean. The rising convection currents bring hot molten mantle materials to the surface, eventually producing the mid-ocean ridges. Some of this material reaches the ridge surface in the form of volcanoes and lava flows, and the rest cools and solidifies before reaching the sea floor.

If convection is assumed to be continuous, the area near the ridge axis (center of the ridge) must always make room for the new materials rising from below. Thus, new ocean floor is continuously created near the center of the ridge. Previously formed ocean floor slowly moves away from the mid-ocean ridge in both directions, at a rate of a few centimeters per year (about as fast as your fingernails grow). This process can explain why ages of rocks increase progressively away from mid-ocean ridges. Older sea floor descends and is consumed into the earth near oceanic trenches. This process is called subduction, and the region around and below the trench is called the *subduction zone.* The fact that no rocks older than 200 million years are found on the sea floor can be explained either by subduction in oceans where trenches are present, or by sea-floor spreading having started within the last 200 million years in oceans where trenches have not yet evolved.

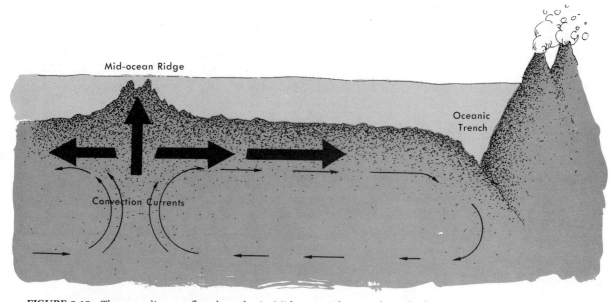

FIGURE 2.12 The spreading sea floor hypothesis. Mid-ocean ridges are formed where convection currents rise, and trenches are formed where they descend into the mantle. New sea floor is created at the ridge and is continuously moved away as convection continues.

The hypothesis of sea-floor spreading has been supported with evidence from two separate sources, the study of *paleomagnetism,* the ancient magnetism of rocks, and *paleontology,* the study of fossils.

Paleomagnetism

Igneous rocks are produced when molten materials such as lava cool near the surface of the earth. They usually contain small amounts of magnetic minerals, which can be thought of as tiny compass needles that line up with the earth's magnetic field when the minerals have cooled to about 600° C. The resulting rock has a weak magnetic orientation, which can be measured using sensitive instruments called magnetometers. The orientation is fairly stable under the temperature conditions existing near the surface of the earth and will not change appreciably even if the earth's magnetic field changes drastically. Sedimentary rocks also possess magnetism. Some of the constituent particles of sediment are magnetic, and when settling to the sea floor, they align themselves with the earth's magnetic field, eventually resulting in magnetized sedimentary rocks.

By the early 1960's, paleomagnetic studies had revealed that the earth's magnetic field has reversed itself many times during geological history. That is, the present north and south magnetic poles of the earth

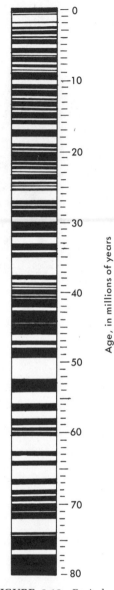

Age, in millions of years

FIGURE 2.13 Periods of magnetic reversals for the past 160 million years. The shaded regions indicate times when the magnetic field was normal (that is, oriented as it is today); and the unshaded areas correspond to times when the magnetic field was reversed. (After Larson and Pitman, 1972.) Figure 2.13 continues on opposite page.

have switched places repeatedly for reasons that are not understood. These 180° reversals are evident from the exact opposite magnetizations of rocks. Figure 2.13 shows the worldwide reversals of the earth's magnetic field for the past 160 million years. Evidence indicates that reversals have taken place throughout most of the earth's history, although the frequency of reversals has apparently increased during the past 50 million years to about double the previous rate. A glance at Figure 2.13 indicates that the reversals are not regular.

The cause of these random changes in the earth's magnetic field is not known. Even the origin of the magnetic field itself is not well understood; it is believed to be caused by convective motions within the earth's core, and hence the reversals of the magnetic field should also be related to variations of these motions.*

The reversals of the magnetic field provide the best evidence for sea-floor spreading, as suggested by British scientists F. J. Vine and D. H. Matthews in 1963. If the sea floor has been spreading, the materials reaching the mid-ocean ridges, when cooled, should be magnetized by the then-existing earth's magnetic field. According to the sea-floor spreading concept, ocean floor is continuously created at the ridges and is pulled away in opposite directions. If the magnetic field reverses, the new rocks formed at the ridges should have a reversed magnetization compared with the older rocks at a distance from the ridge. This should result in ocean-floor rocks that are magnetized in alternating normal and reversed directions, located symmetrically with respect to the mid-ocean ridges, as shown in the theoretical scheme of Figure 2.14.

The magnetism of ocean-floor rocks can be deduced from measurements made by magnetometers carried by airplanes or towed behind ships. Figure 2.15 shows a typical magnetic field pattern produced by the rocks of the ocean floor over a portion of the Mid-Atlantic Ridge, and similar results are obtained from magnetic measurements made over the other mid-ocean ridge systems. Figure 2.16 shows the results of magnetic studies in the northeast Pacific. Such magnetic measurements can be used to determine the rate of spreading of the ocean floor in the past (Fig. 2.17) if one knows the dates of reversals and the distance between corresponding rocks and mid-ocean ridges. The rates obtained range from 1 to more than 8 cm/yr, with an average value of about 2 cm/yr worldwide.

---

*It is generally believed that magnetic reversals require about 10,000 to 20,000 years to complete. This transition time may have significance for life on earth, according to some scientists. The earth's magnetic field is responsible for the **Van Allen radiation belts,** which protect the earth from intense radiation from outer space. If, during reversals, the earth's magnetic field is very weak or absent, the increased radiation reaching the earth may cause mutations of genes. Some scientists claim that a relationship exists between the times of reversals and the extinction or appearance of certain species. More research, however, is required to substantiate this claim, as there is some dispute as to the amount of radioactivity reaching the earth due to the loss of the magnetic field, as well as to the validity of the correlation mentioned here.

Paleontology

As marine organisms have evolved through time and species has replaced species, a fossil record was created in marine sediments that provides a way of determining the age of sediments. By extension, it is also possible to use fossils to test the theory of sea-floor spreading. Almost all volcanic rocks of the sea floor are thought to have been produced on the crests of mid-ocean ridges, and the oldest sediments above them are only slightly younger. If fossils are used to date the basal sediments above the volcanic rocks, one has a very good estimate of the age of these rocks. Comparison of results from this technique and magnetic technique show remarkably close agreement (Fig. 2.17), lending further credence to the theory of sea-floor spreading.

Plate Tectonics

In the late 1960's, the sea-floor spreading theory was modified to take into account all major geological structures of the earth, rather than just the oceans and their features. The new concept is known as *plate tectonics.* The term "tectonics" refers to the origin, movement and deformation of the large-scale structures of the earth. The concept of sea-floor spreading is included in plate tectonics. According to the theory of plate tectonics, the lithosphere of the earth is broken up by several rising convection currents. The resulting pieces are called *plates* (Fig. 2.18). They move laterally, or "drift," over the plastic low-velocity layer (asthenosphere) below. There are about eight large plates and about a dozen smaller ones. Some, such as the Pacific plate, are composed entirely of oceanic lithosphere, while others may contain both oceanic and continental lithosphere. The present plate boundaries are indicated by mid-ocean ridges, oceanic trenches, faults, volcanoes, or, most significantly, abundant earthquakes.

At the mid-ocean ridges, the plates are moving away in opposite directions while new plate materials are continuously being created. As the plates move, they cool and contract, resulting in an increase of ocean depth away from the ridges. The leading edges of many plates, especially in the Pacific Ocean, are consumed into the asthenosphere in subduction zones (trenches). Much of the Pacific Ocean is bordered by trenches, and the rate of subduction exceeds the rate of plate production in this ocean. Thus, the Pacific is a shrinking ocean. In the Atlantic, the rate of plate production far exceeds the rate of subduction, since the Atlantic has only two small trenches within it. Hence, the Atlantic is constantly widening. Some plates, such as the African plate, have no apparent subduction zones. This plate includes both Africa and the ocean floor surrounding it.

Assuming an average rate for sea-floor spreading of 4 cm/yr in the Pacific and a maximum ridge-to-trench distance of 10,000 km, the entire

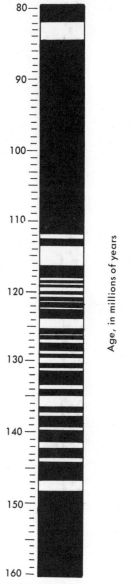

FIGURE 2.13 *Continued.*

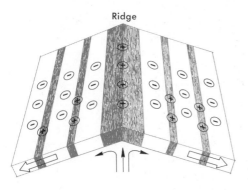

FIGURE 2.14   The normal (+) and reversed (−), magnetizations of the sea floor. Note the symmetry of the magnetizations with respect to the ridge.

floor of the Pacific Ocean can be renewed in about 250 million years. On the basis of this assumption, no present Pacific ocean-floor rock should be older than 250 million years. In fact, no rocks older than 200 million years have been found in any ocean. This lends further credence to the plate tectonics theory.

It is generally believed that episodes of plate tectonic activities must have taken place throughout much of the earth's history. For example, there may have been a single ancient supercontinent that fragmented about 500 million years ago and then reformed as the supercontinent Pangaea. We know most about the present episode of plate tectonics; the breakup of Pangaea and subsequent plate motions. These started about 200 million years ago and have been continuing ever since. Evidence from paleomagnetism and paleontology indicates that the separation of

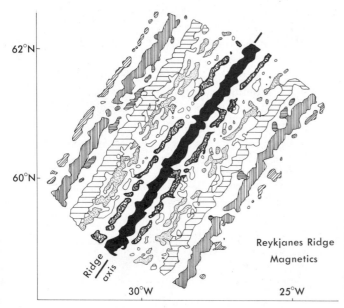

Reykjanes Ridge
Magnetics

FIGURE 2.15   The magnetic field produced by the ocean-floor rocks over the Reykjanes Ridge near Iceland. Note the similarity to Figure 2.14. The outlined areas represent predominantly normal magnetization. The ages of the rocks increase away from the ridge. (After Heirtzler et al., 1966, and Vine, 1968.)

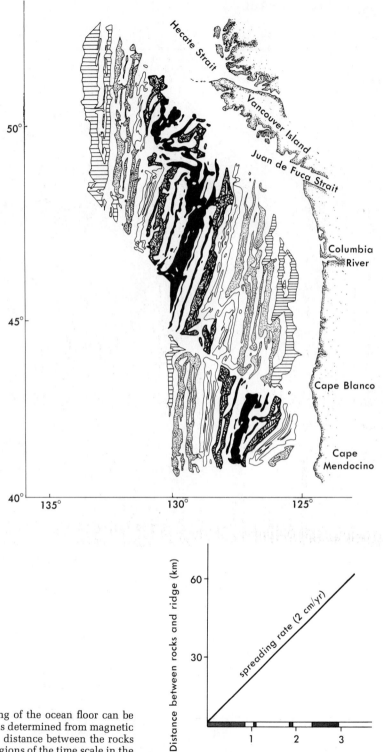

FIGURE 2.16 Magnetic field produced by the rocks on the floor of the northeast Pacific. Note the symmetry with respect to the ridges (the darkest shaded stripes). The ages of the rocks increase away from the ridges. The larger crosshatched areas are about 8 to 10 million years old. The offsets of the patterns west of Juan de Fuca Strait and Cape Blanco are caused by faults or fractures of the lithosphere. (After Raff and Mason, 1961, and Vine, 1968.)

FIGURE 2.17 The rate of past spreading of the ocean floor can be obtained by knowing the ages of rocks (as determined from magnetic reversals or paleontologic data) and the distance between the rocks and the mid-ocean ridges. The shaded regions of the time scale in the diagram indicate normal magnetization.

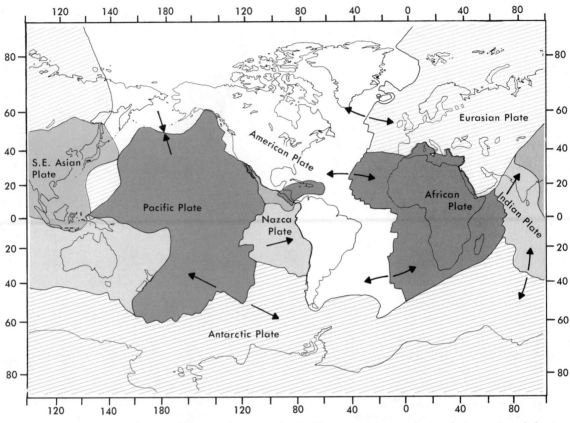

FIGURE 2.18   The major plates of the earth. Arrows indicate relative motion of plates as determined from magnetic data. (Modified after Dewey and Bird, 1970; Isacks et al., 1968; Morgan, 1968; and Vine, 1969.)

Greenland from Europe, for example, occurred only about 60 million years ago. The Gulf of California, the Gulf of Aden, the Red Sea and the East African Rift Valley (not yet a sea) were formed within the last 20 million years. Of these areas, only the Gulf of California and the Gulf of Aden have ridges in them at present, and none have trenches within them. The East African Rift Valley is characterized by numerous shallow earthquakes and volcanic activity (Fig. 2.20A; see also Fig. 2.10) and may well represent an early stage in the formation of a plate—breaking up the lithosphere. The East African Rift Valley may expand and become a new sea millions of years from now if the present activities continue.

The mid-ocean ridges, and hence the plates, are intersected by many fractures (see Fig. 2.10) called *transform faults* (Fig. 2.19). The rocks on both sides of the fault (between a and b in Fig. 2.19) are moving in opposite directions as a result of the general plate movements away from the ridges. In addition to earthquakes beneath the mid-ocean ridges, this differential motion causes earthquakes in the fault zone between the ridges. The San Andreas fault of California is considered to be a trans-

FIGURE 2.19 Mid-ocean ridges are offset by fractures called transform faults. Between points A and B, rocks on both sides of the fracture are moving in opposite directions, while in other parts of the fracture zone the movements are in the same direction. Earthquakes occur beneath the mid-ocean ridges as well as along the fault between A and B.

form fault, and the shallow but powerful earthquakes occurring along it are the result of this differential motion.

Implications of Plate Tectonics

The theory of plate tectonics has far-reaching significance, not only in oceanography but also in geology and biology. The origin of ocean basins, the age of the ocean floor, the origin of the mid-ocean ridges and oceanic trenches, the distribution of deep-sea sediments, and many other features of the ocean floor can be explained in terms of plate tectonics, and no other satisfactory explanations exist. For example, the apparent absence of rocks or sediments on the ocean floor older than 200 million years can easily be explained by this theory. According to the theory, very old oceanic rocks were swept away into the trench system and consumed into the asthenosphere. The progressively increasing age of the ocean floor, as well as the increase in deep-sea sediment thickness with increasing distance from the axes of the mid-ocean ridges, can be explained in terms of continuous movement of sea floor away from the ridges. Thus older rock is carried far from the "young" material at the ridges.

The global implications of plate tectonics have caused a revolution in the earth sciences. For the first time, the cause and distribution of most earthquakes, volcanoes, and mountains can be explained and understood. In addition to the earthquakes and volcanoes along the mid-ocean ridges and rift systems, their presence in other areas, especially along the edges of the Pacific Ocean, can be accounted for. The lithosphere descending into the subduction zone can set up earthquakes in its vicinity. In fact, most deep earthquakes are confined to the subduction zones (Fig. 2.20). These descending plates could melt, at least partially, owing to

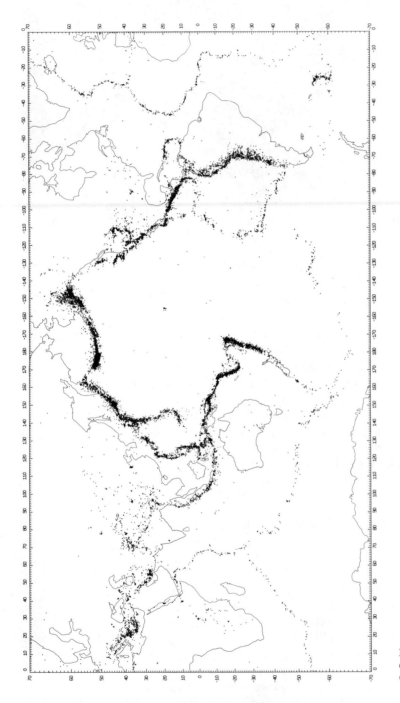

FIGURE 2.20 *A*, Distribution of earthquakes (1961–1967) occurring at depths of 0–100 km.

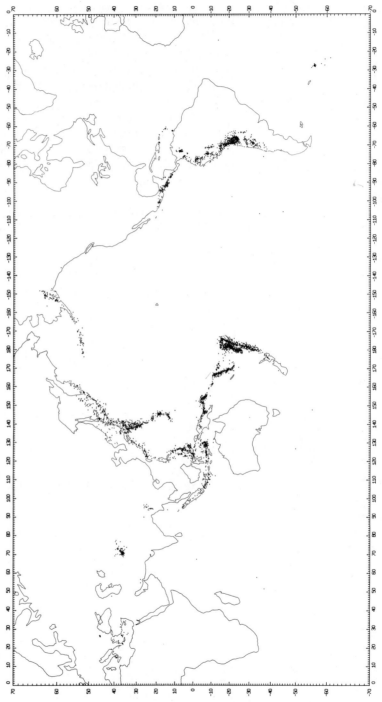

FIGURE 2.20 *Continued.* *B,* Distribution of earthquakes (1961–1967) occurring at depths of 100–700 km. (From Barazangi; and Dorman, 1969.)

friction and could set up volcanic activity landward of the trenches (Fig. 2.21). Eventually, volcanic mountain ranges and volcanic island chains form, as around most of the Pacific Ocean.

The concept of plate tectonics is becoming increasingly important for petroleum and mineral exploration. The geological processes that take place at many plate boundaries apparently concentrate valuable metals there. The location of many metal provinces of the world, such as western South America, may well be determined by ores being brought near the surface by magma rising from the subduction zones. Once this relationship is understood, we may even be able to look for new mineral resources in ancient subduction zones not explored for minerals. In addition, ores may be concentrated by the processes that take place under mid-ocean ridges. Even the origin and distribution of petroleum may be related to plate tectonics. Since most petroleum on earth originated in the oceans, the knowledge of the history of plate movements will help locate areas favorable for the migration and accumulation of petroleum, especially along continental margins.

Submarine petroleum was chiefly formed in stagnant, oxygen-free basins which are not necessarily near plate margins. Plate movements, however, including collisions and breakup of plates, may result in the

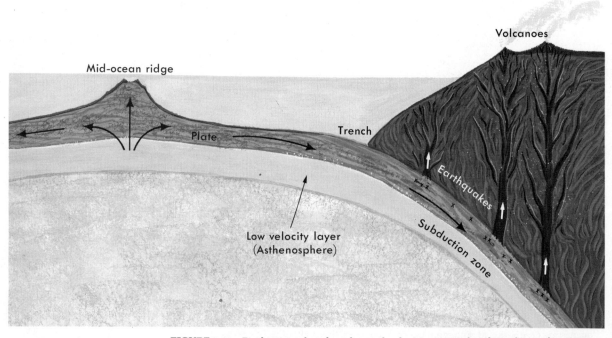

FIGURE 2.21 Production of earthquakes and volcanic activity by plates descending into a trench. The depths of earthquakes increase landward of the trench. Compare with Figure 2.20.

formation and accumulation of oil. Once formed, petroleum may be relocated by plate movements. Hence, knowledge of plate tectonics may be of value in locating areas of petroleum accumulation.

The largest mountains, such as the Alps and the Himalayas, are not volcanic but are composed mostly of folded (wrinkled) sedimentary rock, formed when two plates collided. The Himalayas, for example, were produced when India broke off and moved away from Africa and collided with Asia (see Fig. 2.7). We may use the same explanation for folded mountains that are older than 200 million years, the time at which the breakup of Pangaea started. Certainly, there must have been even earlier episodes of continental drift and collision, resulting in the formation of such old mountain ranges as the Appalachians and Urals.

Collisions, as well as the breaking of plates, may well be of great significance in evolutionary biology. Such events create new environments and destroy pre-existing environments.

Because of the movements of plates, the shapes of the ocean basins and continental margins have been changing throughout the earth's history. As Pangaea broke up and the resulting continents migrated, there was a gradual increase in the number and variety of marine habitats. It is likely that the area comprising shallow marine environments has been greatly expanded, resulting in a greater diversity of habitats. Furthermore, ocean circulation has been greatly modified in conjunction with the changing earth climate, ocean temperature, and distribution of land mass resulting from continental drift.

The study of marine fossils has provided clues to the history of life in the sea and its relationship to plate tectonics. The diversity of species tends to be greatest when there are diverse habitats, reliable food supplies, and a stable environment with regard to temperature, salinity, and oxygen supply. Consequently, diversity generally increases toward the tropics and away from large land masses. In both cases, the stability of food supply and physical environment increases. It has been found that the diversity of marine life increased with time to about 400 million years ago at a time when there were three or four continents. As continents collided and fused (formation of Pangaea), there was a reduction in diversity, and many species became extinct. The least diversity appears during the time when Pangaea was the only continent. Between 400 and 200 million years ago, the number of families of animals living on the continental shores was cut in half. As Pangaea broke up, diversity of life increased again and the number of marine families increased threefold in the next 200 million years. A more detailed investigation of the distribution and diversity of marine fossils will certainly tell a great story of the changing patterns of evolution as affected by the drifting of continents.

During the last 200 million years, populations of terrestrial organisms were split between two continents that were once joined together, resulting in the development of new species. As land masses migrated into different climatic regions, some species became extinct or underwent dramatic adaptations. Quite clearly, the joining of two previously separated land masses, as when India became fused to the Asian land mass, created interactions among species that had never before been in contact with each other.

Many problems concerning the theory of plate tectonics still remain to be solved. The origin and dimensions of the postulated convection currents in the mantle are still unresolved problems in geophysics. One must speculate that some localized source of heat in the mantle could heat the mantle materials, reduce their density, and cause them to rise toward the surface. We have already seen earlier in this chapter that rocks may be partially molten in the low-velocity layer, and convection may very well originate in this zone. Perhaps convection currents extend deeper into the mantle. Issues concerning mantle convection and the driving mechanism of plates are still being debated and remain unresolved.

Origin of Ocean Basins

The preceding discussion of plate tectonics and sea-floor spreading can be used to explain how the Atlantic Ocean was formed after the American plate broke away from the European and African plates. The Atlantic Ocean will continue to grow in size if these plates continue to move. Similarly, we may visualize the Red Sea, the Gulf of Aden, and other such areas as the sites of future large ocean basins if present plate motions continue. The East African Rift may also represent the site of a future ocean. Other oceans and seas, such as the Mediterranean Sea, may decrease in size, and some may even cease to exist, if opposing plates collide.

Many questions still remain to be answered. How and when did the Pacific Ocean originate? An ancestral Pacific must have surrounded Pangaea. The absence of rocks older than 200 million years, can be explained by the rejuvenation of the Pacific floor by sea-floor spreading and the descent of older rocks into its trenches. The Pacific Ocean is probably many times older than 200 million years.

The first ocean basins were formed after the earth's surface cooled and continents began to form. A primitive crust was probably formed by a differentiation process; that is, the separation of lighter rocks from the heavier materials below them. Convection currents, which have been active since the molten beginning of the earth, may have caused this primitive crust to move in a manner similar to the process of plate tectonics (Fig. 2.22). As the earth continued to cool, water vapor condensed and accumulated in the newly formed depressions of the crust. These were perhaps the beginnings of seas. The water was derived from within the earth, and the seas grew into oceans along with the subsequent evolution of oceanic crust and the growth of continents. Since then, plate tectonics have continually modified the shape and distribution of the ocean basins.

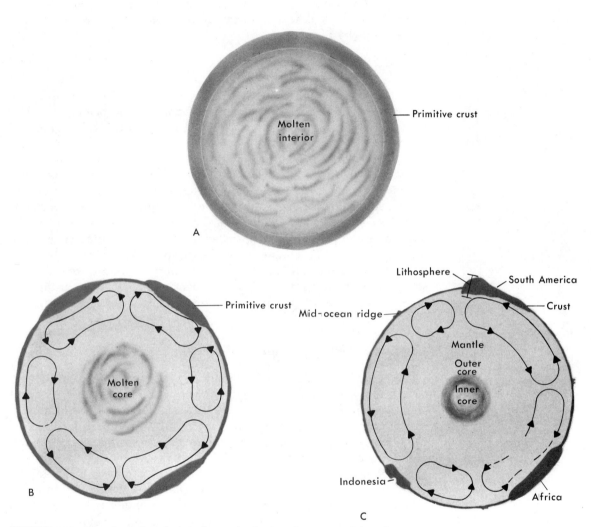

FIGURE 2.22   Hypothetical evolution of ocean basins (not drawn to scale). *A*, formation of a primitive crust. *B*, Convection within the earth causes redistribution of the primitive crust and the formation of continents. Oceans may have begun to form at this stage. *C*, Present earth.

STUDYING THE OCEAN FLOOR

The topography of the vast ocean floor is perhaps not as complex as that of the earth's land surface, but the former is not readily available for direct observation. Therefore, with the exception of submersible dives and drilled cores, almost all of our information about the ocean floor comes to us by indirect means.

Echo Sounding

The old, time-consuming system of determining ocean depth by the use of a weighted wire is not very accurate and has now been replaced by electronic methods developed during and after World War I. Figure 2.23 illustrates the principle involved in *echo sounding.* A transmitter sends a sound wave which is reflected back to the surface by the ocean bottom. A receiver picks up the reflected sound wave. Ocean depth can be calculated if we know the speed of sound in sea water and the time required for the sound to go down to the ocean floor and return. A continuous record (*echogram* or *fathogram*) of the ocean floor profile (Fig. 2.24) can be made by a moving ship. The so-called fish-finder of commercial fishing vessels operates on the same principle, except that the sound waves are bounced back from fish as well as from the bottom. In fact, some fish are so uniform in their schooling habits that they, as well as shrimp-like organisms and squid, produce a false bottom known as a *deep scattering layer.*

Continuous Seismic Profiling

Another method for studying not only the ocean bottom profile but also the sediments and rocks below the ocean floor is called *continuous seismic profiling.* The principle involved is similar to that of echo sounding, but a stronger energy and different frequency source is used. Part of the sound energy is reflected by the ocean bottom, but the other part enters the bottom and is eventually reflected back by rock or sediment layers beneath the bottom (Fig. 2.25). As the ship moves, one obtains a continuous profile of the ocean bottom and the geological structure beneath (see Fig. 2.11). Continuous seismic profiling is limited to the study of ocean sediment to a few thousand meters below the ocean floor, but it can be used in even the deepest water. For deeper penetration below the ocean bottom, explosions or other high-energy sound sources are employed. The *seismic refraction* method (Fig. 2.26) requires two ships, one for sending and one for receiving the sound. This technique is considerably more expensive and time-consuming than the reflection technique discussed above, but it allows for a more precise interpretation of the sub-bottom structure. All of these techniques are extremely important in oceanographic studies and in the detection of possible locations of petroleum-bearing rock structures offshore.

FIGURE 2.23 The principle of echo sounding. A transmitter sends a sound wave, which is reflected back to the surface by the ocean bottom and is picked up by a receiver. By knowing the total time involved and the speed of sound in the ocean (about 1500 m/sec), water depth can be determined. The transmitter may be attached to the hull or may be towed behind the ship by a cable.

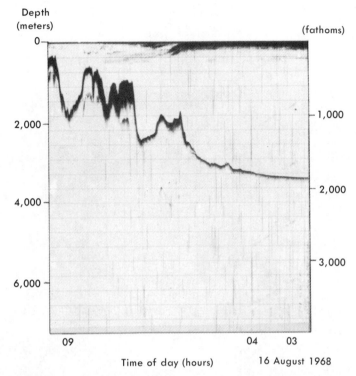

FIGURE 2.24 An echogram, a continuous record of the ocean bottom produced by a moving ship. Note the deep scattering layers near the surface. See also Figure 10.27. (From Lowrie and Eskowitz, 1969.)

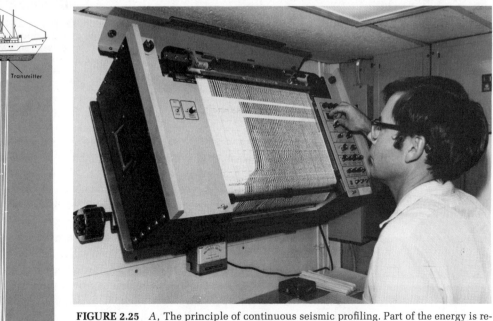

FIGURE 2.25 *A*, The principle of continuous seismic profiling. Part of the energy is reflected back by the ocean floor; the remainder continues into the ocean bottom and eventually is reflected back to the surface by different layers of sediments and rocks. *B*, The monitor of a continuous seismic profiling system on board a ship.

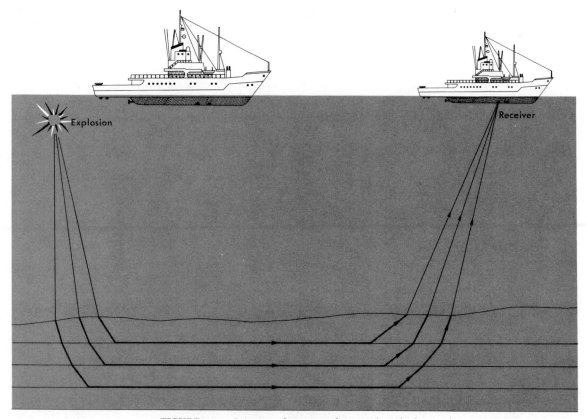

FIGURE 2.26 Seismic refraction technique, for which two ships are required. The ship on the left uses explosives or other high-energy sources to produce seismic waves. The ship on the right records the seismic waves after they have been refracted as shown. The receiving ship may be miles away from the transmitting ship.

Sampling the Ocean Bottom

Ocean-bottom samples can be obtained in various ways. The simplest devices are the *grab* and *dredge* (drag) bottom samplers (Fig. 2.27). The latter is used by a slowly moving ship, primarily to gather sediment and rock samples.

The best technique for deep sampling of the ocean bottom is by *coring.* Most coring devices involve the driving of a pipe or box into the ocean sediments (Fig. 2.28). In order to obtain deeper core samples, especially in rock, drilling equipment very similar to that used in oil production is needed. Unfortunately, drilling requires stable platforms such as those used in offshore oil drilling or onboard a specially constructed drilling ship, such as the *Glomar Challenger* (Fig. 2.29). It is one of the most expensive sampling techniques.

FIGURE 2.27 Some ocean-bottom sampling equipment. *A*, Orangepeel grab; *B*, rock dredge; *C*, anchor bag dredge. (Photos courtesy of Woods Hole Oceanographic Institution.)

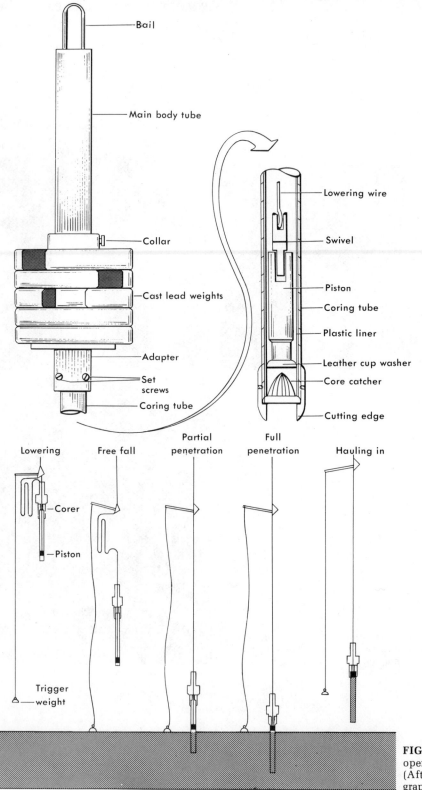

FIGURE 2.28 Structure and operation of the piston corer. (After U.S. Naval Oceanographic Office, 1968, Pub. 607.)

FIGURE 2.29 The drilling ship *Glomar Challenger*. *A*, The ship at sea; *B*, The drilling platform aboard the ship. (*A* courtesy of Deep Sea Drilling Project, Scripps Institution of Oceanography; *B* courtesy of Lamont-Doherty Geological Observatory of Columbia University.)

CONTINENTAL SHELVES, CONTINENTAL SLOPES, AND CONTINENTAL RISES

The *continental shelf* is the relatively shallow flat portion of the sea floor (Fig. 2.30) extending from the shoreline to the *shelf break,* where an abrupt increase in slope occurs. The shelf break is located at an average depth of about 130 meters. The shelves have an average slope of about a tenth of a degree (a drop of about two meters every kilometer) and appear level to the naked eye. The continental shelves constitute about 5 per cent of the earth's surface, or about one third the area of the Soviet Union. The average width of the continental shelves is about 70 km but varies considerably from place to place. For example, practically no continental shelf exists off Chile and portions of California, whereas off northern Siberia its width is about 1500 km. The sediment cover of the continental shelf is made up mostly of sandy materials, except off the glaciated coasts of higher latitudes, where glacially derived gravel-like debris frequently dominates.

Continental shelves are considered geologically to be part of the continents. Many geologists believe that the shelves were produced during the last Ice Age when the sea level was lower, perhaps at the level of the shelf break. During the glaciations of the past two million years, coastal areas could have been worn down to this new lower sea level by stream and wave erosion. When the glaciers last melted, about 15,000

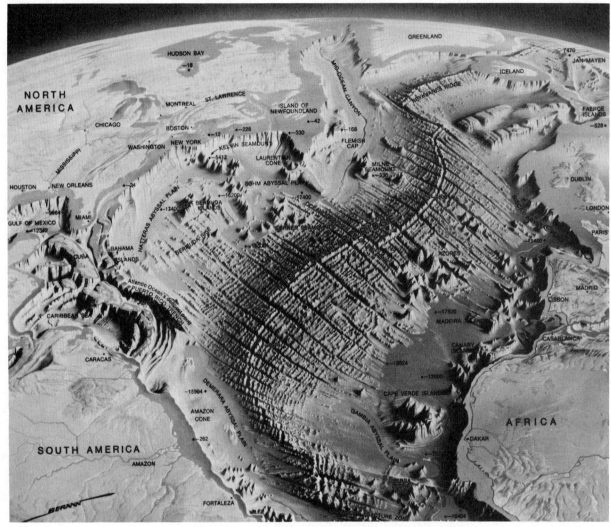

FIGURE 2.30 Artist's conception of the floor of the North Atlantic Ocean. (From a painting by Heinrich Berann, Courtesy of ALCOA.)

years ago, these coasts were submerged by the rising sea to produce the continental shelves. It is possible, however, that not all continental shelves of today were produced in this manner. Some may be the result either of deltaic sediment accumulation from river runoff or of coastal submergence.

The *continental slope* is a band with an average width of about 25 km extending from the shelf break to the deep ocean floor. It has an average slope of about 4° (a drop of about 70 meters per km), but the slope is highly variable. For example, off Santiago, Chile, the slope is approximately 45 degrees. About 60 per cent of continental slope sediment is mud, and the rest is composed of sand, gravel, rock, and organic remains. The origin of the continental slopes is still not well understood, although some may be the result of large-scale movements associated with plate tectonics.

Continental slopes do not always extend to the deep ocean floor. They are often interrupted by broad sedimentary wedges called *continental rises.* Large quantities of sediments transported across the continental shelves and slopes and deposited into the deep ocean by *turbidity currents* (discussed in the next section) as well as by submarine landslides, produce continental rises.

Submarine Canyons

In addition to the numerous submarine gullies, ravines, and troughs found throughout the continental shelves and slopes, there are many large, V-shaped, rock-walled valleys called *submarine canyons.* They are confined mostly to the continental slopes, although many, such as the Hudson Canyon, extend far into the continental shelves. They are almost perpendicular to the coast and usually have many branches. One of the largest, the Monterey Canyon off California, is deeper and wider than the Grand Canyon (Fig. 2.31). Some, such as the Hudson and the Congo Canyon, are found off river mouths.

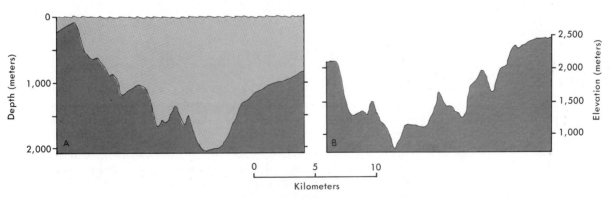

FIGURE 2.31 Comparison of the cross sections of (A) the Monterey Canyon off California and (B) the Grand Canyon of the Colorado River. (After Shepard, 1973.)

The origin of submarine canyons is still unresolved. Undoubtedly, different canyons are produced by different forces. Those found off rivers may be portions of those rivers that must have extended farther out to sea during glaciations, when sea levels were lower, and flooded by the rise in sea level after glaciation. In some cases, sinking of river mouths could be due to large-scale earth movements. However, many submarine canyons are not associated with rivers.

Another explanation for the origin of submarine canyons is erosion by *turbidity currents.* Turbidity currents are dense fast-flowing bodies of sediment-laden water. Their high density results from the large quantities of mud and sand kept in suspension by the high velocity of the current. Because of their greater density, the currents flow downslope along the bottom of the ocean, and the suspended particles, aided by the high speed of the water, are believed to be capable of cutting canyons in hard rocks. It is likely that turbidity currents could maintain as well as deepen submarine canyons that were formed by the drowning of river mouths. Turbidity currents are easy to produce in laboratory experiments, but they have not been directly observed in the oceans. It is thought that earthquakes or other distrubances can dislodge large quantities of sediments resting precariously on many continental shelves and slopes. When the sediments are mixed with water as they tumble down the slopes, turbidity currents could result. The most compelling evidence for turbidity currents comes from breaks in suboceanic telephone cables. In 1929, an earthquake in the Grand Banks area near Newfoundland resulted in the sequential breaking of several transatlantic telephone and telegraph cables (Fig. 2.32). The cables were presumably broken by a turbidity current caused by the earthquake. Analysis of the time of cable breakage (Fig. 2.33) indicates that the turbidity current would have traveled at speeds of up to about 55 knots (knots are nautical miles per hour; 1 knot equals 1.852 km per hour). Drillings in the area confirmed that sand and other shallow-water sediments were carried more than 500 km into the deep ocean. Similar events have occurred in other oceans at other times, for which turbidity currents provide a plausible explanation.

ABYSSAL HILLS AND ABYSSAL PLAINS

Large portions of the deep sea, especially in the Pacific, are characterized by numerous small extinct volcanoes, known as *abyssal hills.* These are usually only a fraction of a kilometer in height. In contrast, other regions of the deep-sea floor may be almost featureless or flat. These are called *abyssal plains.* Seismic profiling studies indicate that abyssal plains are produced by the burial of abyssal hills by sediments, including sediment brought by turbidity currents. Biological sediments, if available in large quantities, may be sufficient to bury abyssal hills and

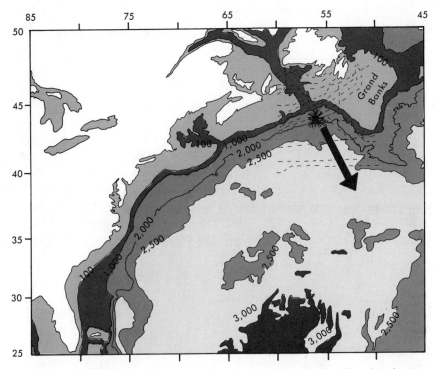

FIGURE 2.32   Breaking of the submarine telephone and telegraph cables after the 1929 Grand Banks earthquake near Newfoundland. The cables (dashed lines) were broken in a sequential manner after the earthquake, which is believed to have triggered a turbidity current, indicated by the arrow. (After Heezen and Ewing, 1952, and U.S. Naval Oceanographic Office: Oceanographic Atlas of the North Atlantic Ocean, Publication 700, 1965.)

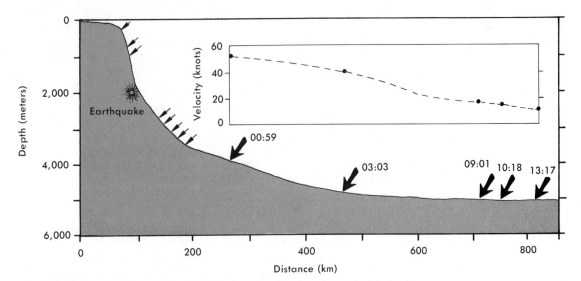

FIGURE 2.33   Cross section of the Grand Banks area, showing times of cable breakage (large arrows) after the earthquake. Small arrows indicate cables broken at the time of the earthquake. Inset shows velocity of the turbidity current as interpreted from the times of cable breakage. (Modified after Heezen and Ewing, 1952.)

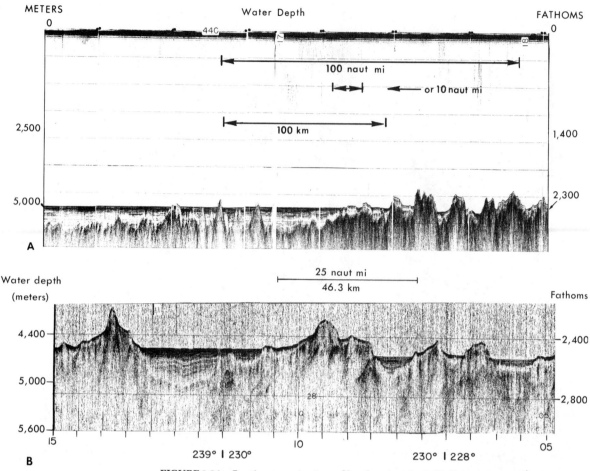

FIGURE 2.34 Continuous seismic profiles showing abyssal hills and abyssal plains along the edge of the Mid-Atlantic Ridge, in water about 5 km deep. Note the burial of abyssal hills to form abyssal plains. (*A* courtesy of Lamont-Doherty Geological Observatory; *B* from Hayes and Pimm, 1972.)

produce abyssal plains. Obviously, abyssal hills exist only in areas where insufficient sediment has been available to cover them. Figure 2.34 illustrates some abyssal hills and abyssal plains.

SEAMOUNTS, ATOLLS, AND GUYOTS

Seamounts are undersea volcanic peaks present in all oceans but most common in the Pacific (Fig. 2.35). Although probably formed in the same manner as abyssal hills, they are larger, reaching heights of more than 1 km. Some, such as the Azores and the Hawaiian Islands, are above sea level. Many seamounts, including the Hawaiian Islands, form chains, with the oldest at one end and the youngest at the other end.

Atolls (Fig. 2.36) are ringlike islands made up of coral reefs sur-

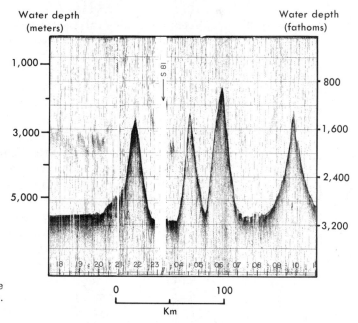

Water depth (meters)

Water depth (fathoms)

FIGURE 2.35 Continuous seismic profile across seamounts near the Mid-Atlantic Ridge. (From Hayes and Pimm, 1972.)

FIGURE 2.36 Atolls of the Tuamoto Archipelago in the Central South Pacific; as viewed from Apollo 7, 100 miles above the earth. (Courtesy of NASA.)

rounding a central lagoon. Drillings into them have indicated that the reefs are built over extinct volcanoes. The reef-building corals live in warm tropical waters (above 20° C), mostly in the upper 20 meters of the oceans (see also Chapter 10).

Fringing reefs (Fig. 2.37*A*) form when corals are attached directly to the shore, as along the coasts of Florida, the Bahamas, and numerous oceanic islands. If a coral-fringed volcano or other land mass subsides, the corals will keep building upward, forming a *barrier reef* (Fig. 2.37*B*). Barrier reefs are separated from the mainland by a lagoon or channel, as in the case of the Great Barrier Reef of Australia. When the entire volcano slowly submerges below sea level, an atoll results (Fig. 2.37*C*). In 1842, Charles Darwin first developed this theory of atoll formation. It is still an accepted explanation for Pacific atolls. Many of the coral reefs of the West Indies and the Great Barrier Reef of Australia, however, appear to have grown upward with the rising sea level following the Ice Age.

Guyots are flat-topped seamounts. Although evidence indicates that the flat surfaces are produced by wave erosion, they are found at various depths. Some are more than 2 km below the sea surface. Apparently these guyots were submerged after the flattened surface was produced. Lowering of sea level during the last glaciation is not a sufficient explanation for most guyots, as their surfaces are at depths far greater than the postulated lowering of sea level during the glaciation. Subsidence due to their own weight may be the answer in some cases. Some guyots may have been formed when an atoll drifted to a colder climate or subsided so rapidly that the corals were killed (Fig. 2.37*D*).

Plate tectonics can explain the origin and distribution of many of the features discussed above. For instance, we have seen that the top surfaces of guyots are found at various depths. Ocean depths increase away from the mid-ocean ridges owing to cooling and subsidence of the lithosphere as it moves away from the ridges (see Fig. 2.30). Guyots, which were formed near the mid-ocean ridges, can be visualized as "riding" on the ocean bottom as a result of sea-floor spreading, to be distributed eventually at various depths. This explanation does not require an additional mechanism to account for their sinking. A similar theory can be extended to the origin of many atolls.

Volcanoes created near the mid-ocean ridges should move away on the spreading sea floor and should become inactive. Subsequently, new volcanoes should be created in their place. The chainlike arrangement of many seamounts and islands, with the youngest and active ones near the ridges and the oldest and inactive ones farther away, may be explained by sea-floor spreading. However, not all such features follow this pattern. For example, the active volcanoes of the Hawaiian Islands are far away from the ridge in the Pacific, constituting a serious weakness in the theories of sea-floor spreading and plate tectonics. The formation of

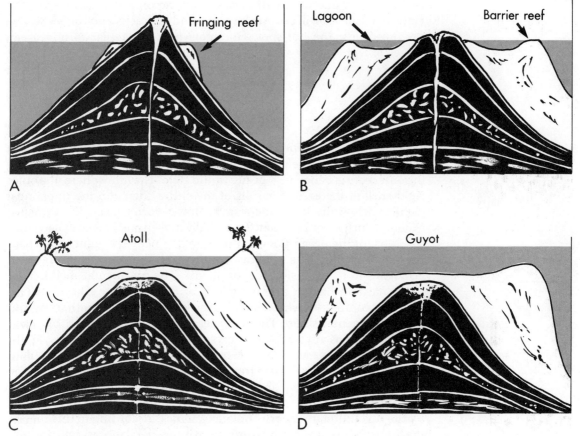

FIGURE 2.37 *A–C*, Origin of an atoll resulting from subsidence of a volcano and the upward growth of a coral reef. *D*, Formation of a guyot by the rapid subsidence of an atoll. (After Shepard, 1973.)

many such volcanoes may be explained in terms of *hot spots,* huge vertical pipelike regions within the mantle, where the materials are believed to be hotter, perhaps molten or partially molten, than in surrounding regions. Several hot spots are believed to exist within the mantle, one of them near the Hawaiian Islands. As the lithosphere moves over the hot spot, volcanoes are produced. As sea-floor spreading continues, the volcanoes move away from the hot spots, and new volcanoes are produced in their places. Thus, according to this theory, the Hawaiian Islands will cease to be active and new active volcanoes will be produced near the hot spot. Hot spots may be similar to mid-ocean ridges except that they are localized rather than extended.

SUMMARY

Seismic studies indicate that the major internal structures of the earth are the crust, the mantle, the outer core, and the inner core. Only the outer core is believed to be in a liquid state. The lithosphere, the outer rigid shell of the earth, consists of the entire crust and the uppermost mantle. Below the lithosphere is the low-velocity layer (the asthenosphere) which is considered to be partially molten and plastic in behavior.

Most of the ocean-floor features are distinctly different from those present on land. The shallow continental shelves are thought by geologists to be extensions of continents. The continental shelf is relatively flat and has an average width of about 70 km. The transition between continental shelves and deep-ocean floor is the narrow but steep continental slope. The deep-sea floor is characterized by many features, such as abyssal plains, abyssal hills, seamounts, guyots, atolls, mid-ocean ridges, and oceanic trenches. The age of the ocean floor progressively increases away from the mid-ocean ridges, but no rocks on the sea floor older than 200 million years have been found.

The origin of most of the ocean-floor features can now be explained by the revolutionary theory of plate tectonics. According to this theory, the lithosphere of the earth is broken up into several pieces, called plates, by convection currents in the mantle rising toward the mid-ocean ridges. The plates have been moving away from the mid-ocean ridges (spreading centers) at an average speed of about 2 cm/yr. The present boundaries of plates are determined based on the presence of mid-ocean ridges, oceanic trenches, mountains, volcanoes, and earthquake zones. The leading edges of plates, if they contain continents, may collide to form mountains. If the leading edge of a plate does not have a continent on it, it may descend deep into the mantle near an oceanic trench. This process is known as subduction and helps explain the absence of ocean floor rocks older than 200 million years.

The major evidence in support of plate tectonics comes from paleomagnetism. Paleomagnetic studies have revealed that the earth's magnetic field has reversed numerous times in the past; that is, the north and south magnetic poles have switched places with each other. These reversals cause alternating normal and reversed stripelike magnetizations of the ocean-floor rocks parallel to the mid-ocean ridges. Such magnetic stripes are unique to the ocean floor. Paleomagnetic studies have enabled scientists to determine the rate of sea-floor spreading and have provided a new technique for determining the age of the sea floor.

IMPORTANT TERMS

Abyssal hills
Abyssal plain
Asthenosphere
Atoll
Barrier reef
Continental rise
Continental shelf
Continental slope
Continuous seismic profiling
Crust
Deep scattering layer
Dredge
Dust-cloud hypothesis
Echogram
Echo sounding
Fathogram
Fringing reef
Gondwanaland
Grab
Guyot
Hot spot
Inner core
Isostasy

Laurasia
Low-velocity layer
Mantle
Mid-ocean ridge
Moho
Mohole Project
Oceanic trench
Outer core
Paleomagnetism
Paleontology
Pangaea
Plate
Plate tectonics
P wave
Rift valley
Sea-floor spreading
Seamount
Seismic refraction
Subduction zone
Submarine canyon
S wave
Transform fault
Turbidity current

STUDY QUESTIONS

1. Why do seismologists believe that the outer core of the earth is liquid and that the inner core is probably solid?
2. What non-seismological evidence can be offered in support of the solid inner core theory?
3. Why was an oceanic site selected for the Mohole Project?
4. Describe the continental drift theory of Wegener.
5. How may the boundaries of plates (Fig. 2.18) be determined? Refer to Figures 2.10 and 2.20. Shade in the Indian and Nazca plates on Fig. 2.20
6. Refer to the age map of the ocean floor (inside jacket cover). Assume a distance of about 4000 km from North America to the Mid-Atlantic ridge. What is the approximate rate of movement (in centimeters per year) of North America in relation to the Mid-Atlantic Ridge?
7. Discuss the implications of plate tectonics for the diversity and distribution of life in the sea.
8. Explain the origin of island chains, such as the Hawaiian Islands, which are far from mid-ocean ridges, trenches, and continents.
9. Discuss the possible role of turbidity currents in the formation of submarine canyons.
10. The flat tops of guyots are found at various depths in the sea. How can this be explained?
11. Describe the origin of atolls.
12. How might an atoll be transformed into a guyot?
13. Compare P waves and S waves.
14. What is the dust-cloud hypothesis?
15. What is the difference between sea-floor spreading and plate tectonics?
16. What is the major proof in support of sea-floor spreading? Explain.

SUGGESTED READINGS

Beloussov, V. V. 1979. Why Do I Not Accept Plate Tectonics? *Transactions of the American Geophysical Union, 60*:207–11.

Bullen, K. E. 1955 (September). The Interior of the Earth. *Scientific American.*

Calder, Nigel. 1972. *The Restless Earth.* New York, Viking/Compass.

Cox, Allan (ed). 1973. *Plate Tectonics and Geomagnetic Reversals.* San Francisco, Freeman.

Emory, K. O. 1969 (September). Continental Shelves. *Scientific American.*

Heezen, Bruce C. 1956 (August). The Origin of Submarine Canyons. *Scientific American.*

RISE Project Group. 1980. East Pacific Rise: Hot Springs and Geophysical Experiments. *Science, 207*:1421–1433.

Rona, Peter A. 1973 (July). Plate Tectonics and Mineral Resources. *Scientific American.*

Sengör, A. M. C. and Kevin Burke. 1979. Comments on: Why Do I Not Accept Plate Tectonics? *Transactions of the American Geophysical Union, 60*:207–11.

Shepard, Francis P. 1973. *Submarine Geology,* 3rd ed. New York, Harper & Row.

Uyeda, Seiya. 1978. *The New View of the Earth: Moving Continents and Moving Oceans.* San Francisco, Freeman.

——————————. 1979. Subduction Zones: Facts, Ideas, and Speculations. *Oceanus, 22(3)*:52–62.

Valentine, James W. and Eldridge M. Moores. 1974 (April). Plate Tectonics and the History of Life in the Oceans. *Scientific American.*

Vine, Frederick J. 1966. Spreading of the Ocean Floor: New Evidence. *Science, 154*:1405–15.

Wegener, Alfred. 1966. *Origin of Continents and Oceans.* New York, Dover Publications. (Translation of the original German edition of 1929.)

Ocean Sediments

The world's oceans hide 71 per cent of the earth's surface from view. If we could drain the oceans, we would see vast areas covered mostly by very fine-grained, muddy sediments. The sediments include the unconsolidated material of the beaches, salt marshes, bays, and ocean floor. They comprise particles that have settled to the bottom. Many sediments are derived from the land and are brought to the oceans by streams, wind, and glaciers. Others are skeletal remains of organisms that lived in the oceans, and a small fraction (meteorite dust) is even extraterrestrial in origin.

Marine sediments are not only the home for a myriad of marine organisms but the site for oil and gas accumulation. Furthermore, sea floor sediments include vast stores of mineral wealth waiting for exploitation.

Sediments provide important clues to the origin and history of the oceans. Fossils in the sediments can be used to determine the age of the sediments, which in turn may tell the age of the ocean floor itself. They also preserve a detailed record of past climatic and chemical history of sea water, a record which is just beginning to be investigated in detail.

The piston corer, a device used to sample seafloor sediments. (Courtesy of Woods Hole Oceanographic Institution.)

3

CLASSIFICATION AND DISTRIBUTION OF SEDIMENTS

Ocean sediments may be classified variously, according to their origin (source), size, chemical composition, or place of deposition. Figure 3.1 shows the general distribution of sediments on the ocean floor. The following is a simple classification of ocean sediments based on their origin and chemical composition:

    (1) Terrigenous (derived from land)
    (2) Pelagic (formed in the sea by marine processes)
      (a) Biological
        (i) Calcareous ooze
        (ii) Siliceous ooze
      (b) Inorganic
        (i) Red clay
        (ii) Chemical precipitates

Terrigenous Sediments

Terrigenous sediments are mostly the result of land erosion. They enter the sea by way of rivers, wind, glaciers, and wave action. Consequently, these deposits tend to be most abundant over the continental shelves, especially near major rivers. Land-derived sediments consist essentially of the same materials found near the surface of the earth. They

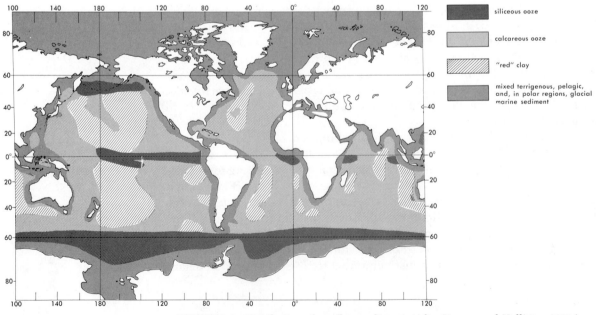

FIGURE 3.1   Distribution of sea-floor sediment. (After Heezen and Hollister, 1971.)

are mostly sand (1/16 to 2 mm in size, usually composed of the minerals quartz or feldspar), silt (1/16 down to 1/256 mm, and the same composition as sand), and clay (less than 1/256 mm and composed chiefly of hydrous aluminum silicates, usually derived from the weathering of igneous rock). In addition, terrigenous sediments include airborne dust and volcanic ash. For example, airborne dust from the 1973 drought in northern Africa was detected as far away as the Caribbean area. Some of the volcanic ash ejected by the 1980 eruptions of Mount St. Helens in the state of Washington has been deposited near the coasts of Oregon and Washington. Some of the finest ash particles have even been carried around the globe.

Because heavy (usually large) particles sink faster than light ones, heavier materials tend to settle to the bottom first and, therefore, closer to the source. Thus, in general, terrigenous materials are coarser near shore and finer away from the shore. Of course, currents can change these distributions. Turbidity currents are capable of transporting even larger particles over greater distances, and the fine sediments on the deep ocean floor may be disturbed by currents moving over them. A current flowing swiftly through a channel picks up the finer particles (leaving the coarser ones) and carries them until the channel widens and the current slows in the less restricted water. Therefore, channels with swiftly flowing water tend to have coarser sediments or exposed rock, whereas sheltered coves frequently become mud flats, characterized by fine sediments. Sediments accumulate and erode more quickly along the beach than in any other part of the ocean environment. In a few areas as much as two to three meters may accumulate in a few months.

Pelagic Sediments

Pelagic sediments, resulting from marine processes, can be either biological or inorganic. Sediments containing 30 per cent or more organic remains are considered *biological sediments.* Most are very fine and are called, appropriately, *oozes.* Some are made of calcium carbonate (calcareous) and others of a glassy form of hydrated silicon dioxide (siliceous).* Calcareous oozes are generally absent in depths greater than about 5000 meters, owing to increased solubility of calcium carbonate under high pressure and cold temperature. Figure 3.2 shows the skeletons of some organisms that constitute most of the biological sediments.

All of the organisms listed in Table 3.1, except the pteropods, are single-celled and quite small. The plant members are frequently less than a tenth of a millimeter in diameter. Siliceous radiolarians and calcareous foraminiferans are usually less than 1 mm in diameter. The calcareous pteropods are larger, about 1 cm in length, but their shells are thin and break into pieces easily.

---

*Siliceous oozes have the same composition as the gemstone opal.

FIGURE 3.2   Skeleta of minute marine organisms accumulate on the sea floor to form sediment. *A*, Radiolaria, *B*, Coccoliths, broken diatom shells and a radiolarian "cage". *C*, Chalk cliffs of Dover, composed of about 90 per cent coccoliths. *D*, Foraminifera. (*A* courtesy of Scripps Institution of Oceanography; *B* courtesy of Susumu Honjo, Woods Hole Oceanographic Institution; *C* courtesy of British Tourist Authority; *D* from Levin, 1978.)

Table 3.1   MAJOR SOURCE AND CHEMICAL COMPOSITION OF BIOLOGICAL SEDIMENTS

| | *COMPOSITION* | |
SOURCE	*Siliceous (SiO₂)*	*Calcareous (CaCO₃)*
Plants		
Diatoms	x	
Silicoflagellates	x	
Coccolithophores		x
Animals		
Radiolaria	x	
Foraminifera		x
Pteropods		x

PLATE 1 Smoking craters on the Hekla volcano, Iceland, on the Mid-Atlantic Ridge. (Photo courtesy Scandinavian National Tourist Offices)

PLATE 2 Volcanic fireworks in Iceland. (Photo courtesy Scandinavian National Tourist Offices)

PLATE 3 Fissure on the Mid-Atlantic Ridge. (Photo courtesy Woods Hole Oceanographic Institution [WHOI])

PLATE 4 Giant tube worms in the area of the Galapagos Rift thermal vents. (Photo by Kathleen Crane, courtesy WHOI)

PLATE 5 "Pillow" lava formations on the Mid-Atlantic Ridge. (Photo courtesy WHOI)

PLATE 6 The Deep Submersible Research Vessel *Alvin.* (Photo by John Porteous, courtesy WHOI)

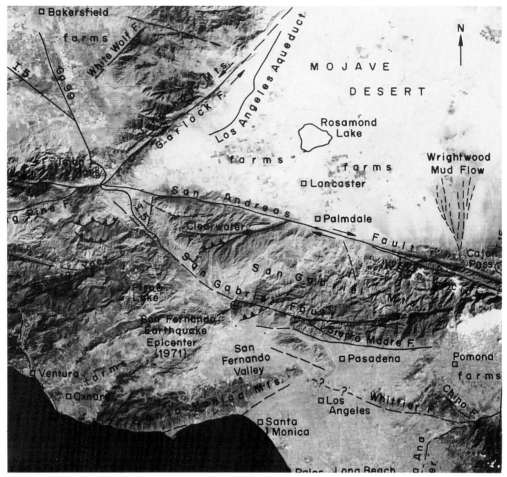

PLATE 7   The San Andreas Fault, in California, the boundary between two lithospheric plates. The Pacific Plate, west of the San Andreas Fault, is moving toward the northwest relative to the American Plate. (Photo courtesy NASA)

PLATE 8   Offshore oil-drilling rig in the Gulf of Mexico. (Photo courtesy American Petroleum Institute)

PLATE 9   The ocean drilling ship *Ben Ocean Lancer.* (Photo courtesy American Petroleum Institute)

PLATE 10   The supertanker Chevron *Nagasaki.* (Photo courtesy Standard of California/American Petroleum Institute)

PLATE 11   A liquefied natural gas (LNG) tanker. (Photo courtesy U.S. Department of Energy)

PLATE 12   A vessel designed for the recovery of spilled oil. The yellow arms direct the oil toward a skimmer in the boat's bow. (Photo courtesy American Petroleum Institute)

PLATE 13   Oil containment with a flexible floating boom. (Photo courtesy American Petroleum Institute)

PLATE 14   An oil spill from the tanker *Ocean Eagle.* (Photo courtesy U.S. Coast Guard)

PLATE 15   A color-enhanced infrared satellite picture of the Gulf Stream. Blue is cooler water; red is warmer water. (Photo courtesy NOAA)

PLATE 16   Artist's rendering of a satellite used to photograph the Earth's surface. (Photo courtesy U.S. Department of Energy)

Diatoms are among the most common of marine plants. They are so important as food for herbivores that they are sometimes called the "grass of the sea." Their siliceous glassy shells settle to the bottom as diatomaceous ooze. In some areas where diatoms formerly were productive, they have formed deposits of *diatomaceous earth* hundreds of meters thick. This rock is used to strengthen concrete, to give a flat finish to paint, and in filters to clarify beer. Thick deposits of diatomaceous earth on land, such as at Lompoc, California, indicate that the area was once covered by the sea. Silicoflagellates also are silica-producing plants. They have the same general distribution as diatoms but are much less abundant. Siliceous oozes tend to be concentrated in highly productive areas that are far enough from shore to avoid masking by terrigenous deposits.

Coccolithophores may be considered plantlike because they contain chlorophyll. Coccolithophore sediments are composed of microscopic (about 0.01 mm in diameter) calcareous buttons, or coccoliths, that once covered the organism's single-celled body and which compose the bulk of most chalk sediments. The white cliffs of Dover, England, are composed of more than 90 per cent coccoliths.

Radiolarians and foraminiferans, both protozoans (single-celled animals), feed by the use of thin projections or *pseudopodia,* onto which dead organic matter (detritus) and bacteria get stuck. The animal then draws the pseudopodia into the cell and digests the food. Radiolarians have internal siliceous skeletons that accumulate to produce siliceous oozes in equatorial oceanic regions of relatively high productivity. Foraminifera also have external calcareous shells and in some areas have produced thick deposits of limestone.

Pteropods are multicelled animals whose shells occur in oozelike sediment. They are closely related to snails and have a calcareous shell. Pteropods keep themselves from sinking by "treading water," actively flapping their winglike feet—thus their common name of "sea butterfly." They may be locally abundant, and when they die, they contribute to calcareous sediment, especially in the shallower mid-ocean ridge areas of the Atlantic Ocean.

In shallow near-shore areas, shells and skeletal remains of clams, snails, sponges, coral, foraminifera, and calcareous algae contribute to the formation of loose sand, which may be cemented together to form sedimentary limestone. Some corals and calcareous algae form large wave-resistant masses called reef limestone.

The *inorganic pelagic sediments* contain less than 30 per cent organic remains and are found in the deepest portions of the ocean floor, and in areas of low biological productivity. Solution in the oceanic water column destroys most calcareous skeletons before they reach the bottom. Such sediments include *red clays,* which in most cases are brown in color. These sediments are very fine, usually between 1 and 50 micrometers* in diameter. Although their origin is not well understood, they

---

*A micrometer is a millionth of a meter.

FIGURE 3.3 Manganese nodules. *A,* Nodules on the sea floor; *B,* The R. V. *Prospector,* a deep-sea mining ship; *C,* Emptying a dredgeload of manganese nodules on board the ship. *D,* Growth rings of a manganese nodule are shown in this cross section, magnified about five times. Scientists at Scripps Institution of Oceanography, University of California, San Diego, believe the rings are created by preserved tubes, built by marine organisms for shelter, and are, therefore, related to biological activity on the nodule's surface during its growth. Scientists have found such tubular structures both inside and on the surface of most nodules studied. Indications are that the nodules may owe their existence to these organisms. (*A* courtesy of Smithsonian Oceanographic Sorting Center; *B* and *C* courtesy of Deep-Sea Ventures, Inc.; *D* courtesy of Scripps Institution of Oceanography.)

include the finest fraction of terrigenous clay (about 1 micrometer in diameter), meteorite dust, and inorganic precipitates. They may even contain very fine but unrecognizable biological sediments.

In addition to the sediments mentioned above, the ocean also contains materials that are formed from chemicals dissolved in water and deposited on the sea floor as *chemical precipitates.* These include submarine *phosphorite deposits* that are found as nodules, sands, mud, or rocks on the continental shelves and slopes. They are especially abundant off the coasts of southern California, southeastern United States, South Africa, and Australia.

It is thought that phosphorite deposits are formed by the precipitation of phosphate produced by the decay of dead organisms falling to the sea floor from highly productive waters above. Many of these deposits are concentrated enough to warrant consideration for mining, because phosphate can be used to produce phosphate fertilizer.

Another important inorganic substance formed in the sea is the *manganese nodule,* found in all oceans. Manganese and phosphorite nodules were initially discovered during the Challenger Expedition in the 1870's. Although manganese nodules form only a very small fraction of ocean sediment volume, they may be of great economic importance in the future. They are chemically precipitated from sea water as coatings of minerals rich in manganese, around objects such as rock particles and shark teeth. These nodules are usually less than 25 cm in diameter (Fig. 3.3). Their exact mode of origin is not well understood. A number of current theories involve biological activity of organisms such as bacteria, foraminifera, and tube-dwelling organisms, which attach to submerged objects such as rock particles and shark teeth. Feces of microcrustaceans, including copepods and shrimps, are rich in trace metals. Some of the fecal pellets sink to the sea floor where they become food for bottom-dwelling life. Bacteria and foraminifera may help to extract metals from the feces and sea water, accelerating their precipitation onto the nodules. Sinking feces may provide a significant route for the transportation of trace metals from the surface to the deep sea.

Manganese nodules grow very slowly, about 2 to 4 mm per million years. This rate is about one thousandth the rate of sediment accumulation. But mysteriously, many nodule deposits remain exposed on the sea floor rather than being buried.

Included in the manganese nodules are copper, cobalt, nickel, and various other valuable substances. The legal aspects of mining these resources in international waters have been discussed in Chapter 1.

CHARACTERISTICS OF SEDIMENTARY PARTICLES

Grain Size, Shape, and Density

In the previous section, sediments were classified on the basis of origin and chemical composition. They may also be classified on the basis of grain size, shape, and density. These characteristics are not only important in describing sediments, but also provide important clues about the history of these particles before deposition. In addition, these characteristics influence the nature of the substrate that bottom dwelling organisms inhabit.

Table 3.2. CLASSIFICATION OF SEDIMENTS BY PARTICLE SIZE

| | DIAMETER | |
PARTICLE	*Millimeters*	*Micrometers*
Boulder	>256	
Cobble	64–256	
Pebble	4–64	
Granule	2–4	2000–4000
Sand	1/16–2	62.5–2000
Silt	1/256–1/16	3.9–62.5
Clay	<1/256	<3.9

Table 3.2 is a simplified classification of sediments by particle size. It is important to note that terms such as sand and clay refer to the size of the particles rather than chemical composition.

The shape of sediment particles may also vary considerably. They may be angular, rounded, or spherical in shape. *Roundness* refers to the curvature of the corners or edges of the particle, and *sphericity* refers to how closely it resembles a ball. Both are the result of wear and tear the particles have undergone before reaching their final destination. These shapes usually imply a history of long-distance travel or, as in the case of some beach sands, repeated attack by wave action. Particle shape determines settling velocity as well as the packing of sediments, and will be discussed later.

Table 3.3 DENSITIES OF SOME SEDIMENTARY PARTICLES

MINERAL	*DENSITY*
Clay	2–2.6
Typical Beach Sands	
Quartz	2.65
Feldspar	2.6–2.75
Diamond	3.5
Black Sands	
Titanium or thoriumrich	
black minerals	4.2–5.1
Magnetite (iron oxide)	5.2
Silver	10.5
Gold	19.5

The mineral particles that make up the sediments differ greatly in their densities. Table 3.3 gives the densities of some sedimentary particles. When particles of varied densities enter the ocean, the settling velocity as well as the distance they travel will depend on their densities. If particles are of uniform size and shape, the heavier ones settle first and the lighter ones settle later, or are carried greater distances. Frequently, especially near shore, this results in an easily discernible sedimentary layering of dark (usually denser) and light-colored particles. Some of these heavy dark minerals (black sands) are rich in iron; others may be sources for such rare elements as titanium and thorium. In certain areas of India and Brazil, these deposits are mined, since they often are concentrated near beach areas by wave action.

Porosity and Permeability

Porosity refers to the amount of interstitial space (open space) within the sediment and is expressed as a percentage of its total volume. *Permeability* is a measure of the ease with which water can pass through the sediment. It is possible that a sediment may have a very high porosity but a low permeability. Low permeability implies that many of the open spaces within the sediment are not interconnected, thus preventing the flow of liquids through it. A good example of a sediment with high porosity and low permeability is mud. Mud deposits consisting of flaky clay particles tend to be highly porous (up to 75 per cent porosity) because the microscopically thin flakes are separated by water-filled spaces thicker than the flakes themselves. The permeability of mud, however, is low due to restriction of the circulation of water through the interstitial spaces. Beach sand, on the other hand, may be highly porous and very permeable.

Porosity is influenced greatly by the shape and degree of sorting, as well as the packing of the particles. *Sorting* refers to the uniformity of particle sizes in a deposit. Sediments composed of uniform size and shape (rounded or angular) have a higher porosity than those having mixed sizes and shapes, because smaller particles may occupy the spaces between larger particles. Similarly, porosity tends to increase with decreasing sphericity. *Packing* refers to the pattern of spatial arrangement of particles (Fig. 3.4). For example, spheres stacked one upon another and side by side results in the loosest form of packing (cubic packing). The same particles can occupy a smaller space and are more stable if they are packed in an offset manner (rhombohedral packing). Most sedimentary particles in nature tend toward rhombohedral packing. The porosity of uniform spherical particles is reduced from 47 per cent to 26 per cent as rhombohedral packing is approached. Beach sands with irregularly shaped particles may have nearly rhombohedral packing, but with a porosity of about 40 per cent.

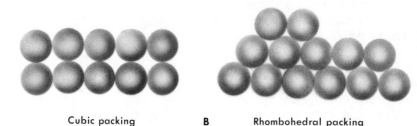

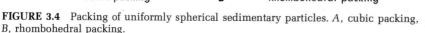

A   Cubic packing   **B**   Rhombohedral packing

FIGURE 3.4 Packing of uniformly spherical sedimentary particles. *A*, cubic packing, *B*, rhombohedral packing.

The interstitial spaces between particles are home to many microscopic organisms, including single-celled protozoans and many-celled, but minute, worms and crustaceans (collectively known as interstitial fauna). The spaces are filled with water, and if the permeability is high, water circulates through the sediment, carrying oxygen and fine organic food to the interstitial organisms. Sediments with low porosity and permeability have little water circulation and generally are oxygen-poor and harbor few species of organisms other than some specialized worms and bacteria. These sediments are frequently black and smell like rotten eggs due to the production of hydrogen sulfide by certain bacteria. Hydrogen sulfide is toxic to many marine organisms, which consequently avoid the black oxygen-poor sediments.

Sediments contain varying amounts of organic matter, including dead and decaying organisms as well as small particles of feces. This organic material provides food for various burrowing organisms, especially worms that ingest the sediment and digest the organic matter therein. Clearly, porosity and permeability of sediments are required for the organisms' movement and respiration. These organisms leave recognizable trails and cause some mixing of sediment. Some even back up to the opening to their burrow and dump their sandy or muddy feces on the sediment surface.

Settling of Sediments

The settling velocity of a particle of sediment depends primarily on its size and density, with larger and denser particles settling at higher rates. The shape of the particle also has an effect, especially for those larger than 1 mm in diameter (Fig. 3.5). For example, a sphere (representing sand), 2 mm in diameter will settle in quiet water at about 30 cm/sec (about 1000 meters per hour). However, disk-shaped particles of the same volume and density will settle at a rate of only about 14 cm/sec. Obviously, sphere-shaped particles may reach the ocean floor soon after entering the sea. On the other hand, particles 0.005 mm in diameter (typical of clays) will settle at a rate of 0.02 cm/sec (1000 meters in two

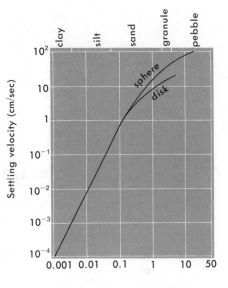

FIGURE 3.5 Settling velocity in relation to particle diameter for spheres and disks. (After Shepard, 1973.)

Particle diameter (mm)

months). This suspended sediment may be carried far out to sea before reaching the sea floor. The 0.005 mm particles could be carried more than 5000 km (more than 3000 miles) by a horizontal current such as the Gulf Stream flowing at 100 cm/sec.

For particles smaller than fine sand, the settling velocity is proportional to the square of the diameter and to the difference between its density and that of the water. This relationship is known as Stokes' Law.* Friction is considered to be negligible for very fine particles. In fresh water, flaky clay particles (less than 4 micrometers in diameter) are negatively charged and tend to remain dispersed. As clay enters the ocean at a river mouth, however, the charges are neutralized by positive charges, such as those of sodium and magnesium, in sea water. The particles tend to clump together and sink as if they were much larger than a single particle. Thus large clumps of clays may settle in coastal areas where relatively strong currents are present, to form mud deposits.

The turbulence caused by the travel through the water of larger and, especially, more irregular particles causes increased friction, and these

---

*Stokes' Law:

$$V = d^2 \, (\rho_s - \rho_w) \, g/18\eta$$
where V = settling velocity
   d = diameter of the particle
   ρ_s = density of the sedimentary particle
   ρ_w = density of the water
   g = acceleration due to gravity
   η = viscosity of the water which increases with decreasing temperature.

particles sink slower than the rate predicted for a sphere. For particles greater than about 0.1 mm in diameter, the settling velocity is proportional to the square root of the diameter and settling does not obey Stokes' Law.

Viscosity (the resistance to flow) of water has an important effect on the settling of very fine particles, although the effect is negligible for particles larger than sand. A 0.005 mm particle would settle at a rate of 0.02 cm/sec at 20° C, but at only 0.01 cm/sec in more viscous 4° C water, the average temperatures of deep ocean water.

SEDIMENTATION

Transportation, Deposition and Erosion

Since the ocean is always in motion, the fate of a sedimentary particle depends on the speed of ocean currents. Figure 3.6 illustrates the relationship between the particle size and erosion, transportation, and deposition of sediments under varying current speeds along the sea floor.

The current speeds necessary to transport ocean sediments increase with increasing particle size. As can be seen from Figure 3.6, the current speeds needed to cause erosion of bottom sediments increase with increasing particle size. For cohesive materials smaller than sand, however, the current speed necessary to erode the bottom sediment increases with *decreasing* particle size. This latter relationship results from the greater cohesive force that exists between smaller particles, especially the flaky

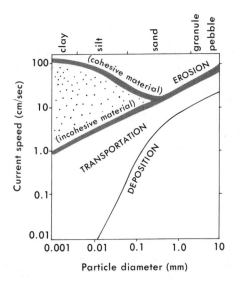

FIGURE 3.6   The relationship between particle size and erosion, transportation, and deposition of sediments under varying current speeds. (After Heezen and Hollister, 1971.)

FIGURE 3.7   Ripple marks on the deep sea floor indicating erosion by deep ocean currents. (Courtesy of NSF-USARP.)

clay particles. Very fine (less than 0.005 mm) non-cohesive particles such as some biologically derived oozes may be eroded and transported by very weak currents, even as slow as 1 cm/sec.

For speeds slower than that required for transportation, particles will remain on the sea floor or become deposited. Tidal and other wave generated currents near the shore transport as well as erode sediments near the sea floor. These currents may have speeds of up to 200 cm/sec (4 knots) or more. Although deeper ocean currents are generally too slow (only a few centimeters per second) to cause appreciable erosion of bottom sediments, bottom current speeds up to 50 cm/sec have been recorded with neutrally buoyant floats and moored current meters. Many parts of the deep ocean floor are marked by ripples and scars (Fig. 3.7); providing evidence of strong bottom currents.

Ripple marks may be formed on the sediment if the bottom current speed is 5 to 15 cm/sec, but currents as slow as 1 cm/sec may cause erosion of the finest marine oozes. Currents over 50 cm/sec can cause large scale erosional features. In water about 500 meters deep at the entrance to the Mediterranean Sea, currents in excess of 80 cm/sec have left the sea floor rocky and mostly free of sediments.

Rates of Sediment Accumulation

Sediments are important to us because they tell much about the history of the oceans (Fig. 3.8). From the kinds of fossils found in various layers, paleontologists may be able to tell not only the age of the layer but

FIGURE 3.8 A sediment core being examined on board ship. (Photo by K. N. Sachs, courtesy of Woods Hole Oceanographic Institution.)

also something about the environment (including climate) that existed when the sediments were deposited. Sediment cores have also revealed that sediment thickness increases away from the axes of the mid-ocean ridges (Fig. 3.9). This finding is consistent with the concept of plate tectonics, which states that the sea floor farther from the ridge is older than the sea floor near the ridge, and thus has had more time to accumulate sediments (see inside back cover). The oldest sediments in the oceans are younger than 200 million years and are found farthest away from the mid-ocean ridges, implying that any older sediments which may have been present were carried deep into the oceanic trenches and recycled. Of course, many parts of the oceans are not bordered by trenches, which indicate that these ocean basins are younger than 200 million years.

Sediments accumulate at rates ranging from less than 0.5 cm to more than 10 cm of thickness per 1000 years. Oceanic red clays accumulate at rates of about 0.2 mm to 2 mm per 1000 years. Many biological sediments tend to accumulate at a rate of about 2 cm per 1000 years (Fig. 3.10), or 20 meters of sediments in a million years. Higher rates of sediment accumulation, however, may occur in areas of high biological productivity, and

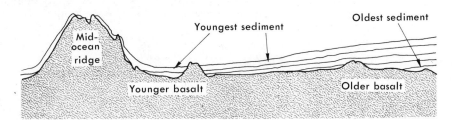

FIGURE 3.9 The thickness of ocean floor sediments increase away from the mid-ocean ridges. The age of the deepest sediments and the rocks upon which they lie also increase away from the ridges.

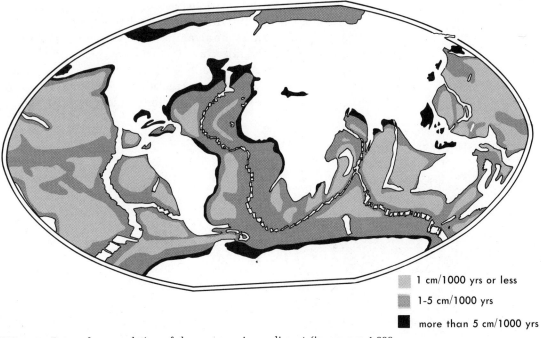

1 cm/1000 yrs or less

1-5 cm/1000 yrs

more than 5 cm/1000 yrs

FIGURE 3.10 Rates of accumulation of deep-sea marine sediment (in cm per 1,000 years). (After Heezen and Hollister, 1971.)

this rate of production of marine life may have varied tremendously over periods of millions of years. Petroleum geologists make use of this concept in estimating the past productivity of an area as an indication of the potential accumulation of petroleum.

CORAL AND OTHER REEFS

Geological features of the shallow-sea floor may be constructed by living organisms. Some of these, such as reefs built by some tube-dwelling worms (Fig. 3.11), use sedimentary particles as a source of building material. Others, such as coral reefs, are formed from dissolved chemicals in sea water. These reefs are a source of sedimentary particles after erosion by wave action and activities by marine organisms such as boring sponges, worms, snails, and parrot fish.

Coral reefs (Fig. 3.12) are the result of chemical precipitation, largely of calcium carbonate, by the living coral. The reef-building corals somewhat resemble sea anemones embedded in cement. They are carnivores, feeding on minute animals that drift by. Reef corals also harbor algae within their tissues. The algae use inorganic nutrients in the presence of light to make organic molecules, some of which leak to the coral and help nourish it. Although the relationship between the coral and algae is almost universal in reef-building coral, the benefits of the relationship are not clearly understood. The algae may receive nutrients and a place to live from the coral, and the coral in turn may receive some organic chemicals, some food, and perhaps vitamins from the algae. Because the algae need light to make necessary compounds, the coral grows only in the well-lighted surface waters, usually limited to the upper 20 meters of

FIGURE 3.11 Reef built by worms that cement sand grains together to form tube-like homes. Each dot in the photo represents the opening of one tube. (Courtesy of H. Gray Multer.)

A

FIGURE 3.12 Coral reef constituents. *A*, Living coral off Key Largo, Florida; *B–E*, Coral skeleta. (*A* courtesy of NOAA, *B–E* from Villee, Walker and Barnes, 1973.)

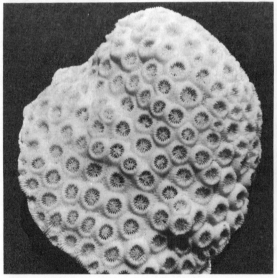

B

C

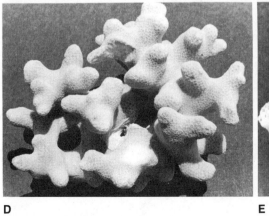

D

E

the tropical ocean. The living coral is found only as a thin veneer exposed to the sea. Deeper down, the rest of the reef is composed of dead skeleta of previous generations of coral.

Also contributing to the coral deposits of the reef are many calcareous plants, including some red and green algae. These plants are very common on and around the coral reef. They extract calcium carbonate from sea water and deposit it in and around their tissues, giving them strength and rigidity.

Coral reefs may become hundreds of meters thick as sea levels rise in relation to the reef. This process can form large geologic features such as the barrier reefs and atolls discussed in the previous chapter. Considerable portions of the Florida peninsula rest upon former reef deposits formed millions of years ago.

Storm waves may erode the reef, producing coral sediment which may range in size from boulders through sand to clay-sized particles. When transported by currents, the suspended coral sediment turns the water chalky in appearance.

Other reef-like structures are formed by tube worms, mussels, oysters, and mangrove trees. For example, some segmented worms (such as *Sabellaria*) form tubes around themselves by cementing together grains of sand. These tubes give the worm protection and in some cases anchor the worm to the substrate. Some tube worms cement their tubes to previously built tubes. A succession of worm generations consequently can build a massive reef, similar in nature to a coral reef. Since the sedimentary particles are held together by an organic cement, which ultimately is biodegradable, these reefs are more prone to erosion and destruction than are coral reefs. Other worms (such as serpulids) form tubes of calcium carbonate secretions. These tubes also may be fused together to form massive reefs, although much smaller than ones built by coral. As with other reefs, the living reef-builder is limited to the outer part of the reef, exposed to flowing sea water. Reef-building worms are filter feeders, consuming small organisms that are carried by the currents.

Mussels and oysters frequently grow attached to one another, forming low-lying reefs. Since they get their food by filtering the water, they benefit by placing themselves above the sediment, which could clog their filter apparatus. When the animals die, their shells contribute to the formation of sands and gravels.

Mangrove trees (Fig. 3.13) have extensive root systems anchored to the sediment in warm, shallow water. The submerged roots act as a net, trapping sediment, forming a reef-like structure. Sometimes mangroves become extensive and trap enough sediment to form swampy islands. About 40 per cent of the land forming today's Florida Keys is believed to have originated in this manner.

FIGURE 3.13 Red mangrove attached to the sea floor by its extensive root system. (From Barnes, 1974.)

RESOURCES OF THE SEA FLOOR

Petroleum

Although the origin of petroleum is poorly understood, most geologists agree on two major points. Petroleum is derived from remains of dead plants and animals, and it is formed exclusively in sedimentary deposits. Plants and animals live and die and the dead bodies are eaten by animals and decomposed by bacteria. When the organic matter is buried under a thickening blanket of fine sediment, lack of oxygen within the sediment results in slowed and incomplete decomposition of the organic matter, composed chiefly of hydrogen, carbon, and oxygen. The oxygen is quickly removed by bacteria, leaving hydrocarbons behind. The presence of free oxygen in the environment would eventually result in the complete decomposition of the organic matter, and the carbon in the oil would be converted to carbon dioxide. Petroleum is produced as heat and pressure gradually increase due to the thickening of the sediment cover. Heavy oils are formed first and then are converted

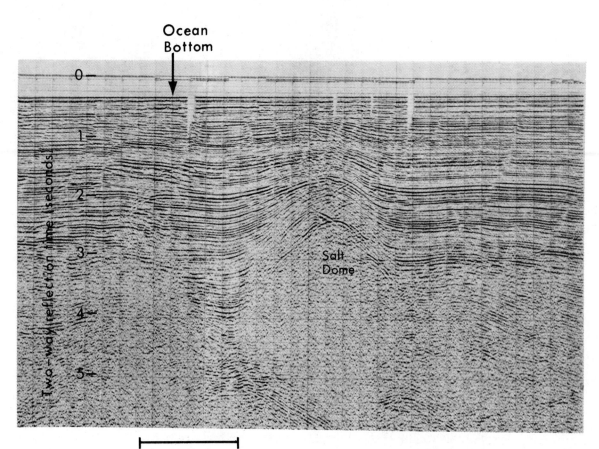

FIGURE 3.14   Continuous seismic profile obtained by a moving ship off the Texas coast. Salt deposited millions of years ago has been forced upward to form a dome, and the consequent tilting of the rocks helps to entrap petroleum. The speed of sound in seawater is about 1.5 km/sec, and that of unconsolidated sediment varies from about 1.7 to 2.5 km/sec. Thus the vertical scale does not correspond directly to depth. (Courtesy Carl H. Savit, Western Geophysical Corporation.)

to lighter, smaller molecules by the heat and pressure. The hydrocarbons migrate from the source beds into reservoir beds such as sand or porous limestone, where they are trapped to form oil and gas pools (Fig. 3.14). Occasionally, oil migrates to the surface along fracture zones in the oceanic sediments to form natural oil seeps.

The composition of crude oil varies, depending on the source of organic matter and the nature of the decomposition process. Pigments such as red hemin from blood and green chlorophyll from plants are altered chemically, but remain as pigments in the petroleum, giving the crude a characteristic color.

OIL IN THE DEEP SEA

In 1968, the drilling ship *Glomar Challenger* drilled through a salt dome on the Sigsbee Knolls area in the middle of the Gulf of Mexico under about 4000 meters of water, striking oil. The hole was filled with cement to prevent oil leakage. The amount of oil is unknown but could be great and someday may be economically retrievable. This was a surprising discovery, since no salt domes or oil were known to occur in the deep sea. Since the discovery is in international waters, the ownership of the oil is uncertain.

The greatest potential for formation of oil and gas reserves is found within sediments having a low diversity of ooze species but great numbers of individual skeletons, as these factors tend to indicate high productivity. Sedimentation in shallow areas with a high productivity and low species diversity will result in entrapment of the decaying organisms before their organic matter is broken down into inorganic salts by bacteria. Many of the productive oil fields of today were this kind of environment millions of years ago. The formation of oil indicates an incomplete utilization of organic matter by relatively few species. A great diversity of species results in a more efficient utilization of organic matter and consequently in less production of petroleum. The energy we release by burning fossil fuels was initially trapped from sunlight by plants millions of years ago. It takes very little time to release this trapped energy that was accumulated over millions of years by plants and animal life. Consequently, reserves of fossil fuels are being rapidly depleted, and soon it will be necessary to turn to other energy sources.

Minerals

In addition to petroleum, the sea floor contains vast resources of minerals (Fig. 3.15). Some such as sulfur, gold, and diamonds on the

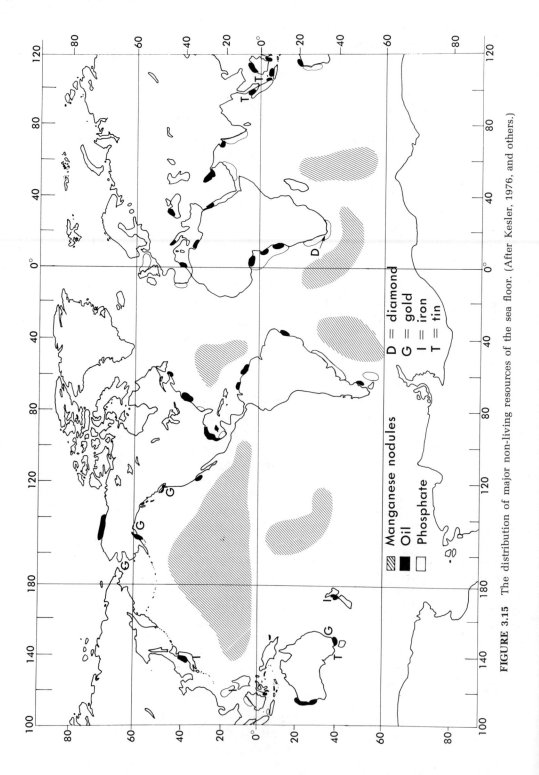

FIGURE 3.15 The distribution of major non-living resources of the sea floor. (After Kesler, 1976, and others.)

continental shelves, are now being exploited profitably. Others, such as nodules of manganese and phosphorite, have great potential, but we are now only in the stage of developing technology for their retrieval.

Mineral particles deposited far from their place of origin are referred to as *placer deposits.* They may be deposited on beaches and continental shelves in relatively shallow and accessible water. Included are deposits of diamond, iron, tin, gold, and other heavy minerals such as ilmenite and rutile, which contain titanium.

The ocean floor also contains *lode deposits* in the rocks where they were initially formed. For example, there are deposits of tin and copper off the coast of Cornwall, England, which are probably extensions of known deposits on land.

Manganese and phosphorite nodules are formed by precipitation from sea water at the sea floor. They are locally abundant and offer great potential for exploitation as soon as legal, political, and technological problems are overcome. Since manganese nodules are frequently found on the international sea bed, questions arise as to who owns them or has the right to take them.

Manganese nodules are concentrated on the deep-sea floor, especially in the Pacific. The mineral content of deep-sea manganese nodules varies greatly from place to place; the most valuable ones are in the eastern tropical Pacific. In addition to manganese, these nodules also contain valuable amounts of such other metals as cobalt, nickel, and copper, which in the future may prove to be more valuable than manganese. Resources of copper in manganese nodules are estimated to be as extensive as those on land. As mineral resources become further depleted on land, those of the sea floor will become increasingly important.

SUMMARY

Ocean sediments are valuable as sources of minerals and petroleum. They also tell us much about the history of the ocean basins and the sea itself.

Sediments are classified as terrigenous (land derived) or pelagic (formed in the sea). Pelagic sediments include those that are inorganically and biologically formed. Many marine organisms have skeleta or shells that are made of silicon dioxide (SiO_2) or calcium carbonate ($CaCO_3$) and remain intact after the organism dies. These organic remains as well as terrigenous sand, silt, and clay settle onto the sea floor and accumulate.

Small lightweight particles are carried great distances by currents and hence may be found far from shore. Large heavier particles tend to become deposited near their point of origin. Fine-grained sediments tend to be tightly packed and low in porosity. Consequently, water circulates through the sediments very slowly, so they tend to have low permeability.

In some locations, production of marine life was so rapid that their remains accumulated on the sea floor faster than they could decompose. Much organic matter was thus trapped within the growing sedimentary blanket. Some of that organic matter eventually was converted into petroleum.

Some ocean-floor features, especially in the tropics, are formed directly by marine organisms, including coral, worms, and mangrove trees. These features frequently grow to massive size and are known as reefs.

IMPORTANT TERMS

Biological sediment
Calcareous
Chemical precipitate
Clay
Coccolithophore
Coral reef
Diatom
Diatomaceous earth
Foraminifera
Manganese nodule
Ooze
Packing
Pelagic sediment
Permeability

Phosphorite
Placer deposit
Porosity
Protozoa
Pteropod
Radiolaria
Red clay
Sand
Siliceous
Silt
Sorting
Stokes' law
Terrigenous sediment

STUDY QUESTIONS

1. List four sources of pelagic sediments.
2. Where are most terrigenous sediments found? Why?
3. Briefly describe the origin of manganese nodules.
4. Discuss the relationship between biological production and sedimentation.
5. Which are more likely to be found in the deep sea: siliceous or calcareous sediments? Why?
6. Discuss the relationships among sorting, packing, and porosity.
7. How are porosity and permeability related to life in the sediment?

8. Discuss the relationship of transportation, deposition, and erosion of sediment to grain size and current velocity.
9. Why does the thickness and age of sediments increase with distance from the mid-ocean ridges?
10. Discuss the economic significance of mineral resources on the sea floor.
11. Discuss the origin of petroleum in the sea.
12. What factors control the settling velocity of sediments in the sea?

SUGGESTED READINGS

Arrhenius, G. 1963. Pelagic Sediments. In: *The Sea*, vol. 3, 655–727. New York, Wiley-Interscience.

Davin, E. M., and M. G. Gross. 1980. Assessing the Seabed. *Oceanus, 23(1)*: 20–32.

Edgar, N. T., and K. C. Bayer. 1970. Assessing Oil and Gas Resources on the U.S. Continental Margin. *Oceanus, 22(3)*:12–22.

Heezen, B. C., and C. D. Hollister. 1971. *The Face of the Deep.* New York, Oxford University Press.

Komar, P. D. 1976. *Beach Processes and Sedimentation.* Englewood Cliffs, New Jersey, Prentice-Hall.

Lisitzin, A. P. 1972. *Sedimentation in the World Ocean.* Tulsa, Oklahoma, Society of Economic Paleontologists and Mineralogists.

Margolis, S. V., and R. G. Burns. 1976. Pacific Deep-Sea Manganese Nodules. *Annual Review of Earth and Planetary Sciences, 4*:229–64.

Multer, H. G. 1977. *Field Guide to Some Carbonate Rock Environments,* 2nd ed. Dubuque, Iowa, Kendall/Hunt.

National Research Council. 1975. *Mining in the Outer Continental Shelf and in the Deep Ocean.* Washington, D.C., National Academy of Sciences.

National Research Council. 1975. *Mineral Resources and the Environment.* Washington, D.C., National Academy of Sciences.

Pettijohn, F. J. 1975. *Sedimentary Rocks,* 3rd ed. New York, Harper & Row.

Shepard, F. P. 1974. *Submarine Geology,* 3rd ed. New York, Harper & Row.

Chemistry of the Oceans

4

Knowledge of the chemistry of sea water is important for an understanding of factors that control the distribution and production of ocean life, movement of ocean water, and pollution. In addition to the resources below the sea floor, sea water itself contains a valuable supply of chemicals and, of course, fresh water, which may be obtained through desalination processes.

THE NATURE OF WATER

All matter, including water, is made up of atoms. A molecule of water, for example, consists of two atoms of the element hydrogen (H) and one atom of the element oxygen (O). All atoms consist of negatively charged electrons that spin around a positively charged nucleus which gains its charge from positively charged protons. The nucleus of most kinds of atoms also contains neutrons, without charge but contributing to the weight of the atom. In Figure 4.1 are some diagrams of the atoms of common elements showing the atomic structures. Note that the hydro-

A rosette water sampler. A cable closes the pod sampling bottles on command from the ship. Water samples are collected at various depths for biological and chemical data. (Photo courtesy of National Oceanic and Atmospheric Administration.)

gen atom is the simplest, consisting of a single electron spinning in an orbit around a proton. An ordinary oxygen atom consists of two electrons orbiting in an inner shell, six electrons in an outer shell, and a nucleus containing eight protons and eight neutrons, as shown in Figure 4.1. The positive charge on the nucleus equals the total negative charge of all the electrons. No atom has more than two electrons in the inner shell, but the outermost shell can contain up to eight electrons. Table 4.1 presents examples of the structure of a variety of elements. All inert (non-reactive) elements have eight electrons in the outer shell, except helium, which has two electrons in its single shell. In these atoms the outer shell is considered complete.

Most elements exist in nature in more than one form, or *isotope*. Isotopes of the same element have different numbers of neutrons, but the same number of protons, in the nucleus. Consequently, they have different atomic weights, which are approximately equal to the total number of protons and neutrons. Of about 90 known elements present in nature, almost 70 have two or more isotopes. Oxygen, for example, includes isotopes with atomic weights of approximately 16, 17 and 18; with eight, nine, or 10 neutrons, respectively, plus eight protons making up the nucleus. Most of these isotopes are stable but some are unstable. The unstable atoms are *radioactive* and break down, releasing electrons, protons, and neutrons.

Since, in all atoms other than inert gases, the outer orbit is incomplete, atoms can combine to form *molecules*. Thus, two hydrogen atoms combine with an oxygen atom to form a molecule of water (Fig. 4.2). Note that the outer shell of oxygen is completed (two electrons are shared with hydrogen atoms) and now has eight electrons, the maximum allowable for the outer shell.

If the water molecule is split, or dissociated, it will form positively and negatively charged *ions* (ions are atoms or groups of atoms that are electrically charged). Thus, some of the water molecules may form H^+ and OH^- ions. When table salt, whose chemical name is sodium chloride (NaCl), dissolves in water, it forms the charged ions Na^+ and Cl^- (Fig. 4.3).

In nature, water is not just a collection of freely moving single molecules. Some of them are bound together into chains, or *polymers*. Cold water and ice have more chains of water and fewer free molecules than warm water. It is the free or exposed molecule of water that is capable of acting as a solvent. The solubility of most substances, including sodium chloride, increases with increasing water temperature. A peculiar exception to this rule is calcium carbonate, which, as already mentioned, becomes increasingly soluble as the water temperature drops.

A water molecule is polar. The hydrogen atoms form a 105° angle with respect to the oxygen atom and provide the positive side of the

Hydrogen (H)

Oxygen (O).

Sodium (Na)   Chlorine (Cl)

FIGURE 4.1 Diagrams of the atomic structures of hydrogen, oxygen, sodium and chlorine. The arrow indicates the site at which sodium and chlorine may combine to form sodium chloride.

Table 4.1   ATOMIC STRUCTURE OF SOME IMPORTANT ELEMENTS

ELEMENT AND SYMBOL	ATOMIC NUMBER (PROTONS)	NUMBER OF NEUTRONS	ATOMIC MASS	ELECTRONS IN VARIOUS SHELLS OR LEVELS
Hydrogen (H)	1	0	1	1
Helium (He)	2	2	4	2
Carbon 12 (C)	6	6	12	2-4
Carbon 14 (C)*	6	8	14	2-4
Oxygen 16 (O)	8	8	16	2-6
Oxygen 17 (O)*	8	9	17	2-6
Oxygen 18 (O)*	8	10	18	2-6
Sodium (Na)	11	12	23	2-8-1
Magnesium (Mg)	12	13	25	2-8-2
Aluminum (Al)	13	14	27	2-8-3
Silicon (Si)	14	14	28	2-8-4
Chlorine (Cl)	17	18	35	2-8-7
Argon (A)	18	22	40	2-8-8
Potassium 39 (K)	19	20	39	2-8-8-1
Potassium 41 (K)*	19	22	41	2-8-8-1
Calcium (Ca)	20	20	40	2-8-8-2
Manganese (Mn)	25	30	55	2-8-13-2
Iron (Fe)	26	30	56	2-8-14-2

*Less abundant isotopes of some of the elements shown above.

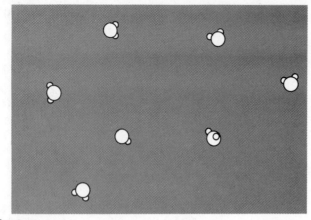

A. Gas

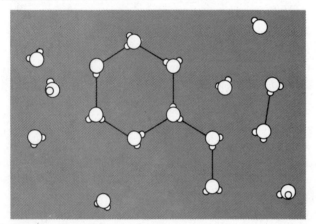

B. Liquid

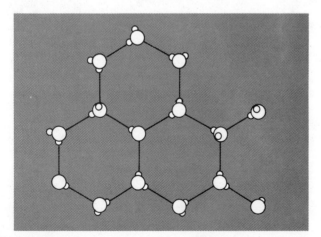

C. Solid (ice)

FIGURE 4.2 The molecular structure of water, illustrating the increased polymerization of water molecules with change in phase from gas to solid (ice).

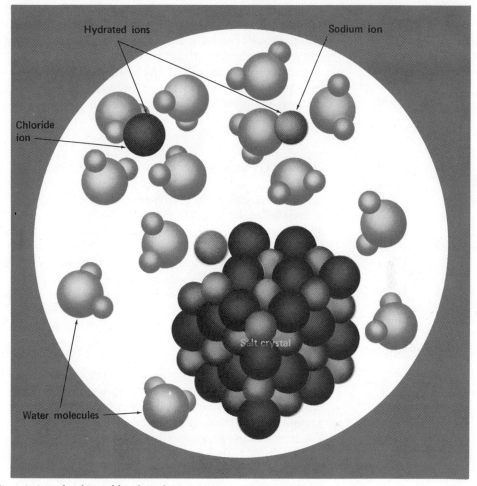

FIGURE 4.3 The dissociation of sodium chloride to form its component ions, Na$^+$ and Cl$^-$. (From Jones et al., 1980.)

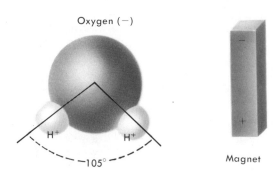

FIGURE 4.4 The polar nature of water.

molecule. The remaining portion has a negative charge (Fig. 4.4). When salt crystals are added to water, the water molecules, acting like magnets, pull the ions of sodium and chlorine out of the crystals. Positive poles of the water molecules attract the negative chlorine ions. Negative poles of the water molecules attract the positive sodium ions. The salt dissolves because the force of attraction between the water and the sodium or chlorine ion is greater than the attraction of sodium and chlorine for each other.

ORIGIN OF SEA WATER

There are many hypotheses regarding the origin of sea water, and all are related to the origin of the atmosphere. These explanations are highly speculative, however, since a time scale of up to five billion years is involved.

One clue to the origin of sea water lies in the study of the distribution of certain inert (noble) gases (neon, argon, krypton, and xenon). The weight of one molecule of water is less than that of these gases. An argon atom, for example, weighs more than twice as much as a water molecule. These gases are found in very small proportions in the earth's atmosphere as compared with their proportions in the atmosphere of the sun and of other stars. If one assumes that the sun and the earth evolved from a similar source, it is implied that large quantities of these inert gases must have escaped from the earth's atmosphere. The ease with which a gas will leave the atmosphere depends on the weight of its molecules, the gravitational attraction of the earth, and the atmospheric temperature. The tremendous gravitational attraction of the sun is capable of retaining essentially all of its atmosphere, in spite of its very high temperature. The relatively low concentrations of inert gases in the earth's atmosphere imply that the early earth's atmosphere must have been very hot, causing a great loss of gases. Because the weight of a water molecule is less than that of these inert gases, most of the water vapor in the early atmosphere was probably also lost. Thus, the present waters of the oceans must have accumulated only after the atmosphere had cooled considerably.

Essentially all of the oceans' waters must have had their origin within the earth,* coming to the surface via hot springs and volcanoes and being released as rocks crystallized from a molten state. Because this process is continuing today, the volume of the oceans should continue to increase in the long run (hundreds of millions of years). All of the waters of the oceans (1.4 billion cubic km) could have been derived in this manner at an average rate of only about ½ cubic km of water per year during the last 3 billion years.

---

*It is estimated that the present atmosphere of the earth could hold only a fraction of one per cent of the total volume of the oceans; also, a great deal of water apparently escaped before the earth cooled.

Although the oldest known salt deposit on earth (evidence of a saline ocean), is only about 800 million years old, at least slightly salty oceans probably existed as far back as 3 billion years ago, because fossils of lime-secreting algae of that age have been found. Lime ($CaCO_3$) is one of the salts of today's ocean. However, this does not indicate the actual degree of saltiness of the ancient oceans. This problem remains a mystery.

How and in what form the dissolved materials in sea water were added to the oceans is not well understood. Most of the dissolved materials were probably initially derived from within the earth through volcanic and hot spring emanations as well as from the weathering (decomposition) of igneous rocks. For example, most of the sodium in sea water could easily be accounted for by rock weathering. On the other hand, the chlorine in sea water could be derived from volcanoes and hot springs. Sodium and chlorine make up about 86 per cent of all the dissolved materials in sea water.

Has the salt concentration of the oceans remained much the same as it is today, or has it been drastically different during past geological time? One can only speculate about this question. A precise answer would require knowledge of the rate of increase of the ocean volume, the rate of chemical addition through volcanic and hot spring activities, the rate of rock weathering, and the rate of removal of these substances from sea water by deposition. Estimates for some of these factors can be made, based on the study of ancient rocks and from a knowledge of the physical and chemical properties of water suggesting that the saltiness of the oceans was not drastically different than it is today, throughout all but its earliest history.

ORIGIN OF LIFE

The early (more than 4 billion years ago) atmosphere of the earth probably contained no free oxygen, as that gas would have been consumed in the oxidation of exposed materials such as iron. It is thought that most of the gases present in the early atmosphere came from volcanic activity and included methane (CH_4), ammonia (NH_3), water (H_2O), nitrogen (N_2), and lesser amounts of carbon dioxide (CO_2) and carbon monoxide (CO). A mixture of hydrogen, methane, ammonia, and water vapor, when exposed to electric sparks, will form amino acids, the building blocks of proteins. Other organic molecules have been synthesized from simple gases in the presence of sparks, ultrasonics, light, heat, and ultraviolet radiation. These include many of the components of living organisms, such as sugars, proteins, and even the constituents of DNA (deoxyribonucleic acid) and RNA (ribonucleic acid). DNA and RNA are the chemical carriers of the genetic information that is necessary for life.

The early atmosphere, subjected to high-energy ultraviolet radiation, heat, and possibly lightning, would have produced many organic compounds that would have dissolved in the water collecting at the surface of the earth. The first living organisms were certainly very simple, dropletlike affairs but were capable of self-duplication (probably using RNA as a genetic material). The sea probably was like a dilute salty broth, with the organic molecules serving as food for the first anaerobic bacterialike organisms that lived in the absence of oxygen.

As organic food became relatively scarce, there was an environmental pressure toward evolution of the process of photosynthesis. The next stage, perhaps, was the evolution of animals that could eat the photosynthetic plants.

It has been estimated that life has existed on earth for as much as 4 billion years and photosynthetic processes about 2 to 3.5 billion years. The development of the sexual reproduction process may have occurred 1 to 2 billion years ago. As oxygen continued to accumulate in the atmosphere, a high-atmosphere ozone (O_3) layer evolved, which shielded the earth from much ultraviolet radiation, the very radiation that was probably responsible for creating the first life. The reduction of ultraviolet light intensity eliminated the need for shielding of organisms by sea water (which permits only very limited penetration of ultraviolet light) and eventually allowed the invasion of land by living organisms. The oldest known marine organisms (3 billion years) are 6 to 7 times older than the oldest known land organisms (400 to 500 million years), suggesting that colonization of the land had to await the evolution of more complex organisms and a suitable atmosphere.

COMPOSITION OF SEA WATER

Sea water is only about 96.5 per cent water; the rest (3.5 per cent) consists of dissolved solids such as common salt (sodium chloride) and relatively small amounts of dissolved gases such as oxygen. Actual concentrations of dissolved substances, however, may vary from place to place, even within the same area. But, even though absolute concentrations may vary, the proportions of the major dissolved substances remain amazingly constant.* For example, the ratio of magnesium to chlorine (Mg/Cl) is practically the same (0.067) in all oceans. This is due to the fact that the world oceans have been mixing with each other for millions of years.

---

*A. Marcet (in 1819) and J. G. Forchammer (in 1865) reported that the relative proportions of elements in sea water vary only slightly, based on analysis of numerous water samples. In 1884, W. Dittmar came to the same conclusion from the analysis of 77 water samples from various depths collected around the world during the *Challenger* Expedition. The ratios obtained by Dittmar are very close to those obtained today.

Dissolved Solids

Water dissolves everything, at least in minute amounts. However, because some substances are more soluble than others, certain elements are more abundant than others in the sea. For example, sodium chloride is very soluble, calcium carbonate (limestone, clam shells, and so forth) is much less soluble, and silicon dioxide (as in quartz and glass) is almost insoluble.

Most of the dissolved material in sea water is composed of relatively few elements. Table 4.2 shows that about 95 per cent of dissolved solids

Table 4.2 ELEMENTAL COMPOSITION OF DISSOLVED SOLIDS IN SEA WATER*

ELEMENT		AMOUNT
Major Element		*In Parts Per Thousand ($^0/_{00}$)*
Chlorine	(Cl)	19.35
Sodium	(Na)	10.76
Magnesium	(Mg)	1.29
Sulfur	(S)	0.86
Calcium	(Ca)	0.41
Potassium	(K)	0.39
Minor Element		*In Parts Per Million*
Bromine	(Br)	67
Carbon	(C)	27
Strontium	(Sr)	8
Boron	(B)	4
Silicon	(Si)	3
Fluorine	(F)	1
Important Trace Element		*In Parts Per Billion*
Nitrogen	(N)	500
Lithium	(Li)	170
Rubidium	(Rb)	120
Phosphorus	(P)	70
Iodine	(I)	60
Iron	(Fe)	10
Zinc	(Zn)	10
Molybdenum	(Mb)	10
Copper	(Cu)	3
Uranium	(U)	3
Cobalt	(Co)	0.1
Mercury	(Hg)	0.1
Silver	(Ag)	0.04
Gold	(Au)	0.004

*Assuming a salt concentration of about 35.12 $^0/_{00}$ and excluding hydrogen and oxygen, which are mostly in the form of water. Only 26 elements are shown here, because most of the others are present in minute amounts. (After Goldberg, 1965; Culkin, 1965; and Fitzgerald et al., 1974.)

are made up of only six *major elements* (chlorine, sodium, magnesium, sulfur, calcium, and potassium). In fact, 86 per cent of all dissolved salts consists of only two elements, sodium and chlorine. While all of the major chemical constituents are important to marine life, sodium chloride, bromine, and magnesium are economically important as well. For convenience, concentrations are reported in parts per thousand by weight (gm per kg or $^o/oo$) of sea water. Note that the major elements are found in concentrations greater than 0.3$^o/oo$.

Actually, some of the elements in sea water are bound to other elements in the form of complex *ions* (electrically charged atoms or groups of atoms). Such ions behave as if they were composed of single atoms. For example, sulfur is usually bound with oxygen to form the sulfate ($SO_4^=$) ion, which has two negative charges. Dissolved carbon adheres to hydrogen and oxygen to form the bicarbonate (HCO_3^-) ion, having a negative charge.

Table 4.3 shows seven major ionic constituents of sea water. Thus, hydrogen and oxygen can be chemically bound with other elements in forms other than water (H_2O). The contribution of hydrogen and oxygen toward the salt content of sea water may be about 4 per cent or more of the total, when they are bound with other elements.

The *minor elements* (bromine, carbon, strontium, boron, silicon, and fluorine) collectively constitute only about 0.3% of dissolved solids and are generally found in concentrations of 1 to 65 parts per million (0.001 to 0.065 $^o/oo$). Carbon, bromine, and silicon, although present only in small amounts, are essential for the growth of marine plants. Carbon in the sea is mostly in the form of bicarbonate (see Table 4.3), but may also be present as carbonate ($CO_3^=$), carbon dioxide (CO_2), or carbonic acid (H_2CO_3). Silicon as silica (SiO_2) in the form of opal makes up the skeletons of some marine plants and animals, including certain sponges. Bromine, which may be extracted from seaweeds, sea water,

Table 4.3   MAJOR IONIC CONSTITUENTS OF SEA WATER*

CONSTITUENT		AMOUNT ($^o/oo$)	CUMULATIVE PER CENT OF TOTAL
Chloride	(Cl^-)	19.35	55.1
Sodium	(Na^+)	10.76	85.7
Sulfate	($SO_4^=$)	2.71	93.4
Magnesium	(Mg^{++})	1.29	97.1
Calcium	(Ca^{++})	0.41	98.3
Potassium	(K^+)	0.39	99.4
Bicarbonate	(HCO_3^-)	0.14	99.8

*Assuming a total salt concentration of about 35.12 $^o/oo$ are taking into account the fact that sulfur and carbon are generally combined with other elements in the sea. (After Culkin, 1965.)

and terrestrial salt deposits (formed in former seas that are now dry land), has a number of industrial uses, such as in photography, in the manufacture of certain drugs, and in detergent gasoline.

Dissolved solids that are found in concentrations of less than one part per million (ppm) are termed *trace elements.* These include ions of nitrogen such as nitrate (NO_3^-), nitrite (NO_2^-), and ammonia (NH_4^+); iron; copper; and iodine, all of which are essential for the growth of plants. Other elements such as mercury, silver, gold, and uranium are valuable, but they are present in concentrations too small to be extracted economically. Excessive amounts of some elements, such as copper and mercury, may be toxic to marine organisms or can be concentrated by certain forms of marine life and in turn be toxic to man.

Residence Time

Salts are constantly entering the sea from the land and to a lesser extent from the atmosphere and the sea floor itself, yet the salinity of the oceans remains nearly constant from year to year. Most of the salt that enters the sea is removed by sedimentation, precipitation, or by biological uptake. The average length of time that a substance remains in the sea before removal is the *residence time.* It is defined in the following way:

$$T = A/R$$

where T is the residence time in years,
    A is the total amount (in grams) of the substance in the sea,
and R is the rate of removal (in grams per year) of the substance,
      if A is constant.
The residence time for iron is about 140 years, for sodium it is about 250 million years. The residence time of river water in the oceans is 40,000 years, assuming that the volume of the oceans is 1.4 billion cubic km and river input is about 35,000 cubic km per year. The rate of river input of water is approximately in balance with removal through net evaporation (evaporation less precipitation). River water, however, accounts for only about 10 per cent of the total water entering the sea each year. Precipitation accounts for most of the rest and small amounts are derived from within the earth. If total water input is considered, the estimate of residence time for ocean water is only about 4000 years.

Ions of elements that have longer residence times, such as sodium, tend to be more soluble and chemically, or biologically, less active than those, such as silicon, iron, and manganese, which have relatively short residence times. Precipitation and biological uptake are important factors in the removal of these elements.

Silicon, for example, has a residence time of about 8000 years. It enters the sea via rivers. Some settles out near shore, some of the dissolved fraction is absorbed by diatoms and other organisms with siliceous skeleta. When these organisms die, the skeleta sink to the sea floor and contribute to the formation of siliceous oozes. Diatomaceous earth was formed in this way from accumulations of diatom shells.

When an element enters the sea faster than it is removed, its concentration increases. This has been observed with lead (residence time: about 2000 years) which has apparently increased from about 0.02 to about 0.07 parts per billion in surface waters since prehistoric times. This is most likely the result of the development of civilization. Because lead is toxic, there may be unfortunate consequences for marine life in the future. Similar increases in the concentration of other hazardous substances released to the oceans from industrial activities put increased stress on the marine environment.

Measurement of Salinity

Salinity refers to the total amount of dissolved solids in water and is usually measured in gm per kg ($^0/oo$) of sea water. The salinity of ocean water generally varies from about 30 to 37$^0/oo$, with an average of about 35$^0/oo$ salinity. However, some deep-sea brines of the Red Sea have salinities of as high as 270$^0/oo$. Estuarine waters (where rivers enter bays and the sea) have a range of salinity from almost fresh water ($0^0/_{oo}$) to about 30$^0/_{oo}$. The distribution of salinity and factors affecting it are discussed in detail in the next chapter.

Accurate determination of salinity is of great importance to biologists and chemists studying the ocean as well as to physical oceanographers, who are concerned with the relationship of salinity to ocean circulation. Patterns of salinity variations in the oceans may reveal the paths of deep ocean currents and of nutrients and the amount of river runoff.

A simple but rather inaccurate method of salinity determination is the evaporation of a known amount of sea water and the weighing of the resulting dry salt. This procedure is inaccurate because some of the salts escape into the atmosphere in bubbles and because any inorganic matter which was not actually in solution (suspended material, for instance) remains along with the dry salt. In addition, water itself may be bound with certain salts in the process of simple drying.

Chemical analysis of sea water, although time-consuming, has been used to measure salinity more accurately. As stated earlier, the ratios of major constituents of sea water remain virtually constant. This enables one to calculate salinity by knowing the concentration of only one of the major constituents. Before the advent of electronic measuring devices now in use, salinity was in fact calculated by measuring the concentration of a single ion. Since the chloride ion (Cl^-) is the most abundant dissolved constituent of sea water and is relatively easy to measure chemically, the chloride content or *chlorinity** (Cl⁰/oo) was most useful in estimating salinity, as expressed in the following equation:

$$S⁰/oo = (1.80655 \times Cl⁰/oo)$$

The density of sea water is directly related to its temperature, salinity, and pressure. Therefore, if the temperature and pressure are held constant, only salinity affects density (Fig. 4.5), and thus salinity may be estimated by measuring density. This procedure requires a simple hydrometer, a small weighted glass tube that floats higher in the water as salinity increases (similar to those used in testing batteries). Rough estimates of salinity may be achieved quickly in this way, but accurate measurements by this method are very time-consuming and consequently are not usually used in oceanographic work.

The index of refraction† of sea water is directly proportional to the salinity for a particular temperature and pressure (Fig. 4.6). Light is bent

---

*Chlorinity is now defined as the weight in grams of silver required to precipitate the chloride, bromide, fluoride, and iodide in 0.3285233 kg of sea water.

†The index of refraction of sea water is the ratio of the speed of light in a vacuum to the speed of light in sea water.

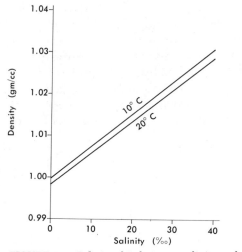

FIGURE 4.5 Relationship between salinity and density of sea water at 10° C and 20° C at atmospheric pressure.

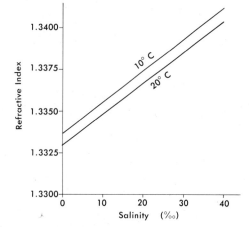

FIGURE 4.6 The effect of salinity on the refractive index of sea water at 10° C and 20° C.

FIGURE 4.7 The refractometer, a device for measuring salinity. Light passes through a thin layer of water and is refracted. The refractive index or the salinity can be read on a scale in the eyepiece by the observer.

(refracted) as it enters a water sample, and this bending is measured using a refractometer (Fig. 4.7), which may be calibrated to give a direct reading of salinity.

The methods described above are either too costly or too inaccurate for widespread use; consequently, most salinity determinations made today are based on measurements of electrical conductivity of sea water. The conductivity of water increases with increases in salinity, temperature, and pressure. Therefore, if the temperature, pressure, and conductivity are known, salinity may be accurately and quickly estimated (Fig. 4.8).

Dissolved Gases

Sea water also contains several dissolved gases, including oxygen, carbon dioxide, nitrogen, and hydrogen sulfide, and the inert gases helium, argon, neon, krypton, and xenon. The inert gases do not ordinarily combine with other elements and are not used by marine organisms. The other gases can be produced or consumed by organisms in addition to being dissolved from the atmosphere. The absolute concentration of each gas in sea water depends to a certain extent on its relative concentration in the atmosphere, its solubility, and the temperature and salinity of the water.

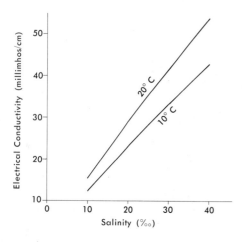

FIGURE 4.8 The effect of salinity on the electrical conductivity of sea water at 10° C and 20° C, at atmospheric pressure.

NITROGEN

The most abundant dissolved gas in the sea is nitrogen (about 15 ppm). Nitrogen as a gas, however, is apparently of little importance to marine life except to a few nitrogen-fixing bacteria and plants. Most of the nitrogen essential to the growth of marine organisms is derived solely from nitrogen-containing salts (about 0.5 ppm in sea water) such as nitrates, which are so rare that they tend to limit the amount of life the ocean can support.

OXYGEN

Oxygen is produced by the following reaction as a result of the photosynthesis of organic matter by plants in the sea as well as on land:

$$CO_2 + H_2O + \text{light (in the presence of chlorophyll)} \rightarrow \text{organic matter} + O_2$$

Oxygen also enters the sea via rivers and the atmosphere. Wave action and diffusion are important processes by which oxygen enters the sea from the atmosphere. Oxygen may leave the sea water by being consumed by marine organisms (animals, bacteria, and plants), as well as through loss to the atmosphere. All organisms, including plants, utilize oxygen during respiration, the process by which they release energy used for growth and activity. In general, plants produce more oxygen than they consume by respiration, and the remainder is used by animals and bacteria. In contrast to photosynthesis, which requires sunlight, respiration can take place day and night. The reaction that takes place during respiration is shown below:

$$\text{Organic matter} + O_2 \rightarrow CO_2 + H_2O + \text{energy}$$

Figure 4.9 illustrates the pathways of oxygen in the sea.

A typical vertical distribution of dissolved O_2 in the sea (excluding polar regions) is shown in Figure 4.10. Oxygen concentrations are gener-

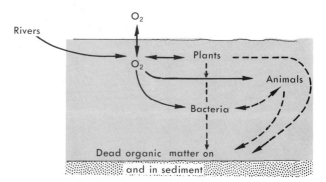

FIGURE 4.9 Oxygen cycle in the sea. Solid arrows indicate flow of elemental oxygen; dashed arrows indicate flow of oxygen in organic matter. Oxygen in CO_2 is supplied to the plants from the atmosphere and water. CO_2 is produced by bacteria, plants and animals during respiration.

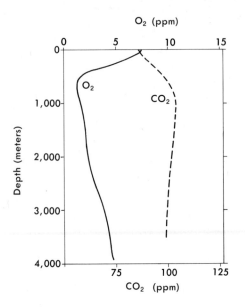

FIGURE 4.10 Typical variation of dissolved oxygen (O_2) and total dissolved carbon (as CO_2) in the North Central Pacific Ocean.

ally at a maximum near the sea surface, as a result of intense photosynthetic activity of plants and the atmospheric exchange of oxygen. This corresponds to the wind-mixed, lighted surface layer. Beneath this layer is the oxygen-minimum zone (at about 700 meters in the figure), in which bacteria and animals consume oxygen by respiration, in the absence of light and plant life. In addition, the lowered oxygen concentrations at these depths is caused by decay of dead organic matter that has fallen from the productive surface layer. Curiously, at greater depths, the oxygen concentration tends to increase with depth. This effect is due to transport of oxygen by deep ocean currents into these depths (Fig. 4.11) from subpolar areas, where cold, dense, oxygen-rich water sinks. (Oxygen is more soluble in cold water than in warm water.)

CARBON DIOXIDE

Carbon dioxide (CO_2) is released when plants, animals, and bacteria respire. Plants consume it, however, during photosynthesis and in

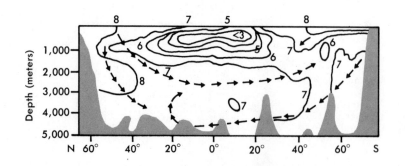

FIGURE 4.11 Vertical distribution of dissolved oxygen (in parts per million) in the Atlantic Ocean. Arrows indicate currents. (Modified after Wattenberg, 1933.)

general consume more carbon dioxide than they produce. In addition, carbon dioxide enters the sea from the atmosphere, and rivers carry it into the sea in the form of carbonates and bicarbonates of sodium, calcium, and potassium and as carbonic acid. The bicarbonate and carbonate salts are produced by weathering of rocks on land. Carbonate is removed from the water by plants and animals which use it in the construction of their shells and skeletons. Some of these hard remains settle to the bottom and become part of the sediment. In cold, deep ocean water (greater than about 5000 meters), the carbonates are more soluble than near the surface, and they dissolve but are eventually recycled back to the surface waters. Consequently, carbonate sediments are typical of warm, shallow water.

Figure 4.12 illustrates the pathways of the carbon dioxide system, discussed above. Notice that the carbon is cycled and recycled through the system. Actually, there is a net input of carbon via the rivers and atmosphere, which is balanced by the deposition of carbonate shells and skeleta onto the sea floor.

The concentration of total dissolved carbon dioxide (including the bicarbonates and carbonates) increases down to about 1000 meters, corresponding to the oxygen-minimum zone (see Fig. 4.10). The lower carbon dioxide concentration near the surface is mostly due to consumption by plants. The increase of dissolved carbon dioxide is related to the decomposition of dead organisms, which results in the conversion of particulate organic matter into nutrients and carbon dioxide.

HYDROGEN SULFIDE

Hydrogen sulfide (H_2S) is a gas that has an odor very much like rotten eggs. H_2S is produced by anaerobic bacteria (bacteria that live in the absence of dissolved oxygen), which may obtain oxygen from sulfur

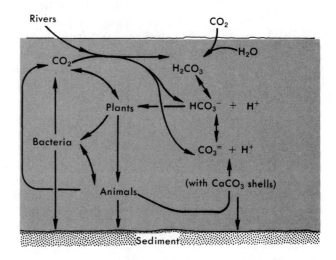

FIGURE 4.12 Carbon cycle in the sea. (H_2CO_3 = carbonic acid, HCO_3^- = bicarbonate, $CO_3^=$ = carbonate, CO_2 = carbon dioxide.)

compounds and ions, and produce hydrogen sulfide as waste during respiration. Other anaerobic bacteria use nitrate (NO_3^-) or nitrite (NO_2^-) as an oxygen source and may produce ammonia (NH_4^+) as a by-product instead of hydrogen sulfide. *Anoxic* conditions (those lacking dissolved oxygen) are found in areas where free oxygen is consumed faster than it can be supplied, as in the deeper water of stagnant basins such as some fiords, in the Black Sea, and in marine sediments such as mud flats. These sediments are characteristically black as a result of the conversion of hydrogen sulfide to other sulfur-containing compounds. When the black sediment comes in contact with oxygen (as when a shovelful of mud is turned over or when a marine animal burrows through the mud), oxygen becomes available to it and the black color slowly disappears as the sulfur compounds are consumed by aerobic bacteria (those that require oxygen).

In addition to the preceding rather localized conditions, hydrogen sulfide may be produced in the open ocean when excessive amounts of organic matter are decomposed, resulting in anoxic conditions. Such a condition exists periodically along the Peruvian coast, in a phenomenon known as *El Niño* (see Chap. 8). During these times, changes in the current patterns result in warmer nutrient-poor water entering an area normally fed by cooler nutrient-rich water upwelled from below. These changes result in extensive fish kills. Decay of the dead fish uses up all the available oxygen, and anaerobic bacteria proliferate, causing hydrogen sulfide buildup. Sulfur compounds tend to coat buildings and ships, giving them a black appearance. Hence, El Niño is sometimes called the "Callao Painter," after a Peruvian port city.

NUTRIENT CYCLES IN THE SEA

In addition to contributing to the salinity of sea water, many of the chemical constituents of sea water are required as raw materials for the production of ocean life. All of the major constituents listed in Table 4.3 are required for plant growth and are used in the living tissues of organisms, as are many minor and trace elements, including nitrogen, phosphorus, iron, copper, and iodine. Some organisms, such as diatoms, silicoflagellates, radiolarians, and certain sponges, have skeleta containing silicon dioxide; however, it is not required as a constituent of their tissues.

Plants require nitrogen and phosphorus in rather large quantities. Living tissues contain about 100 atoms of carbon to about 15 atoms of nitrogen and one atom of phosphorus. The ratio of nitrogen to phosphorus (N/P) by weight in living tissue is about seven gm of nitrogen to one gm phosphorus. This is nearly the same ratio as is found in the open sea, indicating that plants tend to absorb these elements in the same relative proportions. The similarity between the N/P ratios of organic matter and of sea water may be due to evolutionary adaptation of

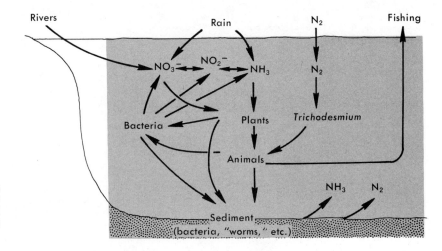

FIGURE 4.13 The nitrogen cycle in the sea. (NO_3^- = nitrate; NO_2^- = nitrite; and NH_3 = ammonia, which forms the ammonium ion NH_4^+.)

plants to the chemical environment, or it may indicate a largely biological source of nitrogen and phosphorus in the sea. When organisms die and decompose, their nitrogen and phosphorus are released back into solution. Near shore and in estuaries, the N/P ratio may vary considerably, owing to differences in composition of nutrient-rich runoff and pollution. Pollution of the coastal ocean is discussed in detail in Chapter 9.

Nitrogen compounds that flow into the sea through river runoff and rain enter the nitrogen cycle (Fig. 4.13) through plants, mainly as the nitrate ion. Plants are consumed by animals and bacteria, which release nitrogen compounds into the water (where they become available to the plants again). Various species of bacteria release nitrates, nitrites, and ammonia and are of great importance in regenerating these nutrients. In warm surface water, a blue-green alga, *Trichodesmium,* is important in fixing dissolved nitrogen and incorporating it into organic matter in areas where other nitrogen sources have been depleted.

Not all nutrients are recycled within the sea. Some dead organisms settle onto the sea floor and are buried by rapidly thickening sediment, and thus some nitrogen compounds as well as other potential nutrients are lost from the sea. Some of this organic matter may eventually be converted to fossil fuels. In addition, some organic matter is carried from the sea as a result of fishing by birds and humans. If the total nutrient content of the sea is to remain stable, potential nutrients lost through sedimentation and fishing must be balanced by input from sources such as rivers and the atmosphere.

The tendency for dead organisms to sink from the well-lit surface water results in transport of nitrogen compounds into the deep sea, where a very small fraction is lost to sediments, and the rest remains until brought near the surface by ocean currents. The concentration of nitrogen compounds, especially nitrates, usually increases with depth (Fig. 4.14), to a maximum at about 1000 meters, because of decomposition of

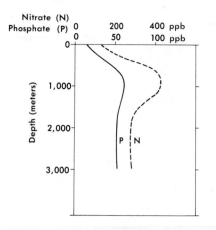

FIGURE 4.14   Typical variations in nitrate and phosphate with depth in the Atlantic Ocean (ppb = parts per billion). (After Sverdrup et al., 1942.)

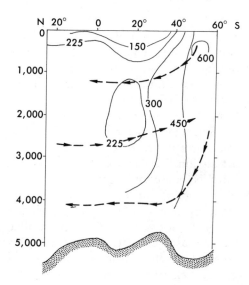

FIGURE 4.15   Vertical distribution of nitrate-nitrogen (in ppb) in the South Central Atlantic. Arrows indicate deep-ocean currents. (Modified after Sverdrup et al., 1942.)

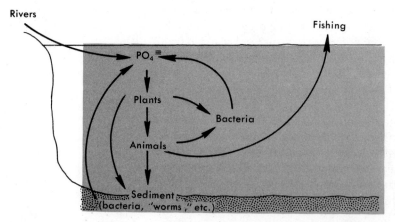

FIGURE 4.16   The phosphate cycle in the sea.

the dead organisms that have fallen from the surface waters. This nutrient maximum is correlated with the oxygen minimum (see Fig. 4.10), which is caused by bacterial metabolism. The nutrient depletion near the surface is most pronounced in tropical waters (Fig. 4.15), where surface heating results in density stratification and a lack of vertical mixing. These waters tend to be low in productivity compared with polar water, where vertical mixing brings back nutrients to the surface.

Phosphates ($PO_4\equiv$) are cycled through the biological system in a manner similar to nitrates. The phosphate cycle is shown in Figure 4.16. Phosphates, like nitrates, enter the sea via rivers and runoff, including rain water running over rocks covered with guano (bird feces). They leave the water through the fishing activities of birds and people. In addition, some phosphates, such as in fish teeth, are trapped in sediment. In some areas of the continental shelf containing heavy concentrations of decaying matter, released phosphates are so highly concentrated that some is precipitated out, forming *phosphorite nodules.* The input and output of nutrients is nearly in balance for the oceans as a whole, but nutrient pollution (an overabundance of nutrients) and over-fishing may upset the equilibrium locally. This may have adverse effects on the biota, especially in promoting the growth of certain undesirable species, such as red-tide organisms.

The distribution of phosphates in the sea is very similar to that of nitrates, both vertically (see Fig. 4.14) and horizontally (Fig. 4.17). The distribution of nutrients reflects their depletion in warm surface water, regeneration in deep water, and transport by currents.

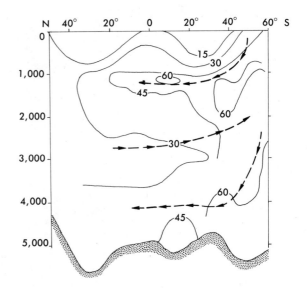

FIGURE 4.17 Vertical distribution of phosphate-phosphorus (in parts per billion) in the Central Atlantic. Arrows indicate deep-sea currents. (Modified after Sverdrup et al., 1942.)

DESALINATION

The fresh water need of the United States in 1960 was about 300 billion gallons a day, has approximately doubled by 1980 and is expected to have tripled by the year 2000. In recent years, the quest for more fresh water has turned our attention to sea water, which is about 96.5 per cent pure water. More than 1000 desalination plants (Fig. 4.18), with a combined capacity of producing over 500 million gallons of water per day, are in operation or under construction throughout the world. The plants in the United States alone have a capacity of more than 60 million gallons per day. The largest desalination plant in the world is in Hong Kong and can produce about 48 million gallons per day.

There are many methods of desalinating sea water. The oldest technique is distillation, whereby sea water is evaporated and fresh water is produced by condensing the resulting water vapor. Julius Caesar is said to have used this method to obtain fresh water for his army. Such stills are used on a small-scale basis even today, especially in desert areas. Other more sophisticated techniques requiring great amounts of energy but capable of desalinating large quantities of sea water are in operation throughout the world.

In spite of the economic considerations, desalination may be an answer to the world's water needs. But it may also create other problems, especially the disposal of the warm and highly concentrated brine, which now is sent back to the sea. This undoubtedly could cause great damage to the ecology of the coastal ocean, especially when many desalination plants are concentrated in one area, for instance, along most of the

FIGURE 4.18 A modern desalination plant at St. Thomas, Virgin Islands. (Photo courtesy of U.S. Department of the Interior.)

Atlantic coast of the United States. No one can foretell what the consequences of brine disposal would be, but research is being conducted on this problem. It has been proposed that the brine be mixed with sewage in such a way that the resulting mixture would sink to greater depths. This proposal, although economically advantageous, warrants further research as it involves many complex oceanographic and environmental factors. Some use for most or all of the brine may be found in the future. For example, metals and other salts might be extracted from it. At present, most of the magnesium and bromine in the United States come from the sea. Although sodium chloride is also extracted from sea water, as in San Francisco Bay, much of the United States' supply currently comes from terrestrial salt deposits.

We should consider other, perhaps safer, alternate sources of fresh water, in addition to conservation and recycling of the presently available water supply. Fresh water can be extracted from the atmosphere. Large cooling towers could be set up in tropical islands to collect the moisture in the air. Cooling can be accomplished by pumping cold deep water from the ocean. Because this water is rich in nutrients, it can be reused for fish farming. Another alternative is to tow icebergs to populated regions and obtain fresh water by melting them. Both of these alternatives may be competitive with desalination of sea water and, most important, may be less damaging to the environment.

SUMMARY

Water is a "universal solvent." It dissolves everything in at least small amounts. This fact is of great importance to the chemical composition of sea water, as well as to the chemical processes that take place in the sea.

The relative abundance of the various constituents of sea water is determined by their solubility as well as their abundance on earth. Seven major constituents make up 99.8 per cent of the dissolved solids in sea water. These are found in nearly the same relative proportions in all parts of the oceans, regardless of the absolute concentration of salts. Less abundant, but important, minor and trace constituents such as nitrate and phosphate are required for the production of life. Quantities vary greatly from one region to another and are dynamically involved in cycles between the living and non-living environments. They are removed from sea water during the production of living matter and returned to the sea upon death.

Rain and river water carry dissolved substances, including nutrients, into the sea. This chemical transport is important to the sea's long-term nutrient balance. It replaces salts lost as organic matter in the form of fish and other marine life removed from the sea by birds and people.

At present, sea water is an important source of sodium chloride, bromine, and magnesium. Desalination, the process by which fresh water is obtained from the sea, is becoming increasingly important in regions where traditional supplies of fresh water are scarce.

IMPORTANT TERMS

Aerobic
Anaerobic
Chlorinity
Desalination
El Niño
Ion
Isotope
Molecule

Nitrogen cycle
Phosphate cycle
Phosphorite nodule
Polymer
Residence time
Salinity
Solvent
Trace elements

STUDY QUESTIONS

1. Discuss the origin of sea water and how it became salty.
2. Why is water such a good solvent?
3. The ratios among the major constituents of sea water vary only slightly from one part of the ocean to another. Why is this so? How is this consistency in the composition of sea water useful in determining the salinity of sea water?
4. Bromine and phosphorus are both required for plant growth. Bromine is almost 1000 times more concentrated in the sea than phosphorus. What are the implications of this fact for the production of marine life?
5. Discuss the significance of the nitrogen/phosphorus ratio in the sea and in marine life.
6. Compare the nitrogen and phosphorus cycles in the sea.
7. How does a salt, such as sodium chloride, dissolve in water?
8. Other than economic considerations, what are the advantages and disadvantages of sea-water desalination? What can be done to minimize any environmental damage that might result from desalination?
9. What are anaerobic bacteria? What part do they play in the formation of hydrogen sulfide? What distinctive feature(s) proclaims the presence of high concentrations of hydrogen sulfide?
10. What is the basic composition of sea water? Define major, minor, and trace elements as they pertain to sea water.

SUGGESTED READINGS

Broeker, Wallace S. 1974. *Chemical Oceanography.* New York, Harcourt Brace Jovanovich.
Davis, K. S., and J. S. Day. 1961. *Water: The Mirror of Science.* Garden City, N.Y., Doubleday.

Edmond, J. M. 1980. Ridge Crest Hot Springs: The Story so Far. *Transactions of the American Geophysical Union, 61*(12):129–131.

Edmond, J. M. 1980. GEOSECS is like the Yankees: Everybody Hates it and it Always Wins. *Oceanus, 23*(1):33–39.

Harvey, H. W. 1957. *Chemistry and Fertility of Sea Water.* New York, Cambridge University Press.

Hunt, Cynthia A., and Robert M. Garrels. 1972. *Water: the Web of Life.* New York, W. W. Norton.

Keosian, John. 1964. *The Origin of Life.* New York, Reinhold.

MacIntyre, R. 1970 (November). Why the Sea is Salt. *Scientific American.*

Popkin, Roy. 1968. *Desalination, Water for the World's Future.* New York, Praeger.

Riley, J. P., and G. Skirrow (eds.). 1965. *Chemical Oceanography.* New York, Academic Press.

Rubey, William M. 1951. Geologic History of Seawater. *Bulletin of the Geological Society of America. 62*:1111–1148.

Sverdrup. H. U., M. W. Johnson, and R. H. Fleming. 1942. *The Oceans.* Englewood Cliffs, New Jersey, Prentice-Hall.

Physical Nature of Sea Water

5

Knowledge of the physical properties of sea water is essential for an understanding of physical oceanographic processes. The circulation of the oceans, the transmission of light (which is essential for life) and the relation of the oceans to weather depend on the unique physical nature of sea water. The sun is ultimately responsible for virtually all physical processes in the sea. A vast amount of solar energy enters the sea and is converted to heat. Some of this energy is used to put water back into the atmosphere through evaporation. The heat reserves in the sea are so great that some may be tapped to produce electricity.

AIR–SEA INTERACTION

The boundary between air and sea is the site of several important processes. It is at the air–sea interface that light enters the sea or is

This strange ship is actually a floating instrument platform. It is towed to sea and flipped into the very stable upright position. *A*, *FLIP* flipping; *B*, *FLIP* flipped. (Photos courtesy of Scripps Institution of Oceanography.)

reflected. Here, water evaporates and also re-enters the sea as precipitation. Gases such as oxygen, carbon dioxide, and nitrogen dissolved in sea water and available for organisms are transferred from the atmosphere to sea water and vice versa at the air–sea interface. It is also at the surface that wind energy is transformed into ocean waves and wind-driven currents. Waves and currents are discussed in separate chapters.

Light in the Sea

When the sea is calm and the sun is directly overhead, about 98 per cent of the light penetrates the sea and only two per cent is reflected. The amount of light reflected increases as the sun approaches the horizon, until essentially 100 per cent reflection occurs. The amount of light that enters the sea varies considerably, depending on a number of factors, including sun angle, sky conditions, sea surface condition, and clarity of water.

The velocity of light is greater in air than in water. Therefore, when light enters the sea, it is refracted or bent downward (Fig. 5.1). Refraction of light follows Snell's Law and is discussed in detail in Appendix B. Because of refraction, the sun appears to a fish or a diver to be closer to the vertical than it really is. As light penetrates into deeper, denser water, it is refracted even more toward the vertical.

As light travels through water, it becomes progressively dimmer, owing to absorption and scattering by tiny particles in the water and absorption by the sea water itself. Sunlight is composed of a spectrum of different wavelengths or colors (Fig. 5.2). Not all colors of light penetrate the same distance into the sea. Most of the red light is filtered out in the upper few meters. Many fish and shrimps that live below the depth

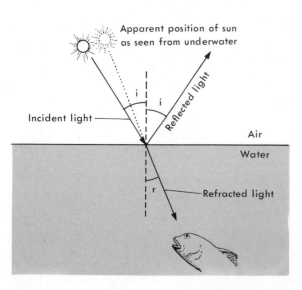

FIGURE 5.1 Reflection and refraction of light at the air-sea interface. Note that because of refraction, the sun appears to be closer to the vertical than it really is.

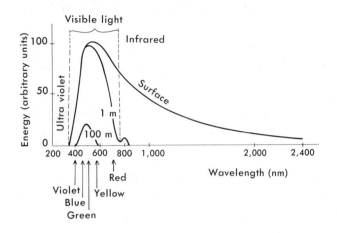

FIGURE 5.2 The spectrum of light reaching various depths in the sea. Wavelengths are given in nanometers (nm) or billionths of a meter. (After Sverdrup et al., 1942.)

to which red light penetrates are themselves red. In the absence of red light, they appear black and are thus camouflaged. In the clearest oceanic water, blue light penetrates furthest (more than 500 meters), and green light penetrates somewhat less. Figure 5.3 shows the total

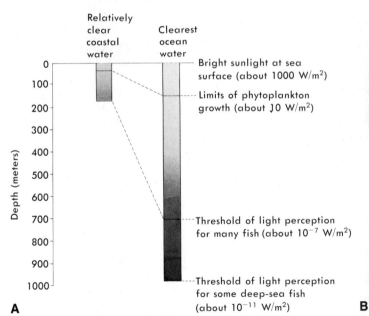

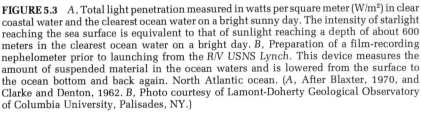

FIGURE 5.3 *A,* Total light penetration measured in watts per square meter (W/m²) in clear coastal water and the clearest ocean water on a bright sunny day. The intensity of starlight reaching the sea surface is equivalent to that of sunlight reaching a depth of about 600 meters in the clearest ocean water on a bright day. *B,* Preparation of a film-recording nephelometer prior to launching from the *R/V USNS Lynch.* This device measures the amount of suspended material in the ocean waters and is lowered from the surface to the ocean bottom and back again. North Atlantic ocean. (*A,* After Blaxter, 1970, and Clarke and Denton, 1962. *B,* Photo courtesy of Lamont-Doherty Geological Observatory of Columbia University, Palisades, NY.)

light penetration in oceanic and coastal waters. At depths of about 600 meters, the intensity of light may be equivalent to that of starlight reaching the surface of the earth. In fairly clear coastal water, selective absorption of blue light by particulate matter and dissolved yellow substances results in relatively greater penetration of green light. About 99 per cent of the total light is filtered out in the upper 50 meters. In very *turbid water* (water that is heavily laden with suspended particulate matter), most of the light is filtered out in the first meter.

As light energy is absorbed in the sea, it is converted into heat, unless the light is absorbed by a living plant, in which case some of the light energy is converted into chemical energy used in the growth of the plant. Since most light energy is absorbed near the surface, this is the region where the greatest warming occurs. Warm water is lighter (less dense) than cold water and tends to stay at the surface, resulting in stratification of warm surface water over cold deeper water. Between these two layers is the transition zone in which temperature changes rapidly with depth, known as the *thermocline* (Fig. 5.4).

The observed color of the sea is due to the color of the light that is reflected from the surface and the light that is back-scattered from within the sea. Because blue and green have the greatest penetration in sea water, they also have the greatest chance for back-scattering; hence the blue or green color of the sea. Light that is absorbed is not reflected so it is not seen from above. The color of the sea varies from deep blue in the clearest open ocean water containing little organic matter to green in relatively clear but biologically rich coastal areas to yellowish or brownish in turbid coastal water.

Near shore, the water tends to look more greenish owing to the additional scattering of light by yellowish substances derived from the decomposition of plants in the sea and on land. These yellow substances that wash to the sea in rivers may even tinge some estuarine

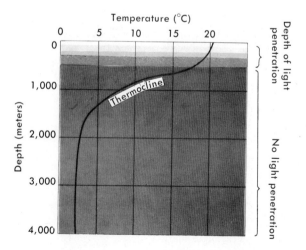

FIGURE 5.4 A thermocline, a zone of rapid temperature change with depth, in relation to light penetration in the open ocean.

water the color of pale beer. Highly turbid water tends to appear brownish because of the reflection of light by brownish suspended sediments, which in turn owe their color to rustlike compounds of iron. Nutrient-rich water may be green, brownish, or even a tomato-soup color as a result of the production of great amounts of highly pigmented microscopic plants.

The overall transparency of water may be measured using the *Secchi disc* method. This is an old technique, which usually involves lowering of a black-and-white disc into the water and noting the depth at which it disappears from view. The depth thus measured indicates the point to which about 18 per cent of surface light penetrates.

Most ocean light measurements today are made with *photometers,* light meters similar to those used in photography, which measure the rate at which light energy in watts (joules per second) reaches a particular point. Frequently, light reaching a particular depth is compared with the light at the surface, giving an indication of the transparency of the water. This technique is, of course, much more accurate than the simple Secchi disc method.

Heat in the Sea

Solar energy and the thermal properties of water largely determine the state in which water exists (solid, liquid, or vapor), as well as the temperature distribution and movement of ocean water. Heat in the sea may someday be an important source of electricity.

SOLID, LIQUID, OR VAPOR

Water is the only substance that may exist as a *solid, liquid,* or *vapor* under ordinary temperatures on earth. Water may be found as ice, liquid, or water vapor (steam).

Water has one of the highest known *heat capacities,* the amount of heat in calories* required to raise the temperature of one gm of a substance by 1° C. Therefore, it has a built-in capacity to resist changes in temperature. For example, the heat capacity of pure water is one calorie per gram (cal/gm), whereas that of iron is only about 0.1 cal/gm, and that of aluminum, nitrogen, and oxygen is about 0.2 cal/gm. Sea water, because of its dissolved salts, has a heat capacity of about 0.93, rather than 1.0 cal/gm but still warms and cools much less rapidly than land or air.

As heat is added or removed, the state in which water is found may be changed (Fig. 5.5). It takes 80 cal/gm to melt ice at 0° C, and 540 cal/gm to boil water at 100° C. Note that, in this case, as calories are added, no

---

*A calorie is the amount of heat required to raise the temperature of one gm (about 1/5 teaspoon) of liquid water by 1° C, from 15° C to 16° C. (1 calorie = 4.186 joules)

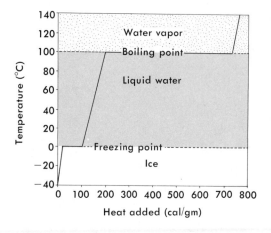

FIGURE 5.5 The changes of temperature and state of pure water as heat is added, starting with ice at −40° C.

change in temperature occurs. However, water may also *evaporate* (be converted from liquid to vapor) at temperatures lower than the boiling point, but more heat is required. For example, at 20° C, 585 calories are required to evaporate one gram of water.

As salts are added to water, the freezing point is lowered (Fig. 5.6). Sea water with a salinity of 37°/oo freezes at about −2° C. Water is one of the few substances known that becomes lighter and floats when frozen; most others contract as they freeze. The fact that ice floats is of great importance to aquatic life. If ice sank, the sea floor might be covered with ice, and the polar seas could freeze solid. Similarly, some lakes in high latitudes would freeze solid in winter, and many would never completely thaw in summer. The distribution of life, especially of bottom dwellers, would be greatly restricted.

HEAT BUDGET

As we have seen earlier in this chapter, *solar radiation* is the most important contributor of heat to the sea. In fact, absorption of light energy accounts for more than 99 per cent of the heat entering the sea. In addition, minute (but measurable) amounts of heat reach the sea from the earth's hot interior. The amount of heat entering the sea is essentially in balance with that leaving the sea. Heat leaves the sea by a variety of means, the most important of which is *evaporation.*

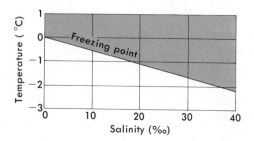

FIGURE 5.6 The relationship between salinity and the freezing point of water.

Evaporation is enhanced when the temperature of the water is higher than that of the air above it and the air is not already saturated with water vapor. Strong winds also enhance the rate of evaporation. As mentioned earlier, a great amount of heat is transferred into the atmosphere when liquid water is converted into vapor. The water heats the air and increases its capacity to retain water. Because water temperatures vary much less than those of the atmosphere, there may be considerable difference in the rate of evaporation during the day and at night and from one location to another. An average of about one cubic meter of water per square meter of surface area evaporates from the sea each year. The greatest evaporation occurs from about 20° to 30° north and south of the equator, the latitudinal belts in which most of the great deserts of the world occur. The reverse process, *precipitation,* is greatest near the equator. These relationships may be due to air-sea temperature differences as well as to trade winds, which carry water vapor toward the equator. When the water is colder than the air and the air is saturated with water vapor, *condensation* of water vapor onto the sea occurs, because the air is chilled and its capacity to hold vapor is decreased. However, this process occurs at a much slower rate than evaporation. In reality, most water is returned to the sea when light, warm, moisture-laden air rises and cools, causing the water vapor to condense onto dust or minute water droplets in the upper atmosphere, forming clouds. The water droplets grow through further condensation and may fall back to the sea or land as precipitation (Fig. 5.7). The rising and

FIGURE 5.7   When moist air rises and cools, condensation of water vapor occurs, resulting in precipitation.

cooling of moist air near the equator and along steep coastal hills and mountains accounts for the increased precipitation in these areas.

Condensation also carries some heat into the sea, but this source is insignificant compared with the loss of heat through evaporation.

Radiation of heat from the sea back into the atmosphere *(back-radiation)* ranks second to evaporation as a means of heat loss from the sea. Loss of heat through radiation depends mainly on the temperature of the water near the sea surface.* Consequently, back-radiation occurs at all latitudes.

Conduction may carry heat into or out of the sea, depending on the temperature of the air and the water. Heat leaves the sea by direct conduction from warmer water, or enters it from a warmer atmosphere or from the earth's interior. Most of the heat thus transferred flows from the sea to the atmosphere. Thus, there is a net loss of heat from the sea due to conduction.

The most important means of heat transfer are as follows:

(1) solar radiation (99+ per cent of incoming heat)
(2) evaporation (55 per cent of outgoing heat)
(3) back-radiation (40 per cent of outgoing heat)
(4) conduction (5 per cent of outgoing heat)

---

*Radiation of heat from an object (or the sea) increases with the *absolute temperature* of the object, expressed in degrees Kelvin: $°K = °C + 273$. Radiation occurs at temperatures greater than absolute zero ($0°K$).

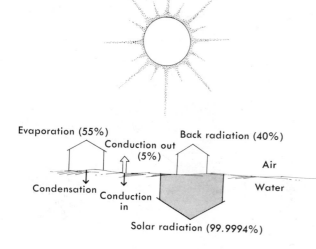

Evaporation (55%)   Back radiation (40%)

Conduction out (5%)

Air

Condensation   Conduction in   Water

Solar radiation (99.9994%)

Heat from earth's interior (0.0005%)

FIGURE 5.8   The heat budget of the oceans. The amount of heat entering the sea is nearly balanced by the heat leaving the sea.

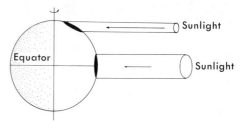

FIGURE 5.9   The amount of sunlight falling on equal areas of earth decreases from the equator toward the poles. The shaded areas on the earth are equal; the light energy reaching these areas is not.

For the oceans as a whole and averaged over a whole year, these factors are related in the following *heat budget equation:*

solar radiation = evaporation + back-radiation + conduction

Thus the amount of heat entering the oceans balances the heat that leaves them (Fig. 5.8), except for minor imperceptible changes from one year to the next. Such changes in the heat content of the oceans have certainly occurred since the origin of the oceans and have probably been related to changes in climate on earth throughout geological time.

More solar radiation reaches a square meter of equatorial ocean than reaches a square meter in polar regions (Fig. 5.9). Consequently, a greater amount of heat enters equatorial water than polar water. Because of the overall oceanic heat balance, a net poleward flow of heat must occur. Within the ocean, this is accomplished primarily by ocean currents. A similar poleward transport of heat also occurs in the atmosphere.

TEMPERATURE OF THE OCEANS

The distribution of temperature in the oceans is one of the factors that controls the distribution of marine organisms and the density of sea water. Because of the great heat capacity of water, the surface temperature of the sea varies far less than land temperatures. Consequently, the sea provides a rather stable environment for marine life and a moderating influence on the coastal climate.

Distribution of Temperature in the Sea

Sea surface temperatures are highest near the equator and lowest near the poles (Fig. 5.10). The average surface temperature is about 16° C, but it varies from about −2° C to about 30° C. Sea ice, however, may have a temperature far below −2° C. The distribution of temperature is partly dependent on current patterns, because some currents transport warm water to high latitudes, and others transport cold water toward the equator. It is because of such surface ocean currents that the surface

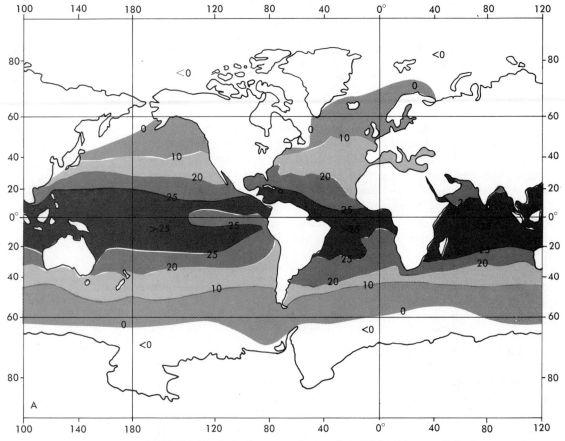

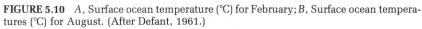

FIGURE 5.10　*A*, Surface ocean temperature (°C) for February; *B*, Surface ocean temperatures (°C) for August. (After Defant, 1961.)

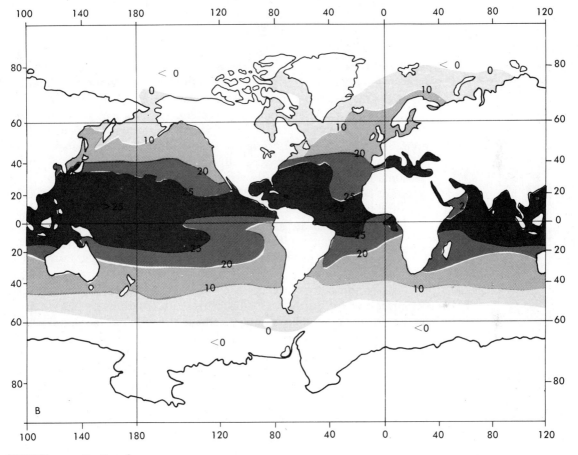

FIGURE 5.10 Continued.

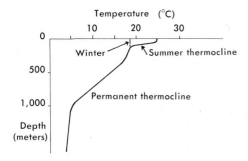

FIGURE 5.11 A double thermocline in mid-latitude regions, where there are appreciable differences between summer and winter air temperatures.

isotherms (lines of equal temperature) do not necessarily parallel lines of latitudes but are displaced poleward in the western Pacific and Atlantic oceans and are closer to the equator in the eastern parts of the oceans.

In the equatorial Pacific, cooler water extending westward from the coast of South America indicates a westward-moving current in this area. Seasonal changes in temperature patterns may also be seen. The isotherms in February are generally south of where they lie in August. Seasonal temperature changes tend to be greater at mid-latitudes than in either polar or equatorial water, especially along the western Pacific and Atlantic oceans. Such variations are at least partly due to the prevailing westerly winds that blow off the land, which is subject to much greater seasonal temperature changes than are found at sea. The eastern Pacific and Atlantic oceans are more constant in their temperature, since the westerly winds blow from the sea with its moderating influence.

The temperature of deep water varies far less than the temperature of surface water. Although day-to-night fluctuations in sea-surface temperatures are usually less than half a degree Celsius, seasonal variations may be much greater. In some mid-latitude regions, a seasonal thermocline may be present near the surface during the summer and fall, owing to seasonal variations in the amount of solar heat reaching the area. In addition, a deeper permanent thermocline may be present in many regions (Fig. 5.11).

In a north–south temperature profile of the Atlantic Ocean (Fig. 5.12) the isotherms are located at greater depths under the equator as compared to polar regions. The 4° C isotherm (representing the overall average temperature of the oceans) occurs at about 1600 meters at the equator, but reaches the surface at about 55° N and S latitudes. Water deeper than 3000 meters is cold at all latitudes. Even at the equator, water near the bottom is very cold, about 2° C at 4000 meters. However, the temperature near the ocean bottom progressively increases toward the North Atlantic, indicating the presence of a northward-flowing deep ocean current. At the poles, there is very little change in temperature with depth. East–west temperature patterns in the oceans are indicated by isotherms that slope gently downward toward the west, as shown in Figure 5.13.

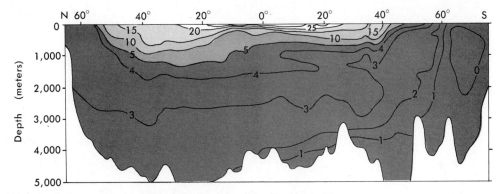

FIGURE 5.12   North-south temperature profile of the Western Atlantic. (After Wüst, 1936.)

Thermal Power

The oceans serve as the largest solar collector in the world. It has been estimated that if all the heat in the Gulf Stream (which flows northward along the southeast coast of the United States and may be responsible for keeping much of northern Europe from becoming icebound) could be transferred to water that is colder by 25° C, tremendous power could be generated (about 75 times the power consumed by the United States in 1970). There are many parts of the oceans where a 20° C to 25° C temperature difference occurs in the top 1000 meters of water (see Figs. 5.11 and 5.12). Cold ocean water could be pumped to the surface to cool the air above warm surface water. The warm water would evaporate and the resulting water vapor could be

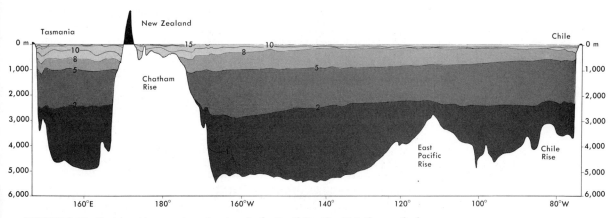

FIGURE 5.13   East-west temperature structure in the South Pacific. Note the gentle downward tilt of the isotherms toward the west. (After Stommel et al., 1973.)

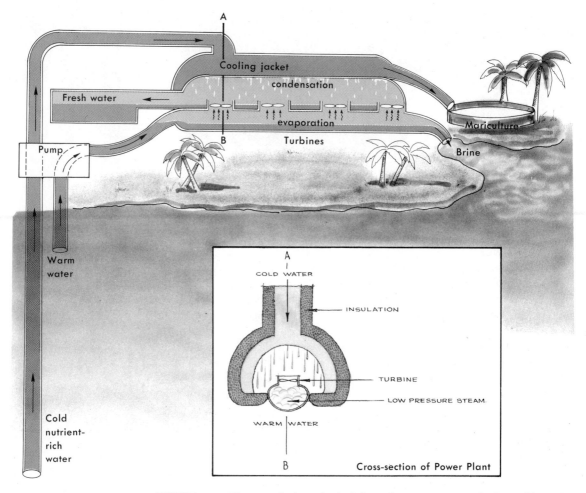

FIGURE 5.14 Diagram of a hypothetical thermal power station, which would use the power of evaporation of sea water to drive steam turbines.

used to drive a turbine and produce electricity (Fig. 5.14). After being warmed, the nutrient-rich water from the deep sea could then be transferred to a mariculture pond. The water vapor, after driving the turbines, would condense on the cold surface above the warm water, providing valuable fresh water.

The idea of using the heat in the sea as a power source was first proposed by D'Arsonval in 1881. In fact, a small power plant using this principle was built by another Frenchman, George Claude, in Cuba in 1929. Although 22 kilowatts of power was thus generated, the operation was not an economic success. If ammonia or propane, which evaporates much more easily than water, were used as a secondary fluid to drive the turbines, a more efficient process would be achieved. In order for this concept to be productive, a tremendous amount of water would have to be pumped, and many millions of dollars worth of equipment would be needed. There are several Ocean Thermal Energy Conversion (OTEC) plants being considered for development. Figure 5.15 illustrates such an OTEC plant. The U.S. Department of Energy has converted a former oil tanker into *OTEC-1* to test heat exchangers for this system. Since ammonia and propane are potentially hazardous, extreme

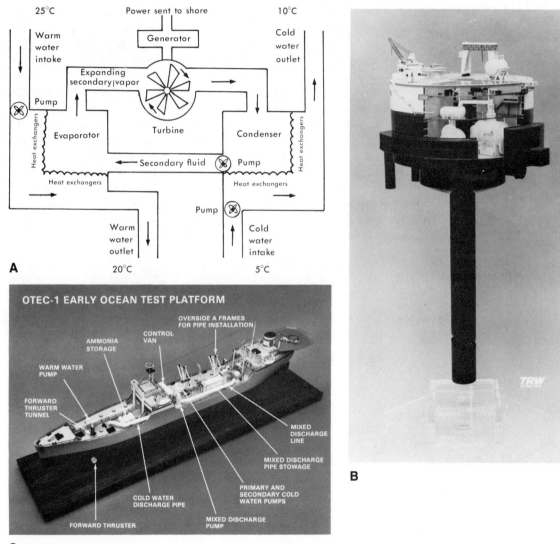

FIGURE 5.15   Ocean Thermal Energy Conversion. (OTEC). *A,* A diagram of a closed-cycle OTEC power plant; *B,* Artist's conception of a floating OTEC power plant; *C,* OTEC-1, a converted oil tanker used to test large heat exchangers in the ocean. (*A* modified from and *B* and *C* courtesy of Dept. of Energy.)

care must be taken in the design and operation of such plants. But the energy is there, and it is renewable. In a few years, we may derive some of our energy from the heat in the sea.

Power thus generated from the sea could be linked to the production of hydrogen or magnesium from sea water and ammonia using nitrogen from the air and hydrogen from the sea. Furthermore, this may bring economical power to remote islands.

Ocean Temperature Measurements

The temperature of surface water may be measured using an ordinary thermometer placed in a sample of water collected with a bucket or a water sampling bottle such as a Nansen bottle (Fig. 5.16). However, the temperature of a deep-water sample will change before it is brought to the surface, so it is necessary to measure its temperature *in situ* (in place). Special thermometers have been developed which may be attached to a Nansen bottle and lowered into the water upside-down and then reversed at the desired depth, automatically registering the temperature at that depth (Fig. 5.17). These thermometers have been devised so that the mercury column remains intact on the way down, and then, when the thermometer is reversed, breaks, isolating a smaller column of mercury which is proportional to the temperature at that depth. After the temperature is read on board the ship, the thermometer is tipped back to the original position. The mercury column is rejoined and is

FIGURE 5.16   Nansen bottle being removed from the wire. (Courtesy of Woods Hole Oceanographic Institution.)

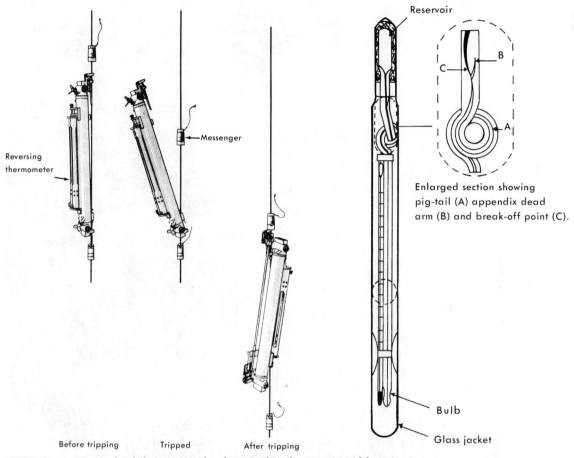

Reservoir

B

C

Enlarged section showing
pig-tail (A) appendix dead
arm (B) and break-off point (C).

A

Reversing
thermometer

←Messenger

Before tripping   Tripped   After tripping

Bulb

Glass jacket

FIGURE 5.17 Nansen bottle being tripped and reversed. A close-up view of the reversing thermometer is shown on the right. (After U.S. Naval Oceanographic Office, 1968. Pub. 607.)

(in reversed position)

ready to be used again. The water collected by the Nansen bottle may be used for various chemical analyses.

The bathythermograph is another temperature-measuring device (Fig. 5.18). It records the temperature and pressure continuously on a coated glass plate, from the surface to the desired depth. Since pressure is closely related to depth, this instrument produces a graph of temperature changes with depth on the glass plate. The conventional bathythermograph has been replaced by the *expendable bathythermograph* or *XBT*. This device consists of a torpedo-shaped probe containing the temperature sensor and a coil of wire that sends the temperature information to the recorder onboard the ship.

Temperature may be measured electronically with a temperature probe, which is a temperature-sensitive device connected to a meter

FIGURE 5.18 The bathythermograph (B-T), an instrument that automatically traces a record of the temperature changes with depth (pressure) as it is lowered into the sea. *A*, Close-up showing B-T, slides, and reader. *B*, B-T being retrieved. (Courtesy of NOAA.)

(Fig. 5.19) on the ship. An electric current is set up in the probe and transmitted to the meter. The current is proportional to the temperature at the depth of the probe. These temperature-measuring units are frequently combined with devices that measure other parameters, such as electrical conductivity, salinity, depth, and dissolved oxygen, all at the same time.

SALINITY OF THE OCEANS

Salinity, the amount of dissolved solids in grams per kilogram of sea water (⁰/oo or parts per thousand), depends on evaporation, precipitation, fresh water input from rivers, and mixing by currents. Salinity in the oceans generally varies from about 32 to 37⁰/oo, except in the Arctic and near shore, where it may be less than 30⁰/oo. The surface salinity in the Red Sea may be greater than 40⁰/oo, due to high rates of evaporation, but the average salinity in the oceans is about 35⁰/oo or 3.5 per cent. Figure 5.20 shows the distribution of surface salinity in the oceans.

FIGURE 5.19   An electronic salinity-temperature-depth (STD) profiling instrument, with remote-tripping water sampling bottles and reversing thermometers. *A*, Instrument being lowered from ship; *B*, Measurements being made aboard the ship. (Photos courtesy of Lamont-Doherty Geological Observatory of Columbia University.)

The surface salinity of the oceans depends on the difference between evaporation (E) and precipitation (P). Figure 5.21 shows the relationship between surface salinity, evaporation, and precipitation in the oceans. Notice that the areas of highest salinity, at 25° north and south latitudes, coincide with the regions where evaporation exceeds precipitation by the greatest amount. The mechanics of evaporation have been discussed earlier in this chapter. For the oceans as a whole, evaporation exceeds precipitation by about 10 cm/yr. Because the ocean-atmosphere-land network is a closed system, approximately this amount of water re-enters the sea via the rivers each year.

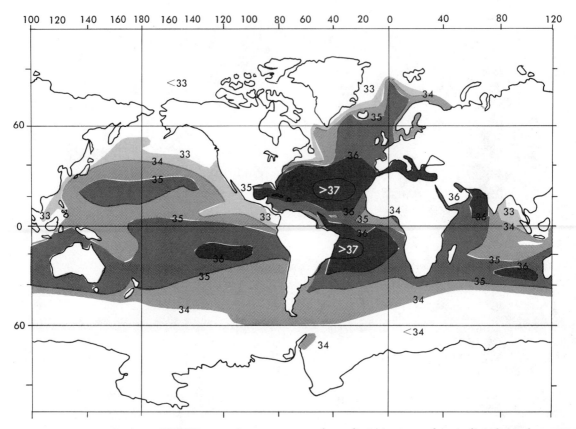

FIGURE 5.20 Average ocean surface salinity (parts per thousand). (After Defant, 1961.)

Surface salinity, evaporation, and precipitation have a well-defined relationship for the oceans as a whole. This relationship is expressed by the equation

$$S = 34.6 + 0.0175 \, (E - P)$$

where S is salinity (⁰/₀₀), and E and P are evaporation and precipitation, respectively, expressed in centimeters per year. It is interesting to note

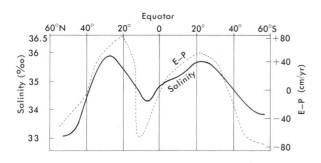

FIGURE 5.21 Relationship between surface salinity (solid) and the annual difference between evaporation and precipitation (E-P) for all oceans (dashed). (⁰/₀₀ = parts per thousand.) (After Wüst, 1954.)

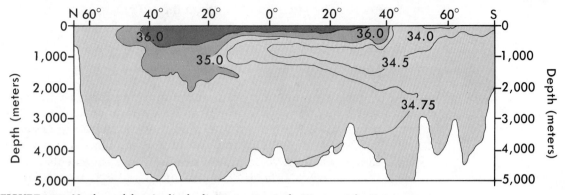

FIGURE 5.22   North-south longitudinal salinity structure in the Western Atlantic in parts per thousand. (After Wüst, 1936.)

that 34.6 represents the average salinity at about 500 meters, implying that the surface waters may be mixed to that depth. Locally, salinity may be raised by freezing of sea water or lowered by river runoff and melting of ice. Extra-saline water from the formation of sea ice is generally the densest water in the ocean, and sinks towards the bottom. The lowered salinity in the Bay of Bengal is due to the tremendous runoff of fresh water from the Ganges River.

In coastal regions receiving large volumes of fresh water from rivers, sharp gradients of salinity changes, or *haloclines*, occur between the fresher surface water and saltier deeper water. In the tropical open ocean, however, the surface water is usually saltier than the deep water, with a halocline frequently existing at about 1000 meters.

Figure 5.22 shows the north–south vertical distribution of salinity in the Atlantic Ocean. Note that there are regions of salinity maxima and minima, which indicate the presence of deep ocean currents originating in high latitudes. These currents are caused by increases in density, which are mainly the result of low temperatures rather than high salinity.

DENSITY OF SEA WATER

Almost all movements in the deep ocean are caused by density gradients analogous to the atmospheric pressure gradients that create winds (which, in turn, drive surface currents in the ocean). Therefore, density is of prime interest to oceanographers in the study of ocean currents. In addition, density is important to life in the sea. Water makes up the bulk of plant and animal cells. Consequently, the density of the biota is nearly the same as the water in which they live, which enables many organisms to be supported in the water and to be more or less neutrally buoyant. However, some plants and animals float on the surface or regulate their buoyancy with the aid of gas-filled bags or oil droplets.

Factors Controlling Density

The density of sea water is a function of temperature, salinity, and pressure. In the open ocean, density increases (from about 1.02 to 1.05 gm/cc between the surface and 5000 meters) with decreasing temperatures and increasing salinity and pressure. Although density at any depth in the ocean can be accurately calculated, measurements of density *in situ* are very difficult. The density of water is usually determined after bringing the water sample to the surface, and discounting the effect of pressure. This method is justifiable for the study of horizontal currents, since one usually compares densities at the same depth (nearly the same pressure) at different localities.

Unlike the uniform effects of pressure on density as a function of depth (one atmosphere per 10 meters of water), the effects of temperature and salinity are localized and form the main driving mechanism of deep ocean currents. Figure 5.23A illustrates the variation of density with temperature, at constant salinity and atmospheric pressure. Figure 5.23B shows the variation in density with salinity, assuming constant temperature and atmospheric pressure. If we know the temperature and salinity of a water sample, its density at the surface (at atmospheric pressure) can be obtained by using a *temperature–salinity (T-S) dia-*

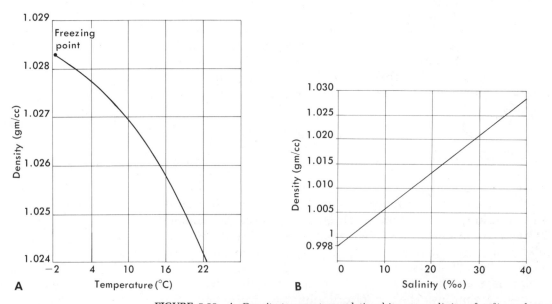

FIGURE 5.23 *A*, Density-temperature relationship, at a salinity of 35⁰/₀₀ and at atmospheric pressure. *B*, Density-salinity relationship at 20° C and atmospheric pressure.

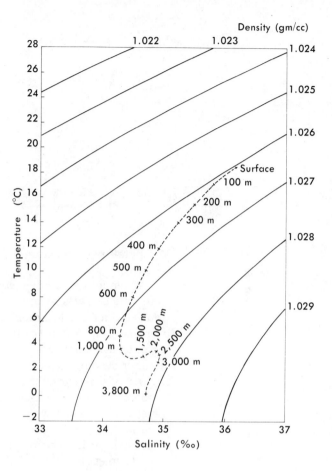

FIGURE 5.24 Temperature-salinity diagram. Density values at the ends of the diagonal curved lines refer to the density of the water at atmospheric pressure. In this case, temperature and salinity values have been plotted for 14 water samples, which were collected at various depths at a sampling station located at about 30° S latitude in the Atlantic Ocean. Depths (in meters) are indicated on the dashed line connecting the data points. (After data of Wüst, 1936.)

gram. Oceanographers use this graph to plot temperature and salinity data from various depths at a particular locality (Fig. 5.24). This data can be used to study the deep ocean currents, because we can identify waters from various localities by their temperature and salinity characteristics.

A uniform increase in density with depth has little effect on organisms. However, in areas where a sharp density change *(pycnocline)* occurs over a small vertical distance—as in regions where very warm and therefore light water overlies a dense, cold layer—several striking phenomena may occur. The sudden increase in density may cause a slowing in the sinking rate of dead organic matter, which will accumulate at this level. An oxygen-minimum zone may be formed owing to the decomposition of this accumulated organic matter by bacteria, which use up oxygen in the process and which may in turn drive away most of the oxygen-requiring organisms that ordinarily would eat the organic matter.

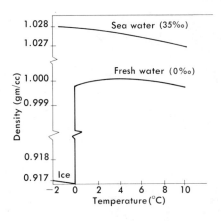

FIGURE 5.25 Temperature-density relationships for fresh water and for sea water of 35‰ salinity.

Temperature of Maximum Density

When ice melts, the resulting fresh water is more dense than the ice from which it was formed (about 0.08 gm/cc greater). The density continues to increase to a maximum at about 4° C, above which the density decreases (Fig. 5.25). For this reason almost all deep lakes in temperate regions have a dense 4° C layer of water near bottom no matter how hot summer becomes, as long as the temperature drops to 4° C for substantial periods in winter. As water is warmed, the molecular movement increases, tending to expand the volume. However, at the same time, the structure of the water changes. Water exists as a mixture of single molecules and *polymers* (two or more water molecules bound together as a single unit). The polymers occur in groups of two, four, or eight united molecules (Fig. 5.26), the size and shape of which change with temperature. The lower density (between 0° C and 4° C) can be credited to the increased concentration of large bulky polymers.

The addition of salt to water lowers the temperature of maximum density (Fig. 5.27) until, at 24.7‰ salinity, the water freezes at the temperature of maximum density. Therefore at higher salinities, sea water freezes before the theoretical temperature of maximum density is

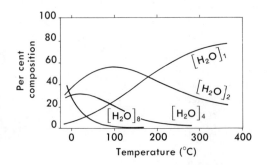

FIGURE 5.26 Relative composition of single water molecules and 2-, 4-, and 8-unit polymers of water at various temperatures. (After Euken, 1948.)

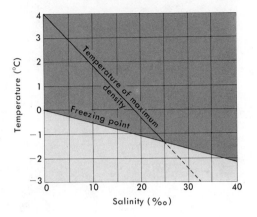

FIGURE 5.27   The relationship of freezing point and temperature of maximum density to temperature and salinity. Note that above 24.7%, the water freezes before the theoretical temperature of maximum density (dashed line) is reached.

reached. As sea water approaches the freezing point, it becomes more dense and tends to sink, causing the water to be well mixed. The oceans do not have a 4° C bottom layer as do lakes but rather have water that may be as cold as 0° C or less. Consequently, the sea surface freezes only when the water is near the freezing point at all levels, from surface to bottom. As the initial ice is formed, much of the salt is excluded, causing the salinity of the water below the ice to increase, further lowering the freezing point. Sea ice, therefore, forms less readily than lake ice.

ICE ON THE SEA

Not all ice on the sea (Fig. 5.28) is formed in the sea. When glaciers move into the sea and break up, they form *icebergs*, perhaps the most spectacular as well as the most dangerous forms of ice in the sea. The Antarctic and Greenland glaciers are the main source of icebergs. Icebergs usually extend no more than about 50 meters above the sea, but five times as much is hidden from view beneath the water. Most of the icebergs in the Antarctic Ocean are *tabular icebergs*, which are relatively flat on top and have nearly vertical sides. They may drift as far north as 35° S latitude in the Atlantic, 45° S latitude in the Indian and 50° S latitude in the Pacific Ocean. Their lengths may be as much as 1500 meters. They are formed as a result of the breakup of *shelf ice* that is attached to and extends from the land mass. Shelf ice is composed of a combination of land ice, snow and ice formed in the sea *(sea ice)*, and may reach about 300 meters in thickness. The Ross Sea and the Weddell Sea regions account for most of the shelf ice in the Antarctic Ocean. In contrast, Arctic and North Atlantic icebergs formed from the breakup of the Greenland glaciers have an irregular topography and may aptly be called *pinnacled* icebergs. Many of these are carried by the East Greenland and the Labrador currents, and may travel as far as 43° N latitude. In the Arctic Ocean, there are many unusually large icebergs

FIGURE 5.28   Ice in the sea.

A, Arctic *pack ice* (which rarely exceeds 4 meters in thickness) north of Alaska. The instrument shown is a buoy that transmits data about temperature, air pressure, and ice movements via orbiting satellites.

B, Pancake ice, a variety of sea ice. Rounded shapes are produced when ice crystals collide against each other owing to motions in the water. An average pancake is about one meter in diameter. Continued cooling may result in the formation of pack ice. The salinity of sea ice rarely exceeds 10⁰/oo.

C, A *pinnacled iceberg* in the Arctic. These icebergs are formed when glaciers reach the sea and are broken.

D, The Ross ice shelf, attached to the Antarctic continent. *Shelf ice* is formed in protected bays such as the Ross Sea from a combination of sea ice, snow, and land ice, and may reach about 300 meters in thickness.

E, Tabular iceberg in the Antarctic. Broken pieces of shelf ice result in tabular icebergs. Similar icebergs called *ice islands* are formed in the Arctic north of Canada and Greenland and may exceed 50 meters in thickness, but may have areas of hundreds of square kilometers.

(*A* courtesy of NOAA; B–E courtesy of U.S. Coast Guard.)

that are called *ice islands.* They were initially discovered by the U.S. Air Force in 1946 and may have lengths of more than 30 kilometers. They are formed from the breakup of shelf ice north of Canada and Greenland.

Sea ice is formed when sea water freezes. The initial ice crystals collide against each other due to wave motion and form rounded ice, called *pancake ice.* An average pancake is about a meter in diameter. Continued freezing of sea water and snow may produce *pack ice* which, in the Arctic, may be as thick as five meters after several years. Essentially the entire Arctic Ocean is covered by pack ice at least part of the year. Most of the sea ice in the Arctic is less than ten years old. Although the initial ice crystals formed are salt-free, as freezing continues, brine may be trapped within the growing mass of ice. Young sea ice may have a salt content of about 5 to 10%. Very old sea ice has little or no salt in it as the denser brine and salt crystals settle out through the ice.

PRESSURE IN THE OCEANS

Organisms that live in the deep sea are subject to tremendous pressure, exerted upon them by the weight of the water that lies over them. A fish at 4000 meters must withstand about 400 times as much water pressure as a fish near the surface. Pressure in the sea increases by about one atmosphere for each 10 meters of depth. Thus, the pressure in a trench at 10,000 meters is about 1000 atmospheres. Life exists at all depths of the sea, and pressure does not pose a great problem for most marine organisms since the pressure inside is the same as that outside the organisms. However, chemical reaction rates may differ under various pressures, and many organisms may be chemically adapted to life at a particular range of depth.

A gas-filled sac such as a fish's swim bladder or the lung of a human diver tends to shrink with increasing depth unless more gas is pumped into the sac to equalize the pressure inside and out. Fish that have swim bladders, for example the many fish that carry on extensive vertical migrations (see also the section on Marine Migrations in Chapter 10), have glands that supply gas during the descent. Similarly, divers may increase their depth range by carrying Self-Contained Underwater Breathing Apparatus (SCUBA), from which they can draw enough air to maintain the balance between internal and external pressure. A diver at 20 meters experiences a pressure of three atmospheres both inside and outside the lungs. When returning to the surface, where there is only one atmosphere of pressure, the diver must continuously exhale to relieve the excess pressure, or else his lungs might rupture.

SOUND IN THE SEA

The propagation of sound waves in the sea is of great importance in oceanography. The human ear can detect sound frequencies from about 20 to 20,000 cycles per second (cps). Sound waves of ultrasonic frequencies, usually above 20,000 cps, are used to determine ocean depths, thickness of sediment, and location of underwater objects such as fish and submarines. Sound transmission in the sea may also someday provide us with a new means of communication. Actually, sound waves are the only effective means of long-range communication in the sea. Light can penetrate a few hundred meters and radio waves can penetrate only a few meters of water.

The velocity of sound in the sea varies from about 1450 to 1550 meters per second, as compared to about 330 meters per second in air, increasing with increasing temperature, salinity, and pressure. Figure 5.29 shows a typical profile of sound velocity and temperature with depth in the ocean. A sound-velocity minimum is generally found at about 1000 meters. Near the poles, however, the minimum is nearer the surface, and in equatorial regions, it may be found at a greater depth. The decrease in sound velocity in the upper 1000 meters is associated

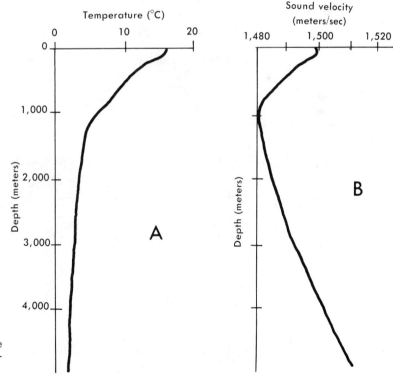

FIGURE 5.29 Typical temperature (A) and sound velocity (B) variations with depth in the sea.

with the decrease in temperature with depth (the thermocline). We know that the gradual increase in velocity below 1000 meters is mainly due to the increase in pressure with depth, because variations of temperature and salinity are generally not appreciable in deep water. However, near the surface, where temperature and salinity variations with depth are significant, irregularities in this pattern may occur, depending on location, and a sound-velocity maximum may be reached at about 100 meters. This effect is of considerable importance in the detection by sonar of underwater objects such as submarines. Figure 5.30 shows the refraction of sound waves near the sound-velocity maximum. (See Appendix B for a discussion of the refraction of waves.) Refraction of the waves upward above the sound velocity maximum and downward below it produces a zone into which the sound waves do not penetrate. This *shadow zone* enables a submarine to avoid detection by a ship's SONAR. This problem can be circumvented by using a *variable depth SONAR,* in which the sound transmitter and receiver are lowered by a cable to any desired depth, thus preventing a shadow zone from interfering with the search for a submerged object. The SONAR discussed above is considered the "active" type because sound is produced by the transmitter. If no sound is sent, but sound produced in the sea is received and located, the system is referred to as "passive" SONAR. Passive SONAR can be used to locate ships, submarines and noise-producing marine animals such as some fish and shrimp.

The sound-velocity minimum shown in Figure 5.29 can aid ships in distress. A depth charge can be made to explode at the depth of minimum velocity. Refraction causes most of the energy associated

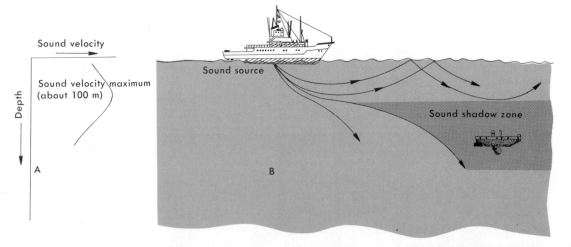

FIGURE 5.30 A sound velocity maximum (A), usually located at a depth of about 100 meters, results in a sound shadow zone (B) when the sound source is near the surface. Sound waves are refracted upward above the sound velocity maximum and downward below it.

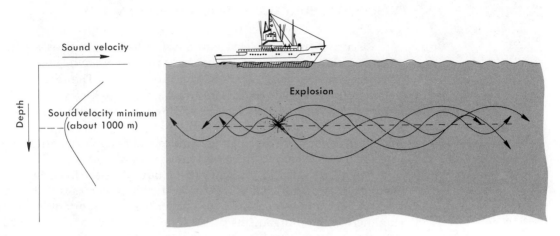

FIGURE 5.31 The sound velocity minimum *(A)*, usually located at a depth of about 1000 meters, results in the Sound (SOFAR) Channel *(B)* in the oceans. Sound waves produced near the sound velocity minimum could travel for thousands of kilometers. Sound energy is refracted toward the sound velocity minimum.

with the explosion to be propagated through this zone (Fig. 5.31), known as the *sound channel* or *SOFAR*. (*So*und *F*ixing *A*nd *R*anging) *channel*. Experiments have indicated that sound waves can be propagated in this manner for thousands of kilometers. The signals can be picked up by listening devices on ships, and the position of the ship in distress can be determined by triangulation, since the signals will arrive at different times at different listening stations. Although this technique is not fully utilized today, it may someday have great practical significance.

Echo sounding (discussed in Chapter 2) often indicates a "false bottom," called the *deep scattering layer,* above the true sea floor. This is caused by reflection of sound waves off fish and other animals in the sea.

Animals in the sea sometimes make noises that can be heard by other animals. Croakers croak, snapping shrimps snap, and even fish and squids swimming through the water set up vibrations in the water that may be sensed by other organisms, usually through their skin (in the case of many fish, a tubular fluid-filled "lateral line" aids in the perception of vibration). Sounds produced by marine animals, including porpoises, and other whales are being studied in the hope that someday we may be able to communicate with them and perhaps train them to respond to orders transmitted through sound. Certain sounds made by animals in the sea may be played back underwater and might serve as a lure. The study of marine biological sounds should, at least, help us to better understand the ecology and behavior of the animals that produce the sounds.

FIGURE 5.32   The oceanic water strider *Halobates* is seen here standing on the water. Tens of thousands of genera of insects are known on land; there is only one oceanic genus: *Halobates*. (Photo courtesy of Scripps Institution of Oceanography, University of California, San Diego.)

VISCOSITY, SURFACE TENSION, AND OSMOSIS

Viscosity, or the resistance of a liquid to flow, decreases with increasing temperature (as syrup becomes runnier when heated). Salinity increases the viscosity of water, but only slightly. The relationship of temperature and viscosity causes tropical water to be less viscous than temperate or polar water, thus providing less resistance to the sinking of tropical plankton. It is interesting that many tropical species of minute organisms have much longer spines and hairs than similar or even the same species found in colder, more viscous water.

Surface tension (the tendency of the liquid surface to resist penetration) decreases with increasing temperature and increases with increasing salinity. Surface tension of sea water is very important in supporting the weight of organisms that rest on the surface of the sea, such as the marine water-strider *Halobates* (Fig. 5.32), which is the only truly oceanic insect. Changes in surface tension due to temperature and salinity probably are not great enough to affect most organisms.

Materials that dissolve in water tend to spread out, or diffuse, until they are evenly distributed in the water. However, biological membranes (and other structures) serve as barriers to restrict diffusion. The membranes that form the boundaries of plant and animal cells are selectively permeable; that is, they allow some molecules, such as water and oxygen, to pass through freely, while others such as the salt ions are restricted in their movement across the membrane (Fig. 5.33). This selective flow of water is called *osmosis.* For example, if a selectively permeable membrane separates salt water from fresh water, there will be a net flow of fresh water into the salt water. The pressure on the membrane, caused by the flow of water, is the *osmotic pressure.*

A freshwater organism placed in the sea tends to shrink as water diffuses through its membranes into the sea, leaving less water in its cells than before. If a strictly marine organism is placed in fresh water, the cells may rupture owing to the diffusion of water into the cells.

Animals such as the green crab *(Carcinus maenas)* that inhabit regions of fluctuating salinity can live in water with a lower salt concentration than is found in their cells. They survive by actively pumping water out of the cell. Their internal salt concentration varies somewhat, but not as much as that of the external water.

SUMMARY

Heat is the driving force for many of the physical phenomena that take place in the sea. The sun is the ultimate source of energy for light and heat in the sea. Evaporation, which transfers heat from the sea to the atmosphere, tends to increase sea-surface salinity, whereas precipi-

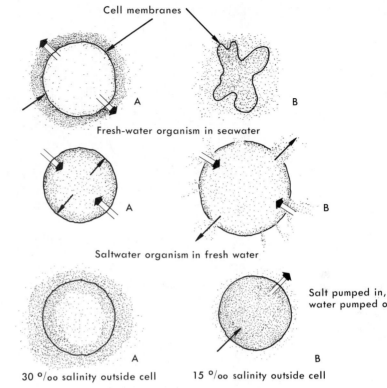

Cell membranes

Fresh-water organism in seawater

Saltwater organism in fresh water

Salt pumped in,
water pumped out

30 °/oo salinity outside cell   15 °/oo salinity outside cell

Salt tolerant organism (green crab)

FIGURE 5.33 The flow of water across cell membranes under various conditions of salinity. *A*, Before exposure; *B*, After exposure. Flow of materials: □ H_2O ⇒; ⊠ salt →.

tation tends to decrease it. Evaporation exceeds precipitation near 20° N and 20° S latitudes, where sea-surface salinity is nearly maximum. Sea-surface temperature ranges from about −2° C near polar regions to about 28° C near equatorial regions. Water temperature near the deep sea bottom, however, is near 0° C everywhere.

The density of sea water increases with increasing salinity and pressure, but decreases with increasing temperature. Fresh water has a maximum density at a temperature of 4° C. The temperature of maximum density of water decreases with increases in salinity until 24.7% salinity, at which it freezes at the temperature of maximum density. The freezing point of water decreases with increasing salinity. Freezing of sea water is further complicated by the fact that after the initial ice crystals are formed, the surrounding residual sea water attains a higher salinity and can freeze only at a lower temperature. This helps explain why sea ice (ice formed in the sea) is only a few meters thick in the Arctic Ocean.

A vast amount of heat is stored in the surface layer of tropical oceans. We may be able to tap some of this energy by building OTEC

(Ocean Thermal Energy Conversion) power plants in tropical oceans. These plants would use ammonia or propane as the working fluid which could be evaporated by using the heat in warm surface water. The resulting vapor could drive turbines. Cold deep water could be pumped to the surface and used to condense the vapor, which could be recycled. OTEC plants using this principle are currently being developed.

Sound travels at about 1450 to 1550 meters per second in water and about 330 meters per second in air. The speed of sound in water increases with increasing temperature, salinity, and pressure. This relationship results in sound-velocity maximum and minimum zones in the sea. Variations in sound speed with depth has important implications for submarine detection and possibly also for communication through the sea.

Two other aspects of the physical nature of sea water are viscosity and surface tension. Viscosity, the resistance of a liquid to flow, is important in retarding the rate at which particles such as plankton sink in the sea. Surface tension, the tendency of a liquid surface to resist penetration, allows some organisms, such as the marine insect *Halobates*, to walk on water.

IMPORTANT TERMS

Back-radiation
Bathythermograph
Condensation
Conduction
Evaporation
Halocline
Heat budget
Ice islands
Iceberg
Isotherm
Osmosis
OTEC
Pack ice
Photometer
Polymer

Pycnocline
Refraction
Sea ice
Secchi disc
Shadow zone
Shelf ice
SOFAR
SONAR
Sound channel
Solar radiation
Surface tension
Thermocline
Turbid
Viscosity
XBT

STUDY QUESTIONS

1. Explain the refraction of light in the sea.
2. Why does the refraction of light vary with salinity?
3. As a diver descends in the sea, red fish become black in appearance and most objects appear blueish or greenish. Why is this so? What happens if the diver shines artificial white light on the submerged objects?

4. At the sea surface, temperature varies from 0° C to 25° C or higher, yet at 1000 meters the temperature is generally less than 5° C. Why is this so?

5. Graph temperature against depth at 20, 40, and 60° S latitude using Figure 5.12 as a source of data. Indicate the thermocline on each graph.

6. How is the depth of light penetration related to the vertical temperature distribution in the sea?

7. Briefly discuss the formation of sea ice.

8. What factors control the distribution of sea surface salinity? Why is it a maximum at about 20° N latitude and 20° S latitude?

9. What is meant by "temperature of maximum density?" How does salinity affect it?

10. How might heat in the sea be utilized to generate electrical power?

11. How is the shadow zone produced in the oceans?

12. What is the SOFAR (Sound) Channel?

13. Describe osmosis. What effect does osmotic pressure have on a freshwater fish which is suddenly placed in a saltwater environment?

14. Discuss the origin of Arctic and Antarctic icebergs. What physical characteristics are typical of each?

SUGGESTED READINGS

Bowditch, Nathaniel. 1977. *American Practical Navigator*, vols. 1 and 2. Washington, D.C., Government Printing Office. (Originally published in 1802.)

Cohen, Robert. 1979. Energy from Ocean Thermal Gradients. *Oceanus*, 22(4): 12–22.

Donn, William L. 1965. *Meteorology*, 3rd ed. New York, McGraw-Hill.

Knauss, John A. 1978. *Introduction to Physical Oceanography*. Englewood Cliffs, New Jersey, Prentice-Hall.

Othmer, Donald, and Oswald Roels. 1973. Power, Fresh Water and Food from Cold Deep Seawater. *Science*, 182:121–125.

Pickard, George L. 1963. *Descriptive Physical Oceanography*. New York, Pergamon Press.

Roll, H. U. 1965. *Physics of the Marine Atmosphere*. New York, Academic Press.

Stewart, R. W. 1978. The Role of Sea Ice in Climate. *Oceanus*, 21(4):47–57.

Sverdrup, H. U., M. W. Johnson, and R. H. Fleming. 1942. *The Oceans*. Englewood Cliffs, New Jersey, Prentice-Hall.

von Arx, William S. 1962. *An Introduction to Physical Oceanography*. Reading, Mass., Addison-Wesley.

Ocean Waves

The most readily observable phenomena in the sea are waves. They range from small ripples to gigantic and highly destructive waves such as tsunamis, which may rise more than 30 meters above normal sea level by the time they reach the shore. Tides, which are caused by the gravitational attraction of the moon and the sun on the waters of the earth, also qualify as waves and will be discussed in the next chapter. Waves are of tremendous importance to us, not only because they make some people seasick and enable others to surf but also because they may erode beaches and destroy artificial structures. Someday electricity may be generated from ocean waves. In addition, waves have many other far-reaching influences on the oceans. For instance, waves breaking on the sea surface make it easier for atmospheric oxygen and carbon dioxide to be dissolved in sea water. Wave action is also responsible for the mixing of the surface water layer. Both of these aspects are of great significance in biological oceanography.

Most ocean waves are caused by wind. However, they can also be caused by submarine earthquakes, submarine landslides, submarine volcanic eruptions, landslides into the sea, ships, and the gravitational attraction of the moon and sun on the earth.

The oil tanker *Argo Merchant* ran aground, broke in two in heavy seas, and sank off Nantucket Island in 1976, causing a major oil spill. (Photo courtesy of U.S. Coast Guard.)

NATURE OF WAVES

Figure 6.1 shows several uniform, ideal ocean waves generated by any of the forces mentioned earlier. The distance between two adjacent crests (or troughs) is called the *wavelength* (L). The vertical distance between the crest and the trough is the *wave height* (H). The time required for two successive crests (or troughs) to pass by a fixed point is the *period* (T) of the wave. The *speed* (C) of the wave is the distance the wave travels in a unit of time such as meters per second, or kilometers per hour. A very useful but simple relationship using the preceding factors for any wave is L = CT. For example, if the period of a wave is 10 seconds, and the wave speed is 15 meters per second, the wavelength is 150 meters.

In reality, ocean waves are never as simple as those shown in Figure 6.1; the real ocean surface is much more irregular. Waves moving in different directions, which may or may not have the same periods, lengths, and heights can interfere with each other, producing a complex sea-surface pattern (Fig. 6.2). The effects of wave interference can easily be seen by dropping two pebbles in a large puddle.

Waves on the sea surface tend to flatten out and may eventually disappear. The force that causes the sea surface to flatten, and return to the initial state of equilibrium, is called the *restoring force.* Very small ocean waves having lengths less than 1.73 cm and periods less than 0.1 second are called *capillary waves,* for which surface tension is the primary restoring force. For waves having lengths greater than 1.73 cm and periods greater than 0.1 second, the major restoring force is gravity and hence these waves are generally called *gravity waves.* The waves discussed in this chapter are gravity waves.

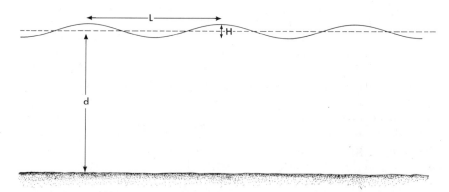

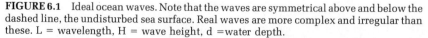

FIGURE 6.1   Ideal ocean waves. Note that the waves are symmetrical above and below the dashed line, the undisturbed sea surface. Real waves are more complex and irregular than these. L = wavelength, H = wave height, d = water depth.

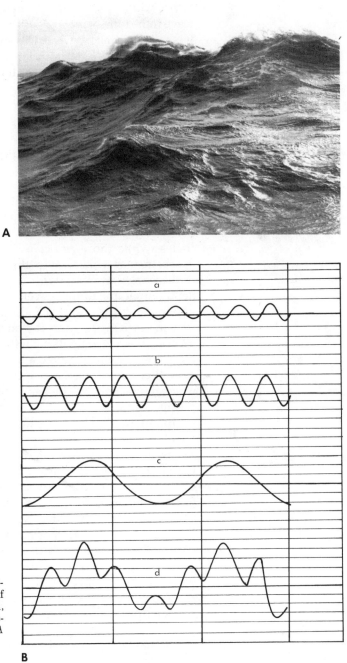

FIGURE 6.2 *A*, Typically complex wave pattern on the open sea. *B*, The superposition of many waves (a, b, and c) of varied wavelength, height, and period, traveling in different directions, produce a complex wave pattern (d). (*A* courtesy of U.S. Coast Guard.)

DEEP-, INTERMEDIATE-, AND SHALLOW-WATER WAVES

Ocean waves are classified into three major categories: *deep-, intermediate-* and *shallow-water waves.* A wave is called a deep-water wave if the ratio of water depth to wavelength (d/L) is greater than 1/2 and it is called a shallow-water wave if this ratio is less than 1/25.* If the ratio d/L is between 1/2 and 1/25, it is called an intermediate-water wave. It is important to note here that the terms "deep" and "shallow" water do not necessarily imply deep and shallow in the usual sense. For example, a tsunami having a wavelength of 150 kilometers will travel as a shallow-water wave in waters less than six km in depth (d/L = 6/150 = 1/25). Very small waves, however, may be classified as deep-water waves even near the beach or in large puddles. Deep-water waves become shallow-water waves as they move into shallow areas in which d/L is less than 1/25.

Many differences exist between deep- and shallow-water waves. Figure 6.3 shows the motions of water particles in these two types of waves. The water particles associated with deep-water waves move in circles, while those associated with shallow-water waves move in ellipses. In deep-water waves, the circular orbital speeds as well as the size of the circles decrease with depth and become almost negligible at a depth equal to half the wavelength (L/2). For shallow-water waves, the horizontal particle speeds and the horizontal dimensions of the ellipses remain unchanged from the surface to the bottom of the ocean. The ellipses, however, get flatter with depth, and at the bottom, the particles move back and forth, almost horizontally. Thus in the case of tsunamis and other large waves which fit the definition of shallow-water waves, water movement can be expected even in the deepest portions of the oceans. These waves can have great influence on shallow bottoms such as continental shelves and portions of the continental slopes, although their effects on the deep-sea bottom are not considered to be significant.

Surprisingly, there is only a very slight forward motion of water in waves until they break near the shore. This effect can be demonstrated by observing a piece of driftwood floating in the ocean. The wood moves forward only slightly with the waves, remaining nearly stationary as the waves pass. The same effect can be observed by watching the movement of a spot of dye or food coloring in a tank of water. The waves will pass by, leaving the dye almost undisturbed.

Another important difference between deep- and shallow-water waves is their respective speeds. The speed of a deep-water wave depends on its length or period, while that of a shallow-water wave depends only on the depth of the water.

---

*Note that "d" refers to the ocean depth from the surface to the ocean bottom at that location and should not be confused with "H," which refers to wave height as defined earlier.

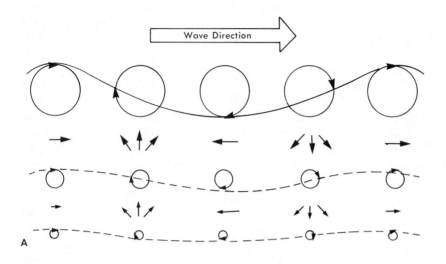

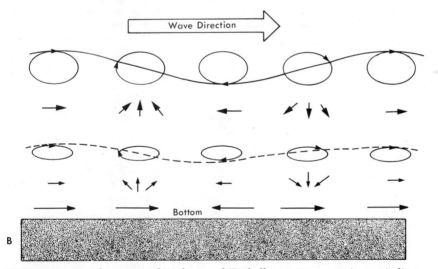

FIGURE 6.3 Particle motions of *(A)* deep- and *(B)* shallow-water waves. Arrows indicate the direction of water particle motion. Note that all water particles in deep-water waves follow circular orbits, decreasing in size with depth. All water particles in shallow-water waves follow elliptical orbits that become flattened with depth, but their horizontal dimensions remain nearly the same.

Figure 6.4 shows the speed of deep-water waves for various values of wavelengths. Figure 6.5 shows the speed of shallow-water waves for various values of depths. These graphs can be used to determine the speed of the waves (excluding capillary waves)* by knowing the wavelength in deep water or the depth in shallow water. The speeds of waves are approximated by the following equations:†

deep-water wave $(d/L > \frac{1}{2})$; $C_{deep} = \sqrt{gL/2\pi}$
Since $L = CT$
$C^2_{deep} = gCT/2\pi$
or $C_{deep} = gT/2\pi$
shallow-water wave $(d/L < 1/25)$: $C_{shallow} = \sqrt{gd}$

where C = wave speed in meters per second; g = gravitation acceleration of the earth (9.8 m/sec^2); L = wavelength in meters; d = depth of water in meters; T = wave period in seconds and $\pi = 3.14$.

---

*The speed of capillary waves can be approximated by the expression:

$$C = \sqrt{gL/2\pi + 2\pi \tau/\rho L}$$

where g = gravitational acceleration, L = wavelength, τ = surface tension and ρ = density of the water. For capillary waves (L less than 0.0173 meter) the speed increases with decreasing wave length. Capillary waves have a minimum speed of about 0.23 meter/sec.

†For intermediate-water wave (d/L between $1/2$ and $1/25$) the wave speed in meters per second is given by the formula: $C_{int} = \sqrt{(gL/2\pi)}\tanh(2\pi\ d/L)$. Values of tanh (hyperbolic tangent) can be found in standard mathematical tables.

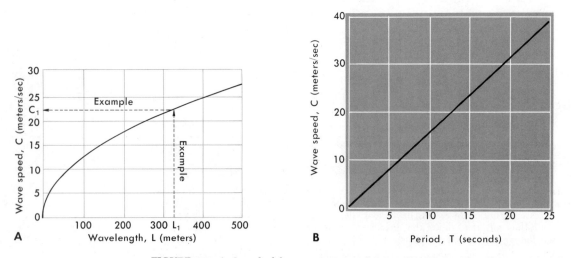

FIGURE 6.4 A, Speed of deep-water waves for various wavelengths. This graph can be used to determine the speed of the waves (C_1) by knowing the length of the wave (L_1). In this case, a deep-water wave with a length of 325 meters moves at a speed of about 22.4 meters per second (about 80 km/hr). B, Deep-water wave speeds for various periods.

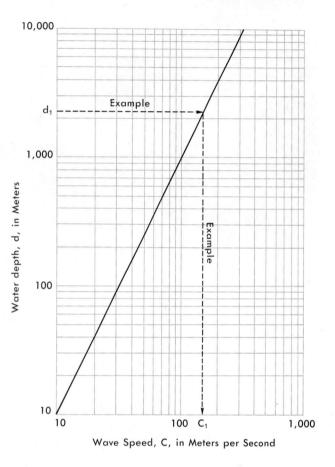

FIGURE 6.5 Speed of shallow-water waves for various depths. This graph can be used to determine the speed of the waves (C_1) by knowing the water depth (d_1). In this case, a shallow-water wave in water 2250 meters deep moves with a speed of 148 meters per second (533 km/hr). Note that both scales are logarithmic.

WIND WAVES

Generation of Wind Waves

As the wind blows over the ocean, very small waves, or capillary waves, are produced by friction between the moving air mass and the sea surface. These waves move and grow as long as the wind blows and die out when the wind stops. If the wind continues in the same direction and the wind speed exceeds about 3.7 km/hr (two knots), more stable gravity waves are formed, which will move forward and across the ocean even after the wind stops. The exact mechanism by which wind energy is transformed into wave energy is not well understood. The ridges of the ripples (crests) present a greater surface area to the wind, making it easier for the wind to pile up the water and push it forward. As long as the wind continues to blow, these waves will grow in size up to a maximum that is

FIGURE 6.6 Waves in the storm area are called sea. (Courtesy of Woods Hole Oceanographic Institution.)

dependent on wind speed. Waves 12 meters high are about the maximum ordinarily present in the open ocean, although waves in excess of 15 meters have occasionally been observed from ships on the high seas. In 1933, the Navy tanker USS Ramapo encountered record high waves. One wave of about 35 meters was observed, with the ship steaming with the waves.

Waves in a storm area (generating area) are called *sea* (Fig. 6.6). The nature of these waves depends on wind speed, duration of the wind, and the distance over which the wind blows (called the *fetch*). The sea waves are choppy and have sharp, short crests which are almost randomly oriented. Away from the storm, the small short-period waves decay and die out, leaving only the more uniform, smooth, long-period waves, called *swell.* However, irregularities in swell may be produced by local winds existing outside the generating area or by interference from swell arriving from different generating areas. Swells can travel for thousands of miles in deep water with little loss of energy.

Forecasting of Wind Waves

Our ability to predict the size, speed, path, and time of arrival of waves is of great importance. Wave forecasting is important to navigation and shore installations as well as to sailing and surfing. Oceanographic cruises require predictions so that scientific observations can be made safely in relatively quiet seas.

Wave forecasting is made empirically, on the basis of numerous past observations at sea and on statistical analysis of data, rather than by pure theory. The wave conditions at any place depend on the following factors:

(1) wind speed;
(2) duration of the wind;
(3) fetch, the distance over which the wind blows;
(4) distance away from the storm area; and
(5) local bathymetric conditions (bottom contours)

The deep-water waves near the generating area (the sea), however, depend only on wind speed, duration, and fetch, information about which can be obtained from meteorological sources. There are many techniques of forecasting deep-water waves, all of which depend on accurate weather forecasting. One of these, developed by C. L. Bretschneider, is shown in Figure 6.7. If one knows the wind speed, the duration, and the fetch, the *significant wave height* can be predicted by using Figure 6.7. The significant wave height is the average height of the highest one third of all the waves that will be observed in the area (generally the height most responsible for damage). Knowing the significant wave height, one can easily compute the average height of all the waves, the average of the highest 10 per cent of all waves, and the most frequent wave height, using the following relationships:

$$\text{average wave height} = 0.63 \times \text{significant wave height}$$

$$\text{average height of the highest 10\% of the waves} = 1.27 \times \text{significant wave height}$$

$$\text{most frequent wave height} = 0.50 \times \text{significant wave height}$$

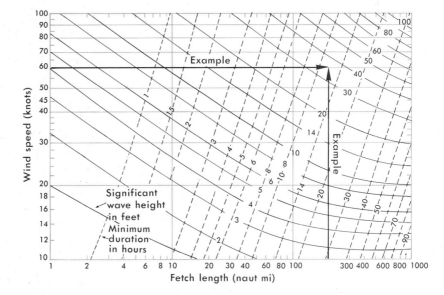

FIGURE 6.7 The Bretschneider deep-water wave forecasting curves. Example: For a wind speed of 60 knots, a fetch length of 200 nautical miles, and a minimum wind duration of 12 hours, the significant wave height is predicted to be 35 ft. (Modified from U.S. Army Coastal Engineering Research Center, 1966.)

In many instances, information about wave conditions near coasts (swell) or in areas far away from the generating area is desired. The nature of these waves will depend, however, on the distance away from the storm area, the interference of waves from other generating areas, local wind conditions, and bathymetry. Forecasting of swell in shallow water is somewhat more complex and involved than forecasting of swell near the generating area, and the reader is referred to the references at the end of the chapter for information on this subject.

WAVE REFRACTION

A casual trip to the beach raises many interesting questions about waves. Why do waves always move toward the beach, usually with crests parallel to it? Why are they more closely spaced (piled up) near the beach than offshore? Why is there more wave action on some parts of the beach than others? To answer these questions, consider a deep-water wave. It will travel constant in direction at a constant speed which depends on its wavelength. When water depth equals half the wavelength, the wave starts "feeling" the bottom and progressively slows down. In addition, it will bend or undergo *refraction* depending on the angle of approach with respect to the shallower bottom contours (Fig. 6.8). Actually, the part of the wave to first feel bottom (in shallow water) slows down, allowing the parts of the wave that are still in deeper water to "catch up" until they, too, feel bottom. Soon the wave becomes nearly parallel to shore. The principle involved in the refraction of water waves (Snell's Law) is similar to that for the refraction of seismic waves, light waves,

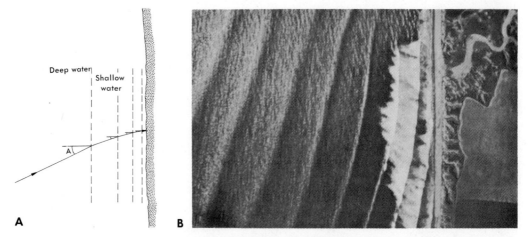

FIGURE 6.8 *A,* Refraction of waves as they pass through regions of decreasing water depth. Dotted lines represent bottom contours. *B,* When swells reach shallow water, sometimes having traveled thousands of kilometers, they may refract, becoming nearly parallel to the shore. (From U.S. Naval Oceanographic Office H.O. Pub. No. 606-e [1950].)

and sound waves discussed earlier. (The refraction of waves is discussed in detail in Appendix B). In Figure 6.8, the arrows show the direction of wave advancement (the ray path) from deep to shallower water. Refraction occurs only when the ray paths are not perpendicular to the bottom contours.

As waves pass through progressively shallower waters, the ray paths become more nearly perpendicular to the bottom contours, as shown in Figure 6.8; that is, the wave crests will be more nearly parallel to the beach. Thus it can be said that waves are "drawn" toward the beach by refraction. (See also Figure 6.9.)

When waves reach shallow water, whether they are refracted or not, their speed and wavelength decrease, their wave height increases, and their periods do not change. In order to keep the same period, the wave crests must get closer together (pile up) and become higher and steeper as they approach the beach.

An important consequence of wave refraction is that the energy contained in waves becomes more concentrated as they approach a headland than when they approach a straight coastline. On the other hand, the concentration of wave energy decreases as the waves enter a bay and spread out (Fig. 6.10), resulting in relatively calm water. The coastline in the illustration has two points or headlands and a bay between them. The bathymetric contours in the shallow-water area, as in many parts of the world, are roughly parallel to the coastline. As the waves enter shallow water, they are refracted. The headlands will undergo greater erosion because of this concentrated wave action, and the bay will receive sediments. These two processes often result in the formation of a straight coastline. Even if the waves approach the shallow water area from a different direction than that shown in the figure, refraction will eventually bend the waves to approximately the orientation shown.

Wave refraction studies are of great importance in choosing locations for new harbors, jetties, and breakwaters. It is essential to know in advance whether the harbor will be filled with sand or will receive too much wave energy, and whether the jetties and breakwaters will be strong enough to withstand the impact of wave action. Another application of wave refraction data is in prediction of the paths as well as the time of arrival of tsunamis (discussed later in this chapter) in various areas so that damage can be minimized.

BREAKERS AND SURF

As waves travel from deep water into shallow water areas ($d/L < 1/2$) they begin to "feel" the bottom and slow down, as the speed is controlled by the water depth. All waves—whether they are sea or swell, deep- or shallow-water waves—will break when the *wave steepness* (ratio of wave height to wavelength [H/L]) exceeds $1/7$. These waves

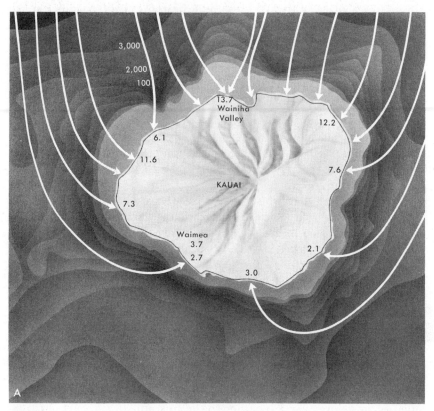

FIGURE 6.9 *A*, Refraction of a tsunami around the island of Kauai, in the Hawaiian Islands. The tsunami originated in the Aleutian area following the 1946 earthquake. Solid lines indicate ray paths, and the arrows show their direction. Notice that the tsunami also arrived on the south side of the island, even though they originated about 4,000 km to the north (top). Tsunami heights along the shore and ocean depths are given in meters. The north-south distance across Kauai is about 40 km.

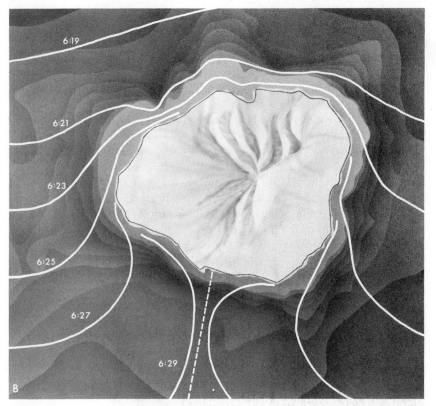

6:19
6:21
6:23
6:25
6:27
6:29
B

FIGURE 6.9 *Continued. B,* Position of wave crest at different times around the island. Note that the waves traveled around the island in less than 10 minutes. (Modified after Shepard et al., 1950.)

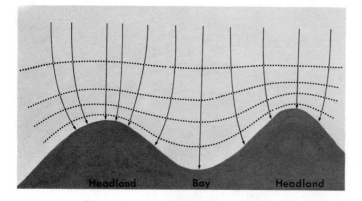

Headland   Bay   Headland

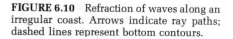

FIGURE 6.10 Refraction of waves along an irregular coast. Arrows indicate ray paths; dashed lines represent bottom contours.

break when the water particles at the crest move faster than the speed of the waves. In shallow water, as the wave slows down, the wavelength shortens and the wave height increases. Studies of breaking waves *(breakers)* have shown that shallow-water waves break when water depth equals 1.3 times the wave height (d = 1.3H). Thus, one can determine the depth of coastal oceans merely by knowing the location and height of various breaking waves. This simple but useful technique was used during World War II landings on many coasts where information on the bottom topography was lacking.

Breakers are classified as follows (Fig. 6.11):

Spilling breakers: break over a large distance and over relatively flat bottom; the crest breaks and "slides" down the face of the wave.
Plunging breakers: front of crest curls and breaks once, over bottom of intermediate steepness, forming a tunnel along the face of the wave.
Surging breakers: do not really break but surge up the slope of very steep beaches or sea walls.

Waves, especially those in the *surf* (zone of breakers), may provide a free ride for surfers. A person may ride a surfboard on the forward surface of the wave and slide downward. Because the wave crest is always moving forward, the surfer may ride the wave onto the beach. By moving at angles to the crest less than 90°, one can adjust the forward motion of the surfboard to the speed of the wave and ride the wave at higher speeds than the wave itself travels. The surfer maintains a relative position on the wave by traversing a path nearly parallel to the wave crest, attaining maximum speed. One might even surf under the crest of a plunging breaker in a sort of aqueous tunnel or cave, called a "tube" by surfers.

FIGURE 6.11 *A*, Spilling breaker; *B*, Plunging breaker; *C*, Surging breaker. (From U.S. Army Coastal Engineering Research Center, 1973.)

The forces involved in surfing are shown in Figure 6.12. The force of buoyancy (B) is always perpendicular to the surface of the water, and the force of gravity (G) is always downward. Thus, the resulting force (F) is directed forward and downward along the wave face, producing a forward motion of the surfboard. The principles depicted in the diagram apply also to surfing without a board—"body surfing."

A

B

FIGURE 6.12 *A*, Forces involved in surfing. B = buoyancy; G = gravity; F = resultant force. *B*, Surfing on (and under) a large wave. (Photos by Dan Merkel, courtesy of *SURF-ING Magazine.*)

TSUNAMIS

Tsunamis or *seismic sea waves* are giant ocean waves usually caused by large submarine earthquakes. They are sometimes called "tidal waves" although they have nothing to do with tides. In the open ocean, tsunamis usually have lengths of about 150 km but may have lengths of up to 1000 km, speeds sometimes exceeding 800 km per hour, and heights of about one meter. The period of a tsunami ranges from about 10 minutes to an hour. Thus, they are unnoticed in the open sea and cause no damage to ships. But when they approach shallow water, their wave heights increase drastically, sometimes running up the shore more than 30 meters above normal sea level. A tsunami may advance like a *bore* (an abruptly advancing ridge of water) in very shallow water, sometimes exceeding 15 meters per second (about 34 miles per hour). Tsunami damage (Figs. 6.13 to 6.15) is usually great near coasts where the offshore slope is very steep. Because of their long wavelengths, they travel as shallow-water wave (C = $\sqrt{gd}$) over most portions of the oceans. For an ocean depth of 4000 meters, C = $\sqrt{9.8 \text{ m/sec}^2 \times 4{,}000 \text{ m}}$ = 715 km/hr (445 mph). Tsunamis can travel thousands of kilometers with little loss of energy and are able to cause tremendous damage to coastal areas far from their point of origin. For example, a tsunami created by the 1964 Alaskan earthquake caused many deaths and much property damage along the Pacific coasts of Canada and the United States. In Crescent City, California, 12 lives were lost and property damage totaled $11 million. The tsunami of May 23, 1960, at Hilo, Hawaii, caused by an earthquake in Chile, resulted in the death of 61 people and about $20 million of damage to property.

A

FIGURE 6.13 *A*, Crescent City, California, before the Alaskan earthquake of March 27, 1964. *B*, Enlarged view of the outlined area two days after the tsunami hit the city. Note the destruction of buildings. (Photos courtesy of NOAA.)

FIGURE 6.14 *A*, The port area at Seward, Alaska, before the March 27, 1964, earthquake. *B*, The same location after the earthquake, showing tsunami damage along the waterfront. (Photos courtesy of U.S. Geological Survey.)

FIGURE 6.15 Tsunami damage at Kodiak, Alaska, following the 1964 earthquake. (U.S. Navy photo.)

Most tsunamis are caused by sudden vertical displacement of the sea floor, as by vertical faulting (Fig. 6.16), submarine landslides, or landslides into water, all of which may be caused by earthquakes located near coastal areas. Most tsunami-producing earthquakes have magnitudes greater than 6.5 on the Richter scale* and originate at depths usually less than 50 km. It is important to note that not all such earthquakes result in tsunamis. For example, the California earthquakes rarely produce tsunamis because the faults associated with them generally move horizontally rather than vertically. It is generally believed that in order for a

---

*Magnitude refers to the energy released by the earthquake at its source. Because the magnitude scale is logarithmic, an increase of one "magnitude" actually involves about a thirtyfold increase in energy release. The energy is doubled for each increase of 0.2 unit of magnitude.

Some examples of magnitudes of well-known earthquakes include the following: San Francisco, 1906 (8.3); Chile, 1960 (8.0); Alaska, 1964 (8.4); and San Fernando (near Los Angeles), 1971 (6.5).

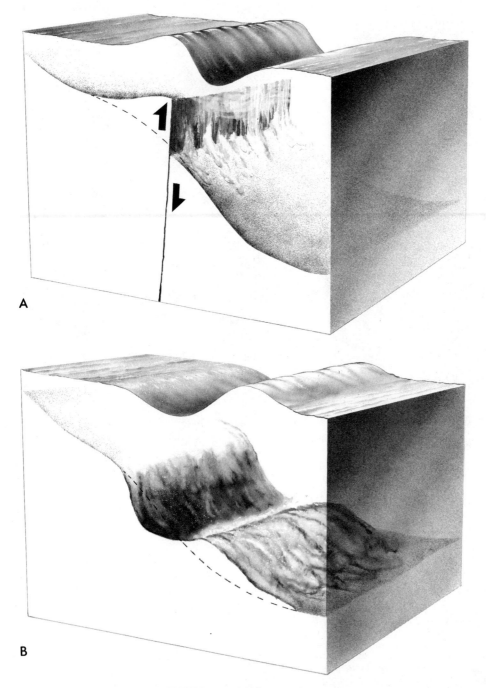

FIGURE 6.16 Mechanics of tsunami generation. *A*, Fault; *B*, Submarine landslide. The dashed line in both cases indicates the previous profile of the sea floor.

tsunami to be produced, vertical displacement must be at least several meters in a large area and the fault length at least 100 km. Earthquakes may trigger submarine landslides (Fig. 6.16) in regions where large quantities of sediment rest on the continental shelves and slopes. This is the mechanism which is responsible for some turbidity currents and the resulting breakage of submarine telegraph cables (see Chapter 2). Tsunamis can also be produced by landslides into water and submarine landslides that are unrelated to earthquakes, and by submarine volcanic eruptions or explosions. The explosion of the volcano Krakatoa (1833) produced large tsunamis in the Pacific, which killed more than 35,000 people. However, landslides into water and submarine landslides generally cause only local tsunamis.

Tsunamis are most common in the Pacific, because earthquakes are more common there, but they have occurred in all oceans. In the 20th century alone, about 200 tsunamis have been reported around the Pacific Ocean. One of the largest tsunamis in history, however, occurred in the Atlantic Ocean after the 1755 submarine earthquake near Lisbon, Portugal. Waves rose up to about 15 meters along the coast of Morocco, and at Tangier the water rushed inland 2.4 km. This tsunami radiated in all directions and was observed as far away as the West Indies.

A tsunami is not a single wave but rather a series of waves which may persist for hours. The first motion may be either an advance or a withdrawal of water. Many tide records indicate an initial rise in sea level about a third as large as the succeeding withdrawal. The initial lowering of water may be like a very low tide, exposing animals and seaweed that are normally covered. People have been known to be lured toward the sea by this lowering of water, only to be killed by the advancing wall of water in the next large wave. The most destructive waves arrive within the first four hours. Figure 6.17 shows the records made at two locations of the tsunami that resulted from the Alaskan earthquake of 1964.

Although tsunamis cannot be prevented, they can be predicted, and death and damage to property can be minimized. The National Oceanic and Atmospheric Administration of the Department of Commerce operates the international Seismic Sea Wave Warning System with headquarters in Honolulu, Hawaii (with a regional center in Palmer, Alaska), serving the entire Pacific coast. The principles involved in the warning system are as follows: (1) knowledge of the exact location of earthquakes in coastal areas, using critically placed seismic stations around the Pacific; (2) detection of waves having tsunamilike characteristics generated after the earthquake at intermediate locations; (3) a communications network to transmit the information; and (4) construction of probable paths and determination of expected arrival time of the tsunami at various coastal areas, using knowledge of the principle of wave refraction. Since we know the major tsunami-producing areas of the Pacific, refraction diagrams and travel times for numerous hypothetical tsunamis can be worked out ahead of time. Figure 6.18 is a wave refraction diagram

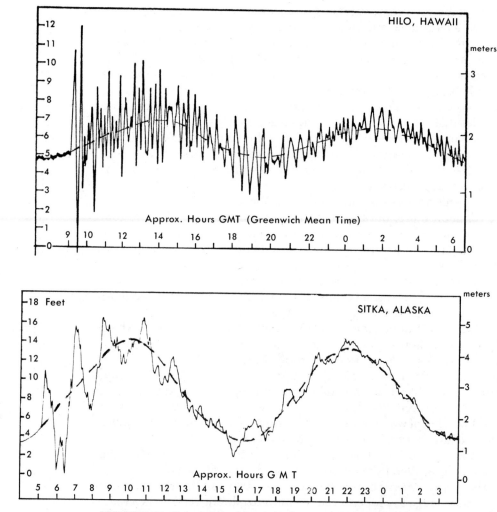

FIGURE 6.17   Tsunami resulting from the 1964 Alaskan earthquake as recorded by tide gages at Hilo, Hawaii, and Sitka, Alaska on March 28–29, 1964. Dotted curves indicate projected tide levels. Note that the tsunami arrived four hours later at Hilo than at Sitka. (After Wilson and Torum, 1968.)

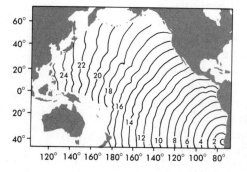

FIGURE 6.18   Wave refraction diagram of a tsunami that originated on the coast of Chile in 1960. Numbers indicate the hours of travel time from the source. (After The Committee for the Field Investigation of Chilean Tsunami of 1960, Report on the Chilean Tsunami of May 24, 1960, as observed along the Coast of Japan, December, 1961.)

for the 1960 Chilean tsunami. The success of any warning system depends mainly on the willingness of people to obey evacuation instructions. Many deaths have been prevented since the inception of the system in 1948. However, the system is ineffective in areas close to the source of the tsunami.

INTERNAL WAVES

Just as wind waves are produced at the air–sea boundary by pressure differences, waves known as *internal waves* (waves at an internal boundary) may be formed between layers of water of different densities. It takes less energy to produce an internal wave between layers of water than to produce a comparable surface wave between the sea and the air. Fresh water from river runoff, ice melt, or a layer of warm water at the surface may be much lighter than the salty or cold water below. This contrast produces a sharp density boundary between the lighter water on top and the denser water below. Waves may be set up on the boundary by forces such as surface waves, tides, earthquakes, or ships' propellers. These waves can have periods from a few minutes to hours or days. Their wave heights vary considerably, and heights of about 100 meters have been reported. Their speeds,* however, are very low compared with ordinary surface waves. Internal waves may also develop in areas where there is a continuous increase of density with depth.

The most familiar internal waves are the short-period waves produced by slow-moving ships in a stratified ocean, where a thin layer of low-density water overlies dense water. Ships are known to lose speed when turbulence from the propellers churns into the density boundary (Fig. 6.19) and sets up internal waves. The ship "sticks" and makes very little progress. This is known as "dead water." At speeds of about two knots or less, much of the ship's energy may be spent in making internal waves rather than in propelling the ship forward. The severity of this problem depends on the propeller design and may be overcome by increasing the ship's speed to about five knots.

Although internal waves do not result in any noticeable waves at the surface, they are often accompanied by parallel slicks at the sea surface, which can be used to track the progress of the internal waves. They may also be observed by noting temperature and/or salinity

---

*The following equation can be used to compute the speed of internal waves in a two-layered ocean:

$$C = \sqrt{\frac{gd''d'}{d}\left(\frac{\rho'' - \rho'}{\rho''}\right)}$$

where C is the speed of the internal wave, g is the acceleration of gravity, d' is the thickness of the surface layer, d" is the thickness of the deeper layer, d is the total thickness of the two layers. ρ' is the density of the surface layer, and ρ'' is the density of the deeper layer (see Fig. 6.19).

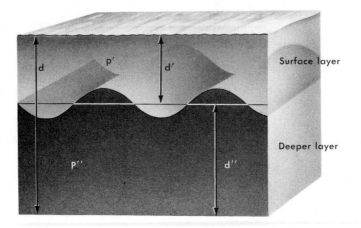

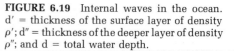

FIGURE 6.19 Internal waves in the ocean. d′ = thickness of the surface layer of density ρ′; d″ = thickness of the deeper layer of density ρ″; and d = total water depth.

fluctuations with time at various depths. Internal waves sometimes cause submarines to rise and fall, and they were suspected of causing the loss of two nuclear submarines, *Thresher* and *Scorpion*. They also may interfere with the transmission of sound in the sea. The phenomenon of internal waves may be important for biological processes in the oceans in that they contribute to mixing of deeper nutrient-rich water with surface water, especially if the internal waves break. Nutrients from deeper waters are thus brought to the surface, where sufficient light is present for primary production by plants.

STANDING WAVES

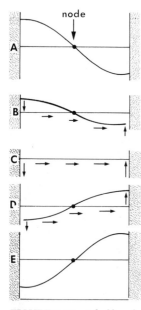

FIGURE 6.20 A half-cycle of standing wave in a closed basin at different stages. Arrows indicate the direction of water movement and are reversed during the other half-cycle.

The waves that we have discussed so far are called *progressive waves,* because they travel away from their source. In closed and semienclosed basins such as lakes and bays, a different type of wave, a *standing wave* or *seiche,* may be initiated. These waves (Fig. 6.20) may be thought of as two progressive waves travelling in opposite directions or as waves that are alternately reflected first off one side of the basin and then off the other. If the wave length is one half, or a multiple of the length of the bay, one or more *nodes,* lines of no vertical motion are set up. Between the nodes the water simply rises and falls. The period of such waves in a closed basin is given by the equation, $T = (1/n)\, 2\zeta/\sqrt{gd}$ where n is the number of nodes, ζ is the length of the basin, *g* is the acceleration due to gravity and *d* is the depth of the basin. These waves may be caused by storms, earthquakes, and tides. A more familiar example of a standing wave is the sloshing of water in a shallow tray, such as a refrigerator pan, as it is carried across the room. The role of tides in setting up standing waves in bays is discussed in more detail in the next chapter. Seiches initiated by storms and earthquakes may be as high as several meters and cause considerable damage to docks and boats in a harbor.

STORM SURGES

Storm surges, also known as *storm waves* or *storm tides*, are caused by the strong and steady winds associated with hurricanes (tropical cyclones) or other severe storms. Unfortunately, storm surges and tsunamis are both popularly referred to as "tidal waves" even though they have different origins and have nothing to do with tides. In storm surges, the winds simply pile up the water against the coast. The rise in water level is greatest in partially enclosed, shallow bodies of water such as the Gulf of Mexico and is negligible around isolated islands. The rise in water level is augmented if it coincides with high tides, especially with the highest tides at new and full moon. Sea-level rises in excess of 10 meters have been reported. Damage from storm surges can be tremendous, especially in low-lying coastal areas (Fig. 6.21).

Hundreds of thousands of people have died as a result of destructive storm surges in places surrounding the Gulf of Mexico, the Bay of Bengal, and the North Sea. The storm surge that hit Galveston, Texas, in 1900 caused thousands of deaths. In 1938, a hurricane and its storm surge hit the New England coast, resulting in the death of almost 500 people and about $300 million in damage. The disastrous storm surge that inundated Bangladesh in 1970 is estimated to have resulted in 500,000 deaths.

In addition to flooding of coastal areas, very strong winds and heavy rain usually accompany storm surges, making evacuation very difficult.

FIGURE 6.21 Storm damage at Scituate Beach, Massachusetts. (Photo courtesy of U.S. Army Coastal Engineering Research Center.)

SUMMARY

Waves include a wide variety of periodic undulations of the sea surface, from tiny capillary waves to tsunamis and tides. Waves may occur at any interface between fluids of distinctly different densities. The waves with which we are most familiar occur at the air-sea interface, but internal waves sometimes occur between layers of water of different density such as at a pycnocline.

Although most ocean waves are caused by wind, they may also be caused by earthquakes, landslides, and volcanic explosions. Tsunamis may be caused by the latter types of disturbances and may be hundreds of kilometers long and travel at speeds of hundreds of kilometers per hour.

As waves travel unimpeded over the sea surface, they grow longer and flatter and eventually disappear. The force that causes them to flatten is termed the "restoring force." This force is surface tension in the case of the very small capillary waves and gravity for the familiar larger waves.

The generation and propagation of waves follow well-defined rules and it is possible to predict their growth, path, and decay with great accuracy.

As waves enter shallow water their particle motion is altered and they grow higher as they slow down. Eventually they break and are destroyed. Breaking waves may cause damage or may provide a recreational ride for surfers.

If waves enter shallow water at something other than a right angle, they bend, or are refracted, toward the beach. Refraction of waves tends to cause headlands to have rough water and bays to have quiet water.

IMPORTANT TERMS

Deep-water wave
Fetch
Intermediate-water wave
Internal wave
Node
Period
Plunging breaker
Progressive wave
Refraction
Sea
Seiche
Seismic sea wave
Shallow-water wave
Significant wave height

Spilling breaker
Standing wave
Storm tide
Storm wave
Surf
Surging breaker
Swell
Tsunami
Wave forecasting
Wave height
Wavelength
Wave speed
Wave steepness

1. Determine the speed of a deep-water wave having a length of 400 meters (use Fig. 6.4). At what depth would this wave become a shallow water wave if its length remains unchanged? At what speed will this wave travel? (Use Fig. 6.5.)
2. Would a shallow-water wave with a length of 20 meters and a height of 3.0 meters break? If so, at what depth in the ocean?
3. If a tsunami travels 3500 kilometers in five hours, what is its average speed? What is the average ocean depth crossed by the wave?
4. It took about four hours for the 1964 tsunami that originated in Prince William Sound, Alaska, to reach Crescent City, California, a distance of approximately 2600 kilometers. Calculate the average speed of the tsunami and the average ocean depth between these two locations.
5. Briefly describe the Seismic Sea Wave Warning System.
6. Describe the forces involved in surfing.
7. Why are waves generally higher near headlands than they are in bays? What are the long-term consequences for an irregular coastline?
8. If a 47-knot wind blows over a fetch of 100 nautical miles for at least eight hours, what is (a) the significant wave height in meters and in feet? (b) the average wave height in meters and in feet?
9. Why do waves pile up near shore?
10. Describe the three most common types of breakers found near shore.
11. Why do tsunamis cause little damage out at sea and great damage along coastlines?

Barnstein, J. 1954 (August). Tsunamis. *Scientific American.*
Bascom, Willard. 1964. *Waves and Beaches.* Garden City, N. Y., Doubleday.
_____ 1959. (August). Ocean Waves. *Scientific American.*
Newman, J. N. 1979. Power from Ocean Waves. *Oceanus,* 22(4):38–45.
Pierson, W. J., G. Neumann, and R. W. Jones. 1955. *Practical Methods for Observing and Forecasting Ocean Waves by Means of Wave Spectra and Statistics.* U.S. Navy, H. O. Pub. 603.
Smith, F. Walton. 1973. *The Seas in Motion.* New York, Thomas Y. Crowell Co.
U.S. Navy. 1944. *Breakers and Surf: Principles in Forecasting.* H.O. Pub. 234.
U.S. Navy. 1944. *Wind Waves at Sea, Breakers and Surf.* H.O. Pub. 602.

Ocean Tides

Tides, the periodic rise and fall of the oceans, are the most predictable of all the motions of the oceans, yet many aspects of the tides are not fully understood. Knowledge of tidal fluctuations as well as tidal currents is important not only in navigation but also in the ecology of the coastal oceans. A lack of understanding of tides caused Julius Caesar's fleet to meet disaster on the beaches of England. His army invaded Kent just before a high tide, which subsequently caused many of his ships to be wrecked upon the shores.

Although the enormous energy of the ocean tides may never be fully harnessed, power is already being produced from the tides, as in the Rance estuary in Brittany, France. Similar facilities may someday be set up in other areas of the world.

TIDES ON AN IDEAL EARTH

Let us consider tides from an ideal point of view, that is, tides on a spherical earth covered by a uniform ocean, the bottom of which is

The tidal power station on the Rance estuary in France. (Courtesy of French Embassy Press and Information Division.)

assumed to be frictionless. Ocean tides are caused by the gravitational attraction of the moon and the sun acting on the waters of the rotating earth, and the centrifugal forces resulting from the mutual revolutions of the earth–moon–sun system. Contrary to popular belief, the moon does not really revolve about the earth; both the earth and the moon revolve about a common center of gravity located between their centers. Similarly, the earth and the sun revolve about their common center of gravity. The earth's own gravity and the centrifugal forces resulting from the earth's rotation on its axis in no way cause oceanic tides, as these forces do not vary with time. If we ignore the effects of the sun for the time being, we can state that every water particle in the oceans is subjected to two forces, the gravitational force of the moon and the centrifugal force due to the earth-moon revolution about each other. The tidal effects of the sun are similar to those of the moon, but only about 46 per cent as great as those produced by the moon.

The gravitational attraction due to the moon is greatest for a water particle directly beneath the moon and least for a water particle on the other side of the earth (Fig. 7.1A). The earth and the moon revolve about their common center of mass approximately every 27 days. This mutual revolution takes place about an imaginary axis passing through the center of mass of the earth-moon system. This axis is located about 1700 km below the earth's surface. As the earth rotates on its own axis, the position of this common center of mass is continually changing. Fig. 7.1B illustrates the mutual revolution of the earth-moon system. Note that during the course of one revolution, the center of the earth describes a circle whose radius is the distance between the center of the earth and the common center of mass (Fig. 7.1C). In fact, every particle of the earth describes a circle with the same radius (neglecting the rotation of the earth). Centrifugal force is proportional to the radius of revolution (about 4700 km) and the square of the angular velocity of the particle (one revolution per 27 days). The centrifugal force that each particle of the earth experiences is equal. The centrifugal forces are always perpendicular to the axis of revolution and are directed away from the moon.

For the earth-moon system, there is an overall equilibrium between their gravitational attraction and the centrifugal force due to their mutual revolution. If this were not true, the earth and the moon would either fly apart or crash into each other. Tides are the result of local imbalances in these forces. Only at the center of the earth are these forces exactly equal and opposite, as illustrated in Figure 7.1A and C. For points on the earth facing the moon, the gravitational forces exceed the centrifugal forces, and for points on the other side of the earth, the centrifugal forces exceed the gravitational forces.

The gravitational and centrifugal forces acting on a water particle always oppose each other, resulting in the net forces shown in Figure 7.2A. These forces are symmetrical with respect to the earth's axis of rotation when the moon is above the equator and are distorted when

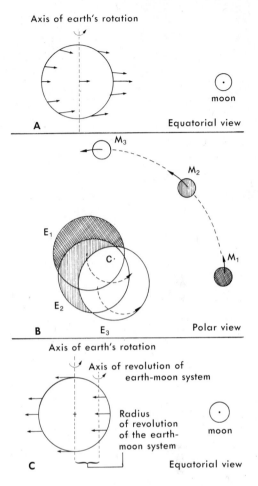

FIGURE 7.1 *A*, Gravitational attraction between the moon and selected water particles. *B*, The mutual revolution of the earth-moon system, where C is the center of mass. E_1, E_2, and E_3 are positions of the earth at times corresponding to the position of the moon at M_1, M_2, and M_3 respectively. The dotted arrows indicate the paths of water particles on the earth's surface. Note that all particles describe circles with the same radius. *C*, Centrifugal forces resulting from the earth-moon revolution on selected water particles. The centrifugal forces are all equal, parallel and directed away from the moon.

the moon is north or south of the equator. The *declination* (the angular distance of the moon north or south of the earth's equator) of the moon may be as much as $28\frac{1}{2}°$ north or south of the equator. The interval between northern and southern declinations is 13.7 days. Figure 7.2*B* shows the horizontal components of the tide-producing forces. These are the forces acting parallel to the sea surface, causing two equal bulges of water (Fig. 7.2*C*). The vertical components simply cause a very small decrease in the earth's gravity, since they are oppositely directed. Because of the earth's rotation and the moon's revolution around the earth, most places on an ideal earth would experience two equal high tides (bulges) and two equal low tides during a period of 24 hr 50 min, or one *lunar day*. A lunar day is the time between two successive appearances of the moon over the same longitude. If the earth did not rotate, these tides would occur only twice each lunar month.

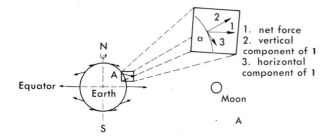

FIGURE 7.2 *A*, Net gravitational and centrifugal forces on water particles; *B*, Horizontal components of the tide-producing forces; *C*, Equal bulges produced by the forces shown in *B*.

The gravitational effect of the sun and the moon on water particles varies inversely with the square of the distance of these celestial bodies. The net tide-producing forces similar to those shown on Figure 7.2*A* vary inversely with the cube of the distance of these celestial bodies from water particles on earth. The sun, although nearly 30 million times more massive than the moon, is about 490 times farther away from the earth than the moon. The sun's tidal effect is, in fact, only 46 per cent of that of the moon. When the sun, the moon, and our ideal earth are on a line (new moon or full moon), the resulting two tidal bulges will be larger than those produced by the moon alone (Fig. 7.3*A*). These tides, called *spring tides,* would be repeated every two weeks during full moons or new moons. Spring tides produce the highest high tides and the lowest low tides. The *tidal range* is the difference between two successive high and low tide levels and reaches a maximum during spring tides.

When the sun, the moon, and the earth are at right angles to each other, *neap tides* develop (Fig. 7.3*B*). This occurs during the first or third quarter of the moon (about one week after spring tide). Because the sun's weaker tidal effects are directed perpendicular to those of the moon, the net effect is a reduction in the size of the bulges produced by the moon alone. There would still be two high tides and two low tides during a

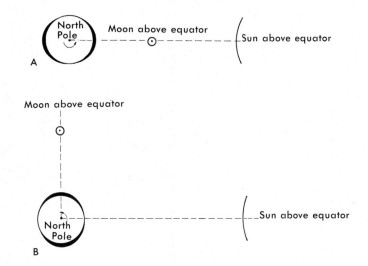

FIGURE 7.3 Tidal bulges produced by various orientations of the earth, moon, and sun. *A*, Spring tide; *B*, Neap tide, seven days after spring tide. Note that the moon could be on the opposite side of the earth for *A* or *B*, yet the bulges would remain the same.

lunar day, but the tidal range would be at a minimum. Between spring and neap tides, a more complex situation should develop.

TIDES ON THE REAL EARTH

Tides on the real earth are more complex than those expected on an "ideal" earth. Some of these differences and their causes are discussed below.

Types of Tides

Figure 7.4 shows the tides observed at various locations throughout the world. At some locations (Fig. 7.4*A* and *B*) there are two approximately equal high tides and two low tides every day. These are *semidiurnal* (semidaily) *tides*, and their periods are about 12 hr. 25 min. (one half lunar day). Such tides are found along the Atlantic coast of the United States. In some areas (Fig. 7.4*C*, *D*, and *E*), however, two unequal high tides and/or two unequal low tides may occur each lunar day. These are called *mixed tides*, and they are common along the Pacific coast of the United States. *Diurnal* (daily) *tides*, with only one high and one low tide each lunar day, predominate in areas such as the Gulf of Tonkin (Fig. 7.4*F*) and portions of the Gulf of Mexico. Notice that no tide is purely semidiurnal nor purely diurnal.

Diurnal Inequality

The twice-daily high and low waters for the semidiurnal and mixed tides are rarely equal in height. Only when the moon is above the equator

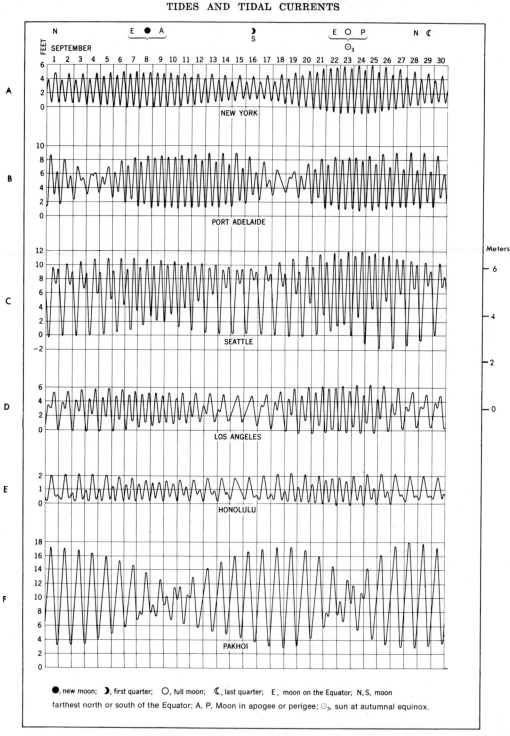

FIGURE 7.4 Tidal curves for various localities. Pakhoi is in the People's Republic of China, on the Gulf of Tonkin. Note that the meter scale on the right does not apply to Honolulu, as its vertical scale is exaggerated. (From Bowditch, 1962.)

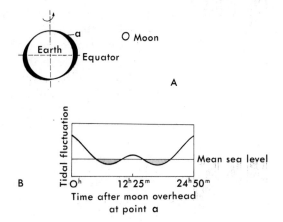

FIGURE 7.5 *A*, Tidal bulges when moon is not over the equator; *B*, Resulting diurnal inequality of the tidal curve for point a.

can these tides be expected to be equal. At all other times, a *diurnal inequality* exists (Fig. 7.5). During these periods, the tide-producing forces are more diurnal than semidiurnal. This situation is reflected on many of the tidal curves shown in Figure 7.4 during maximum declinations of the moon.

THE MYSTERIOUS GRUNION

A mysterious relationship exists between the tide and the grunion (*Leuresthes tenuis*), a small fish (about 15 cm in length) that swims up the beaches of Southern California to spawn during the peaks of the high spring tides of spring and summer. The grunion spawn for three to four nights following the full or new moon. This ensures that the eggs will remain undisturbed for at least 10 days, the time necessary for development before hatching. The female burrows tail first into the sand just after the high tide and lays her eggs. The male arches himself around the head of the female and releases the sperm into the moist sand. After fertilization, the eggs develop, and the next high spring tide washes away the sand and releases the young fish, which then swim toward the sea. Many of the young grunion are eaten by other fish before they mature. Some of the adults are captured by birds, eager Californians, and tourists who flock down to the beach as the grunion stream up the beach. However, enough of the young grunion survive to continue the species, and some of the adults scramble back to the sea and eventually may return to spawn again.

Time Lags

Regardless of type, tides more or less repeat themselves at each location on successive lunar days, although there may be a time lag of up to a few hours between the appearance of the moon overhead and the occurrence of a high tide. This time lag is constant for each locality and is caused by friction between the ocean bottom and the water and by the fact that continents block the free transfer of water around the globe. It can also be seen from Figure 7.4 that spring and neap tides do not exactly coincide with the alignment of the earth, the moon, and the sun discussed under ideal tides. That is, spring tides, for example, do not always coincide with full or new moons. There is usually a few days' delay, which is also constant for each locality. The reasons for this delay are not well understood. Perhaps it, too, is related to friction and continental blockage.

Other Tidal Complications

Tidal ranges and periods not only vary from day to day, but drastic differences in tidal ranges and periods are sometimes observed over short distances. For example, the tidal range at the head of the Bay of Fundy (Fig 7.6) may exceed 15 meters, whereas at the mouth of the bay the tides range only about three meters. Tides along the northern shores of the Gulf of Mexico (Fig. 7.7) are predominantly diurnal in nature, and the tidal range is less than one meter, whereas along the east (Atlantic) coast of Florida, the tidal ranges are nearly two meters and are of the semidiurnal type. We will now investigate why tides vary in this manner.

Local weather conditions such as wind and barometric pressure can alter the tidal range and the timing of successive tides. Storm surges resulting from hurricanes and other tropical storms may cause tides to be several times higher than predicted. Still another factor is the size and shape of the ocean basins. All of these elements actually control the tides locally, although their ultimate causes are the tide-producing forces described in our discussion of ideal tides.

Tides also vary because the earth-moon and the earth-sun distances vary with time. The orbits of the moon and the earth are not circular, but are elliptical as shown in Figure 7.8. Once a month the moon is closest to the earth (perigee) and two weeks later is farthest from the earth (apogee). The tide-producing forces are greatest at perigee and least at apogee, resulting in the biweekly inequality of the tides (see Fig. 7.4). Similarly, the earth's orbit brings it closest to the sun on January 2 (perihelion) and farthest from the sun on July 2 (aphelion), resulting in a semiannual inequality.

RESONANCE

The water in a rectangular enclosed basin such as a bathtub can be set in motion by application of some external force. The water will rock back and forth (Fig. 7.9) with a period that depends on the length of the basin and the water depth. This is the *natural period* of the basin. There

FIGURE 7.6 Tides in the Bay of Fundy. *A*, High tide; *B*, Low tide. (Photos courtesy of National Film Board of Canada.)

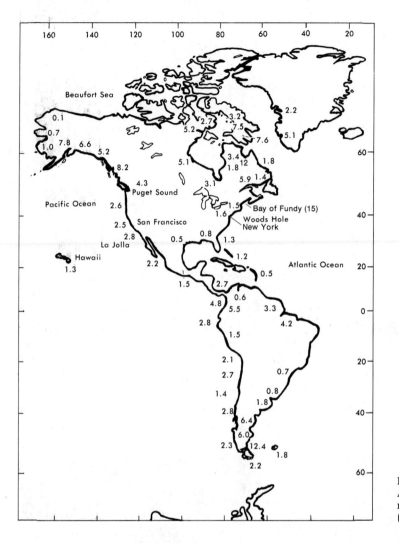

FIGURE 7.7 Map of North and South America, showing maximum tidal ranges in meters at selected locations. (After Doty, 1957, and others.)

will be no vertical motion in the middle of the basin, the *nodal line.* The period (T) for such a closed basin is given by the equation:

$$T = \frac{2\zeta}{\sqrt{gd}}$$

where ζ is the length of the basin, g is the gravitational acceleration and d is the basin's depth. The waves in the basin are *standing waves* produced by reflection from opposite sides of the basin. In standing waves, the water surface simply moves up and down, and there is no forward motion of the waves. Most of the waves discussed in Chapter 6 are called

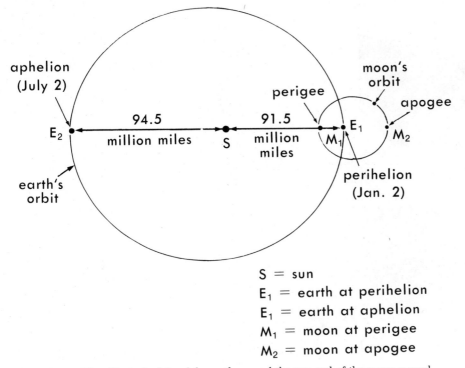

S = sun
E_1 = earth at perihelion
E_1 = earth at aphelion
M_1 = moon at perigee
M_2 = moon at apogee

FIGURE 7.8 The elliptical orbits of the earth around the sun and of the moon around the earth.

progressive waves, in which the wave form advances in some direction. If the period of the external force and the natural period of the basin coincide, *resonance* will be set up, causing amplification of the rocking motion. A good example of resonance is that of pushing a swing; each push is good for only a small amount of motion, but if the pushes are properly timed, the swing goes higher and higher. For a basin that is open on one end, such as the Bay of Fundy, the situation is very similar to that of the enclosed basin, except that a single nodal line is located at the mouth of the bay. The period of such a basin is twice that of the enclosed basin of the same length and depth, and is given by the following equation:

$$T = \frac{4\,\zeta}{\sqrt{gd}}$$

Thus the period increases with increasing length and decreasing depth of the basin. The natural period of the Bay of Fundy (about 12 hours) is very close to the period of the ocean tide entering the bay. Thus resonance is

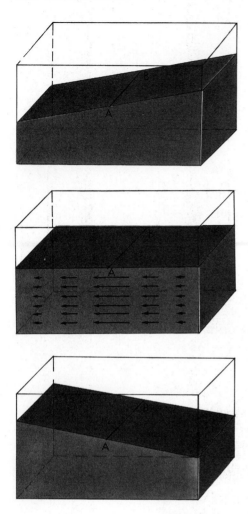

FIGURE 7.9 Resonance set up in a rectangular (closed) basin for half a tidal cycle. Arrows indicate resulting current. Note that no tidal fluctuations occur along AB, the nodal line.

set up in the bay. This explains why the tide is at a maximum at the head of the Bay of Fundy and at a minimum at its mouth (the nodal line), where the tides are those of the nearby Atlantic.

A similar explanation can be offered for the predominantly diurnal tides found in certain large basins such as the Gulf of Tonkin and portions of the Gulf of Mexico. As seen earlier (Fig. 7.5), during the northern and southern declinations of the moon, the tide-producing forces of the moon are more diurnal than semidiurnal. Similarly, except during equinoxes, the sun is either north or south of the equator, and hence the sun's tide-producing forces are also more diurnal than semidiurnal. Thus, if a basin has a natural period of about 24 hours, it will respond preferentially to the diurnal forces rather than to the semidiurnal forces. This is exactly what happens in the Gulf of Tonkin and portions of the Gulf of Mexico. As can be seen from Fig. 7.4*F*, not only are the tides in the Gulf of Tonkin diurnal during the northern and southern declinations of the moon, but also the tidal ranges are very great due to resonance.

EFFECT OF THE EARTH'S ROTATION

In addition to the resonance mentioned above, tides in wide bays and estuaries are modified by the earth's rotation, namely, by the effect of *Coriolis force* (Appendix C). Coriolis force tends to deflect all moving objects near the surface of the earth, including moving water particles, which are deflected to the right in the northern hemisphere and to the left in the southern hemisphere. Coriolis force increases from zero near the equator to a maximum at the poles. Figure 7.10 illustrates the effect of Coriolis force on semidiurnal tides in a bay in the northern hemisphere, starting with high tide at the head of the bay. As the tide begins to retreat, Coriolis force deflects the water toward the right, causing high tide in the bay to eventually move in a counter-clockwise direction around the bay. Note that there is no tidal fluctuation at point A, in the middle of the bay. This is similar to the nodal line of Figure 7.9, except that the line has been reduced to a point—the *amphidromic point*—by the effect of Coriolis force. It is evident from this illustration that the tides do not always occur at the same time at different points even at the head of the bay.

TIDAL BORES

In many shallow, steep, funnel-shaped rivers, the tide advances as a single roaring wall of water, called a *tidal bore* (Fig. 7.11). Bores may move at speeds of up to 25 km/hr and may have heights up to 8 meters. The generation of bores is essentially the same as that of breakers in shallow water. Spectacular bores occur in the Tsientang River in the People's Republic of China, in the Amazon, in the Bay of Fundy, and in many English and French rivers.

Tides In The Open Ocean

Only in the Antarctic Ocean can tides be expected to move completely around the globe from east to west, as seen in ideal tides. The other oceans and seas can be thought of as gigantic basins, and the tides present in them are determined by the natural periods of the basins, modified by Coriolis and frictional forces. The natural period of the Atlantic Ocean "basin" is about 12 hours and hence it responds to semidiurnal tide-generating forces. In addition, the semidiurnal tides from the Antarctic Ocean travel northward into the Atlantic. The resulting tides in and around the Atlantic Ocean are semidiurnal. In contrast,

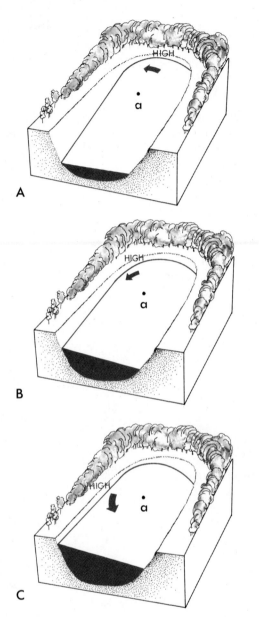

FIGURE 7.10 The effect of Coriolis force on semidiurnal tides in a bay in the Northern Hemisphere. The tides move in a counter-clockwise direction about point a, the amphidromic point. Only a quarter of the tidal cycle is shown. The tidal conditions shown will be repeated 12 hours later.

mixed tides are found in the Pacific and Indian Oceans, because the dimensions of these oceans are such that they respond to both diurnal and semidiurnal forces.

Because there are no fixed points that can be used as reference points, it is difficult to measure tides in the open ocean. They can be computed using data from numerous tidal observations along the coasts and considerations of the size and shape of the ocean basins. It is esti-

FIGURE 7.11 Tidal bore in the Bay of Fundy. (Courtesy of NOAA.)

mated that the maximum tidal range in the open ocean is about one meter. Tides may be considered as a type of wave with a period of about 12 or 24 hours, which travel as shallow-water waves because the ratio of the wavelength to ocean depth is very large (see Chapter 6 for definition and discussion of shallow-water waves). Therefore, tidal ranges are generally amplified near the coasts, similar to amplification of height of other ocean waves. Figure 7.12 shows the computed tidal map of the oceans for an assumed position of the moon directly above the equator on the Green-

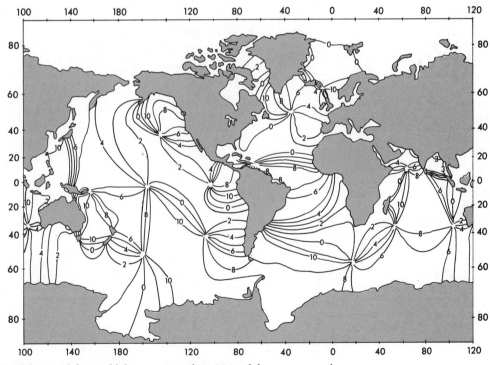

FIGURE 7.12 Cotidal map of the world for an assumed position of the moon over the equator, at 0° longitude (the Greenwich meridian). Numbers indicate times of high tide (in hours) after the moon has crossed the Greenwich meridian. Note the many amphidromic points, where no tide will be produced by the moon. Tides rotate around these points. (After Cartwright, 1969.)

wich meridian (0° longitude). The effects of the sun are not considered. The lines shown are called *cotidal lines* and represent points connecting simultaneous high tides for each hour after the moon has crossed the Greenwich meridian. Low tides would occur about six hours before or after high tides. Tides move around the amphidromic points, counter-clockwise in the northern hemisphere and clockwise in the southern hemisphere because of Coriolis force. No lunar tide should occur at the amphidromic points shown. For this reason, at islands such as the Solomon Islands, which are close to an amphidromic point of lunar tides, the tides follow the sun rather than the moon. High tides, for example, would occur each day at the same time in these places.

SEA LEVEL

Elevations and depths are referred to some datum or reference level. Topographical maps in the United State give elevations above *mean sea level,* which is the average height of the sea surface for all stages of tides over a period of 18.6 years. The concept of sea level as used in mapping in the United States is based on the determinations of mean sea level at 26 tidal stations along the Pacific Ocean, the Atlantic Ocean, and the Gulf of Mexico. This is referred to as the "sea level datum of 1929." Tides can vary considerably over short distances, depending on the geometry of the ocean basins. Therefore, mean sea level as indicated on topographical maps may not agree with the true mean sea level at a particular coastal location.

Ocean depths and tide tables are referred to other reference levels. In the United States, *mean low water* (the average height of all low tides at a place each day over a period of 18.6 years) is used for charts and tide tables of the Atlantic Ocean and the Gulf of Mexico. For the Pacific Ocean, *mean lower low water* (the average height of the lower of two low tides each day over a period of 18.6 years) is the standard. Land features on nautical charts, however, are usually referred to by *mean high water* (the average height of all high tides at a place each day for a period of 18.6 years).

Tidal Currents

Tides move large quantities of water. The water piled up along the coast during high tide must return to the sea, thus creating a current. Knowledge of tidal currents is important in navigation, especially in bays and estuaries. Tidal currents are capable of cutting channels in such bodies of water, which further affect navigation. Tidal currents are influenced by the same forces that govern tides, and their patterns can be complicated. In the open ocean, tides can travel as progressive shallow-water waves, and hence the water particles travel in elliptical paths. Although the theoretical horizontal velocities of these currents should be the same from top to bottom, frictional forces reduce them considerably at depths. Maximum currents (usually a fraction of a knot) will be found associated with the crests and troughs of these tide waves. Tidal currents in the open ocean and in large basins are influenced by Coriolis force. As

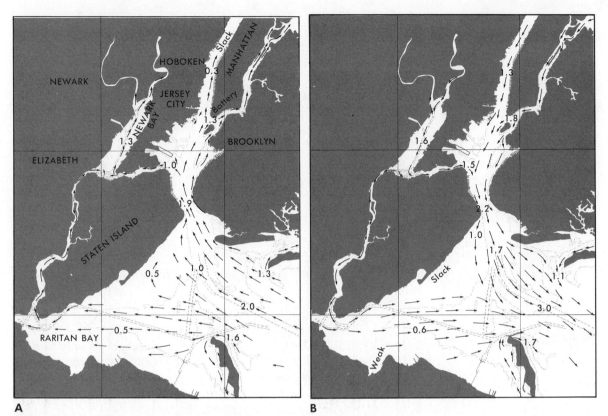

FIGURE 7.13 Tidal current charts for New York City. *A*, Four hours after low water, *B*, Four hours after high water, at the Battery. Tidal current speeds are shown in knots. (After NOAA: Tidal Current Charts, New York Harbor, 7th ed. Washington, D.C., 1956.)

a result, clockwise-rotating currents will be formed in the northern hemisphere (one complete rotation during one tidal cycle).

In many estuaries and bays, as well as in areas where tides exist as standing waves, maximum currents occur below the nodal line (see Fig. 7.9) when there is no tidal fluctuation in the bay. The current velocities at other locations are maximal between high and low tides. Theoretically, no currents should exist at high or low water, as they should begin to flow just after the turn of the tide. However, the actual tidal currents observed may vary considerably from this prediction, depending on the configuration of the basin (Fig. 7.13).

MEASUREMENT AND PREDICTION OF TIDES

Measurement

Near coastal areas, tides are commonly measured by a continuously recording mechanical device known as a *tide gage.* In principle, it con-

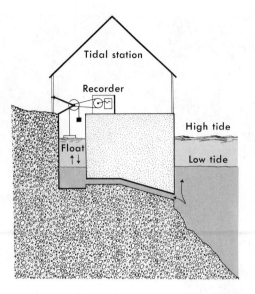

Tidal station

Recorder

High tide

Float

Low tide

FIGURE 7.14 A tide gage station.

sists of a well or shaft connected to the sea far below the lowest possible tide level (Fig. 7.14). The water level in the well responds only to tidal fluctuations and not to waves or other short-period fluctuations such as those produced by a passing ship. A float in the well rises and falls with the tides. The float is connected by wire to a pulley, which moves a pen and produces a continuous tidal record.

Some modern electronic tide-recording devices have been developed that measure the differences in pressure at a point on the sea floor caused by surface fluctuations. Such devices can be used to measure tides even in the open sea.

Measurements of tidal current are considerably more difficult than measurements of tidal height. In addition, these measurements are complicated by nontidal currents existing in the area. Numerous current-measuring devices are available, the most common ones being current meters. Nontidal currents and various current-measuring devices are discussed in the next chapter.

Prediction

Tides at any place cannot be accurately predicted by knowing only the positions of the moon and the sun relative to that place, since they also depend on other factors such as the size and configuration of the basin. Accurate tidal predictions, therefore, require actual tidal observations for at least a year and, preferably, much longer to establish these factors. Such predictions may be made using a mechanical instrument

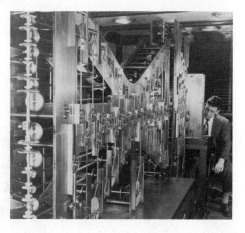

FIGURE 7.15   A tide predicting machine. This method of predicting tides has been replaced by computers. (Courtesy of NOAA.)

called a *tide predicting machine* (Fig. 7.15), originally invented by Lord Kelvin in 1872. In the United States, however, electronic computers have been used to predict tides since 1966. Tables for the prediction of tides and tidal currents are published annually for many representative coastal locations throughout the world. In the United States, they are published by the National Oceanic and Atmospheric Administration (NOAA) of the Department of Commerce and can be purchased from map and marine-supply dealers throughout the United States.

Tides do not repeat themselves from one year to the next, because the relative positions of the earth, the moon, and the sun are approximately repeated only once every 18.6 years. Thus, if tides were to be predicted statistically, at least 18.6 years of tidal observation would be required.

POWER FROM TIDES

The world's total retrievable energy from tides and currents is less than the total energy that could be obtained from wind, geothermal sources, or rivers. However, in certain areas with a great tidal range, such as the Bay of Fundy, the potential exists for producing economical tidal power.

From the seventeeth to the nineteenth century, numerous small tidal-powered mills were operated in estuaries from New Brunswick to Georgia. Usually the flood tide was allowed to pass through a tide gate in a dam. When the tide began to ebb, the gates slammed shut and the water turned a wheel or turbine which powered the saws or millstone. Water also powered a generator that provided electricity to light the work area. Some of these mills operated successfully well into the twentieth century.

FIGURE 7.16   The tidal power station on the Rance estuary, in France. (Courtesy of French Embassy Press and Information Division.)

They were eventually replaced by electrically powered mills, the operation of which was not limited to the time provided by the outgoing tide. In addition, the mills could be placed on land, closer to the improving land transportation.

Currently, there are only two tidal power plants in operation in the world. One of these is a Soviet pilot plant on the White Sea. The other is the hydroelectric power plant on the Rance estuary in France.

The maximum tidal range in the Rance estuary is about 13 meters and approximately 170 million cubic meters of water enter and leave it during a semidiurnal tidal cycle. In 1960, an electric power plant was built in a dam near the mouth of the Rance (Fig. 7.16). The blades of the turbines are turned by the tidal flow in and out of the estuary. When in full operation, this power plant is capable of producing about 670 million kilowatt-hours of electricity annually. This is roughly half the amount produced by an average hydroelectric power plant in France.

Tidal power plants may be built in many parts of the world. A joint United States–Canada tidal power plant has been proposed for many years to be built in the Passamaquoddy Bay near the mouth of the Bay of Fundy. The tidal flow in this area is more than 10 times that involved in the Rance estuary. No firm decision, however, has been made. Far more power could be generated by a power plant built in the Bay of Fundy itself. Although tidal power can account for only a small portion of the world's energy needs, it is pollution-free and dependable on a long-term basis. The major obstacles are economic feasibility and possible ecological damage from the changes it would cause in the normal tidal flow.

SUMMARY

Tides are waves with periods of about 12 or 24 hours. They are caused by the gravitational attraction of the sun and the moon on the earth and the centrifugal forces that result from the mutual revolutions of the earth–moon–sun system.

Even though the sun is much more massive than the moon, its tidal effect is only about 46 per cent of that of the moon, because of the sun's greater distance from the earth. When the earth, moon, and sun are lined up (new or full moons) the tides are usually the greatest (spring tides). When the positions of the earth, moon, and sun form a right angle (first or third quarter of the moon) the tides are at a minumum (neap tides) since the effects of the sun partially cancel those of the moon.

Ocean tides are modified by the shapes of ocean basins, rotation of the earth, declination of the moon, and friction with the ocean bottom. These factors determine whether the tides at a particular location are semidiurnal (12 hours) or diurnal (24 hours) as well as determining tidal ranges and tidal irregularities.

Although the nature of tides is complex and varies greatly from place to place, they are highly predictable. Tide tables published in the U.S. are the product of electronic computers capable of predicting local tides years in advance. Such predictions are based on years of observation of real tides.

The rise and fall of the sea causes tidal currents. In some restricted areas such as the Rance estuary in France, power is extracted from this flowing water by turbines placed in dams.

IMPORTANT TERMS

Amphidromic point
Aphelion
Apogee
Cotidal line
Declination
Diurnal inequality
Diurnal tide
Lunar day
Mean sea level
Mixed tides

Natural period
Neap tide
Nodal line
Perigee
Perihelion
Resonance
Semidiurnal tide
Spring tide
Tidal bore

STUDY QUESTIONS

1. Why are there two high and two low tides each day along many coasts?
2. How and when are spring tides and neap tides produced?
3. What causes the diurnal inequality of tides?
4. How does Coriolis force affect tides?
5. What is meant by "the natural period of a basin?"
6. Explain the unusually great tidal ranges in the Bay of Fundy.
7. Why are there only one high and one low tide each day in some areas?
8. Calculate the natural periods for the following open basins:

	Length (ζ)	Depth (d)	Natural Period
A.	300 km	60 meters	_____
B.	10 km	10 meters	_____
C.	160 km	90 meters	_____
D.	5 km	8 meters	_____

9. High tide appears earlier in Southern California than it does in Northern California. Why is this so?
10. Discuss the prediction of tides.
11. Discuss the advantages and disadvantages of tidal power.

SUGGESTED READINGS

Bowditch, Nathaniel. 1977. *American Practical Navigator.* Washington, D.C., Government Printing Office.
Clancy, Edward P. 1968. *The Tides.* Garden City, N.Y., Doubleday.

Defant, Albert. 1960. *Ebb and Flow.* Ann Arbor, University of Michigan Press.

Goldreich, Peter. 1972 (May). Tides and the Earth-Moon System. *Scientific American.*

Macmillan, D. H. 1966. *Tides.* New York, American Elsevier.

McDonald, James E. 1952 (April). The Coriolis Effect. *Scientific American.*

National Oceanic and Atmospheric Administration. 1976. *Our Restless Tides.* Washington, D.C., Government Printing Office.

Ryan, Paul R. 1979. Harnessing power from tides: State of the art. *Oceanus.* 22:64–67.

Ocean Currents

8

Although not as perceptible as the surface waves, water from the surface to the bottom is constantly in motion from ocean currents. Surface currents are caused primarily by the force of wind against water, whereas deep ocean circulation is related to density differences of the water within the ocean. Currents are of great significance to all of us and to ocean life. Knowledge of ocean circulation is important to sailors who wish to take advantage of surface currents in their voyages. It is useful in the prediction of weather. Currents also affect the movement of nutrients and pollutants. Ocean circulation is responsible for bringing nutrients to surface waters in the biologically productive regions of the oceans. In fact, the oceans would be nearly lifeless if it were not for their circulation.

A current buoy in the Chester River, Maryland. (Photo courtesy of National Oceanic and Atmospheric Administration.)

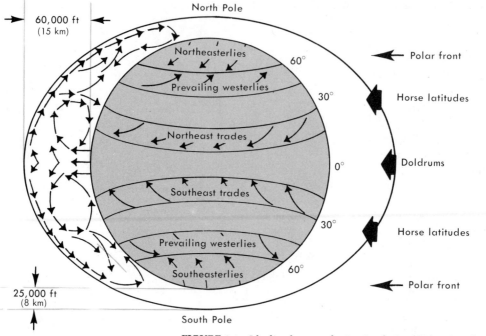

FIGURE 8.1 Idealized atmospheric circulation. (After Bowditch, 1977.)

SURFACE CURRENTS

Atmospheric Circulation

The primary cause of surface ocean currents is wind. The ultimate driving force of the winds over the earth is solar energy. As seen earlier (Chap. 5), the equatorial regions of the earth receive more solar energy per unit area than does the rest of the earth. Figure 8.1 shows an idealized atmospheric circulation pattern of the earth. The equatorial air masses rise as a result of increased heating and eventually move poleward at higher altitudes. Consequently, low pressure belts known as the *doldrums* are created along the equator. In order to compensate for the rising air, air masses move toward the equator near the surface. The poleward-moving equatorial air gradually cools and sinks at about 30° N and S latitudes (the subtropical high-pressure belts known as the *horse latitudes*), and upon reaching the earth's surface, part of the air spreads poleward while some moves toward the equator. The poleward-moving air meets the cool, dense polar air moving toward the equator at about 60° N and S latitudes, forming the *polar front*.

Because we are dealing with a rotating earth, *Coriolis force* (see Appendix C) must be considered. This force tends to deflect all moving objects toward the right in the Northern Hemisphere and to the left in the Southern Hemisphere. The equatorward flow of wind near the earth's surface, for example, is deflected by Coriolis force toward the west, but because of the effects of friction, the deflection of air masses is not complete. Thus, in the Northern Hemisphere, surface winds tend to blow from the northeast and in the Southern Hemisphere from the southeast. Figure 8.2 shows average wind patterns over the oceans during winter and summer. Except for distortions caused by the presence

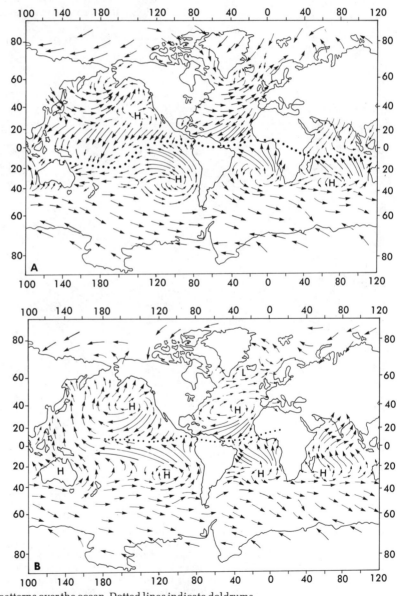

FIGURE 8.2 Average surface wind patterns over the ocean. Dotted lines indicate doldrums. *A*, Northern winter; *B*, Northern summer. (After Bowditch, 1977.)

of land masses, the wind patterns of both the Northern and Southern Hemispheres are roughly symmetrical with respect to the equator. The most prominent of all features shown are the *trade winds* (at about 20° N and S latitudes), which generally blow from the northeast in the Northern Hemisphere and from the southeast in the Southern Hemisphere. Between about 30° and 60° latitudes are the *prevailing westerlies.* They blow from the southwest and northwest in the Northern and Southern hemispheres, respectively. Poleward of the prevailing westerlies are the *polar easterlies*. Before attempting to explain the relationship between atmospheric and oceanic circulation, let us first look at the surface current patterns.

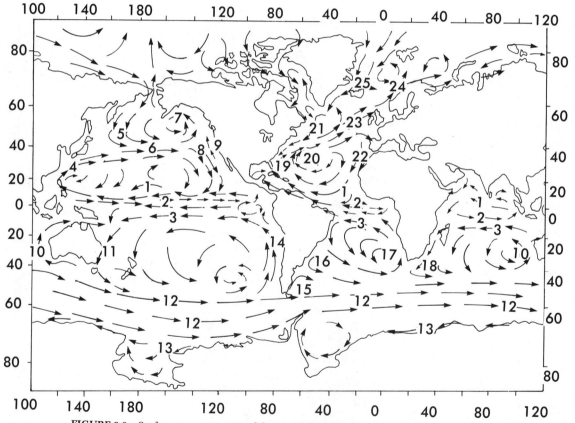

FIGURE 8.3 Surface ocean currents of the world for February. 1. North Equatorial Current. 2. Equatorial Counter Current. 3. South Equatorial Current. 4. Kuroshio. 5. Oyashio. 6. North Pacific Current. 7. Alaska Gyre. 8. California Current. 9. Davidson Current. 10. West Australia Current. 11. East Australia Current 12. West Wind Drift. 13. East Wind Drift. 14. Peru Current. 15. Falkland Current. 16. Brazil Current. 17. Benguela Current. 18. Agulhas Current. 19. Florida Current. 20. Gulf Stream. 21. Labrador Current. 22. Canary Current. 23. North Atlantic Current. 24. Norway Current. 25. East Greenland Current.

Surface Current Patterns

One of the striking features of global surface circulation (Fig. 8.3) is that most of the water flows in closed loops, or *gyres*. Most of these gyres flow in a clockwise fashion in the Northern Hemisphere and counterclockwise in the Southern Hemisphere. Notable exceptions, however, occur in high latitudes in the North Pacific and North Atlantic, where the currents may form a number of small counterclockwise gyres. In general, surface currents are restricted to the upper few hundred meters of the sea, although parts of the Gulf Stream can be detected as deep as 1500 meters. A few of the major surface ocean currents are discussed below.

EQUATORIAL CURRENTS

The currents in the equatorial regions consist of the westward-flowing *North* and *South Equatorial Currents*, which are separated by the eastward-flowing *Equatorial Counter Currents*. In the Atlantic, the Equatorial Counter Current is highly developed in the east, near the coast of Africa. These currents generally have speeds of about 25 cm/sec (0.5 knot) and may flow at speeds of up to 125 cm/sec (Table 8.1). The direction of the North Equatorial Current in the Indian Ocean is reversed during the Northern Hemisphere summer (when the monsoon winds blow from the southwest) and flows eastward, as does the Equatorial Counter Current.

Table 8.1   SPEED AND TRANSPORT OF SELECTED MAJOR SURFACE CURRENTS*

CURRENT	MAXIMUM SPEED (cm/sec)	AVERAGE SPEED (cm/sec)	TRANSPORT (million m³/sec)
Gulf Stream	300	100	100
Kuroshio	300	90	50
Florida	250	35	25
California	—	40	—
Brazil	—	—	10
Peru	100	—	20
West Wind Drift	50	—	100–200
East Wind Drift	150	—	—
North and South Equatorial	125	20–35	45
Pacific Undercurrent (Cromwell)	150	—	40

*Data compiled from various sources, including Bowditch (1977), Defant (1961), Warren (1966).
100 cm/sec = 3.6 km/hr = 2 knots.

WESTERN BOUNDARY CURRENTS

The warm westward-flowing equatorial currents are deflected poleward near the western boundaries of the oceans (the east coasts of continents). These currents are known as *western boundary currents* and include some of the most intense ocean currents (sometimes exceeding 200 cm/sec). These boundary currents are the *Kuroshio* and the *East Australian Currents* in the Pacific Ocean, the *Florida, Gulf Stream,* and *Brazil Currents* in the Atlantic Ocean, and the *Agulhas Current* in the Indian Ocean. Some western boundary currents are quite extensive vertically, flowing at depths as great as 1500 meters or more, considerably deeper than most surface currents.

The northward-flowing Gulf Stream is not easily distinguished by temperature from the warm *Sargasso Sea* along which it flows. However, the current flows along a boundary between the warm Sargasso Sea water on its right and cold water from the *Labrador Current* on its left and acts as a barrier, preventing the mixing of the two waters. Its flow is not straight but is deflected eastward toward Europe and is further characterized by a number of meandering loops along its margins, which help mix the waters laterally (Fig. 8.4). In addition, the current pattern varies with time. The pattern of the Gulf Stream is similar to that of the Kuroshio and other western boundary currents. The Labrador Current flows southward along the coast of Labrador, bringing cold water to the east coast of North America. It is a rather sluggish current (about 25 cm/sec or 1 km/hr). As it flows southward between the Gulf Stream and land, it gradually mixes with the Gulf Stream, but the cooling influence may sometimes extend as far south as Cape Hatteras.

FIGURE 8.4 Infrared image of the East Coast of the United States. The warm Gulf Stream appears as a dark meandering band, with several clockwise rotating eddies. Colder coastal water appears lighter, and the clouds appear white. Photo taken on May 1, 1977. (Courtesy of NOAA.)

OTHER MAJOR SURFACE CURRENTS

In the Pacific Ocean, the North Pacific Current carries most of the water from the Kuroshio toward the east, until it eventually splits into northward- and southward-flowing branches. The southward-moving current, the *California Current*, completes the major clockwise gyre in the North Pacific Ocean when it joins the North Equatorial Current. The northward-flowing water contributes to the smaller counterclockwise *Alaskan Gyre*. The *Davidson Current* flows northward during the winter between the California Current and the coast of Oregon and Washington. The southwest winds of winter confine fresher waters toward shore, and river effluent is deflected toward the north. In the drier summer, north winds and the resulting upwelling shift the surface water seaward and the diminished fresh water mixes with the ocean water more freely.

In the Atlantic Ocean, the *North Atlantic Current* carries water eastward. The southward-flowing branch of this current, the *Canary Current*, completes the major clockwise gyre in the North Atlantic. At the same time, water is deflected northward from the North Atlantic Current to form several smaller counterclockwise subpolar gyres.

The surface circulation in the Southern Hemisphere is virtually a mirror image of that in the Northern Hemisphere, except for the absence of the minor subpolar gyres. In the South Pacific, the counterclockwise gyre is completed by the globe-circling *West Wind Drift* and the northward-flowing *Peru Current* (also known as the *Humboldt Current*). Similar current patterns exist in the South Atlantic and South Indian oceans. Between the West Wind Drift and the Antarctic continent is a narrow westward-flowing current, the *East Wind Drift*.

Causes of Surface Currents

EKMAN TRANSPORT

In 1902, V. W. Ekman, a Swedish oceanographer, provided the mathematical basis for understanding wind-driven ocean currents. When a wind blows over the ocean, the water particles near the surface tend to move in the same direction as the wind. Once this motion is initiated, Coriolis force begins to act, deflecting the particles to the right in the Northern Hemisphere and to the left in the Southern Hemisphere. The motion of a thin layer of surface water will be about 45° to the right of the wind in the Northern Hemisphere. The wind energy, however, is transmitted even below the surface layer. Because of friction, the water at the surface tends to pull the underlying water with it but at a lower speed. Coriolis force further deflects this deeper water. This reaction continues to some depth at which the frictional forces and the water movement due to this process become negligible. The vertical

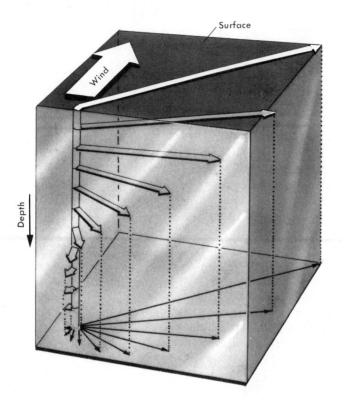

FIGURE 8.5 The Ekman spiral (the effect of wind on water near the surface) in the Northern Hemisphere. (After Baker et al., 1966.)

pattern of the horizontally moving water forms a sort of spiral called the *Ekman spiral* (Fig. 8.5). The water at some depth may even be moving in an opposite direction to that of the wind at the surface. The overall effect is that a surface layer of water may have a net motion 90° to the wind. This is known as the *Ekman transport*. The depth of this layer may extend downward to several tens of meters.

UPWELLING

Because of the Ekman transport, an interesting phenomenon occurs along some coasts. Assume that a wind is blowing parallel to a coast in the Northern Hemisphere, as shown in Figure 8.6. With the coast to the left of the wind, the net transport, 90° to the right of the wind, is away from the coast. Deeper water must move toward the surface along the coast to replace the water that has moved offshore (Fig. 8.7). This process is known as *upwelling*. Upwelling may transport water from as deep as 400 meters, although depths of 100 to 200 meters are more common. It is actually a rather slow process, involving rates of flow of a few tens of meters per day. The cold, deeper water is generally rich in nutrients, and when it is upwelled, surface productivity is enhanced.

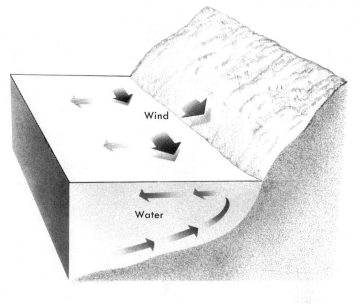

FIGURE 8.6 Upwelling in the Northern Hemisphere, with wind blowing parallel to the coast. The deeper water replaces surface water transported offshore.

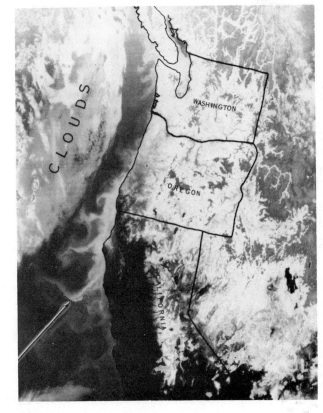

FIGURE 8.7 Infrared image of upwelling along the west coast of North America. The lighter areas (arrow) indicate cold, upwelled water reaching the surface. Photo taken September 11, 1974. (Courtesy of NOAA.)

Many of the world's most productive fisheries are in upwelling areas, for example, portions of the west coasts of North and South America (see also Chap. 12). Because wind conditions near the coasts are usually variable, upwelling is frequently a seasonal process. If the wind were to blow in the opposite direction to that shown in Figure 8.6, a reverse circulation would be established. This would tend to draw nutrient-poor water from offshore toward the coast and might eventually cause a decline in productivity.

Along the Peruvian coast, upwelling has produced one of the world's greatest fisheries. Occasionally, warm nutrient-poor water inundates the area from the north and the resulting decrease in productivity causes the death of many animals such as anchovies and the guano birds that eat them. This results in anoxic conditions and the production of hydrogen sulfide gas by anaerobic bacteria. Hydrogen sulfide discolors the hulls of ships and the exteriors of homes and is sometimes referred to as the *Callao Painter*, after a Peruvian port city. The phenomenon usually occurs around Christmas, so it is also known as *El Niño* ("the child"). Recent occurrences of El Niño have been in 1976, 1972, 1965–6, 1957–8, 1953, and 1941.

Attempts are now being made to predict episodes of El Niño. They usually follow known changes in atmospheric pressure and wind patterns that affect the ocean currents. If the trade winds are strengthened for about 18 months or more and then weaken, the water that they piled up in the western Pacific sloshes eastward, strengthening the Equatorial Counter Current. Eventually, warm nutrient-poor water is pushed toward Peru, contributing to El Niño.

GEOSTROPHIC BALANCE

In the preceding discussion, we have considered the effects of wind, friction, and the earth's rotation on the ocean surface. Although there is a good correlation between surface wind patterns (Fig. 8.2) and the large-scale surface currents (Fig. 8.3), a total agreement is lacking, especially when Ekman transport is considered. For example, the trade winds in the equatorial North Atlantic blow from the northeast, yet the surface current (North Equatorial Current) flows almost due west. Near the South American coast, however, the current is deflected toward the northwest by the land mass. The presence of land masses plays an important role in determining the configuration of many surface currents.

Ocean currents are also influenced by the presence of irregularities of the sea surface, the *sea-surface topography*. The sea surface contains "hills" and "valleys," especially near the central portions of the oceans. The surface of the Sargasso Sea in the north Atlantic, for example, forms a "hill" consisting of warm light water surrounded by colder dense water. The sea surfaces of the western Atlantic and Pacific also are higher than those in the east. Sea-surface topography is primarily due to spatial variations in sea-water density. Light, usually warm, water stands higher than dense, usually cool, water. Wind patterns over the sea also contribute to the formation of these topographic features. Because the topographic variations frequently involve only a few tens of centimeters, these features are generally deduced by indirect means from the distribution of sea-water density. Sea-surface topography is now being studied directly with radar altimeters aboard satellites. The data thus generated tend to agree with the more detailed topography generated by indirect means.

Figure 8.8A shows a sea-surface profile across the equator in the Atlantic or Pacific Ocean. It show a topographic high or "hill" centered about 5° N latitude and topographic lows or "valleys" north and south of it. Ekman transport due to the southeast trade winds that blow across the equator in the Pacific and Atlantic causes a divergence of surface water at the equator (Fig. 8.8B). This brings colder upwelled water toward the surface, displacing warmer surface water northward and southward, and results in a topographic valley at the equator.

The North and South Equatorial Currents flow westward along sloping surfaces, as shown in Figure 8.8A. These currents can be

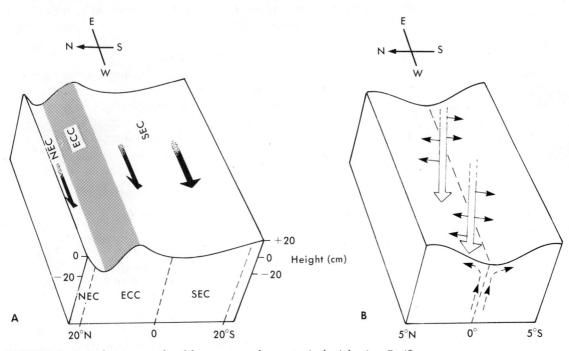

FIGURE 8.8 *A*, Surface topography of the sea, across the equator in the Atlantic or Pacific Ocean. NEC, North Equatorial Current; ECC, Equatorial Counter Current; SEC, South Equatorial Current. *B*, A topographic valley is produced when the southeast trade winds (double arrows) cross the equator. The small arrows show the poleward Ekman transport of surface water. The dashed arrows show the resulting upwelling of cool, dense subsurface water at the equator. (*A*, after Defant, 1961, and Wyrtki, 1979.)

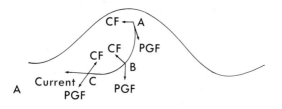

FIGURE 8.9 Geostrophic balance. Deflection of a current by Coriolis force, directed to the right of the current, and the pressure gradient force, directed downhill, on a sloping sea surface in the Northern Hemisphere. Note that the Coriolis force and pressure gradient force shown at points A, B, and C are balanced only at point C. (CF, Coriolis force; PGF, pressure gradient force.)

explained only if the Coriolis force acting on the current is balanced by an equal and opposite force. This opposing force is the *pressure-gradient force* (Fig. 8.9), the horizontal force due to the decrease in pressure between the two points, caused by the sloping sea surface. Any water that is transported up the slope would flow back down the slope owing to gravity and eventually would be deflected to the right in the Northern Hemisphere and to the left in the Southern Hemisphere. This type of flow, in which the Coriolis force and the pressure-gradient force are exactly balanced *(geostrophic balance)*, is called a *geostrophic current*. A purely geostrophic current is assumed to be frictionless and unaccelerated, because no forces acting in the direction of the current are considered. Of course, wind and friction are quite significant factors, as mentioned earlier. The east-flowing Equatorial Counter Current also can be accounted for by geostrophic flow.

Although wind is the primary driving force, most ocean currents, especially surface currents, tend to flow around hills of light water, as in the clockwise-flowing gyre around the Sargasso Sea "hill" (Fig. 8.10), or around valleys of denser water, as in the counterclockwise Alaskan gyre. All of the major ocean currents that we have discussed are considered to be in geostrophic balance.

EQUATORIAL UNDERCURRENTS

In addition to the surface currents mentioned earlier, another type of current, although not at the surface, is important. These currents are the *equatorial undercurrents*, initially reported in the Atlantic in 1876 by Buchanan, one of the investigators on the *Challenger* expedition. This phenomenon was largely ignored until 1951, when an undercurrent was observed in the East Pacific moving under the westward-flowing South Equatorial Current. Tuna-fishing gear was observed to be carried eastward below the westward-flowing surface currents. This undercurrent is known as the *Cromwell Current* after Townsend Cromwell, an Americal physical oceanographer, who first studied it. Since that time, equatorial undercurrents have been confirmed for the Indian as well as for the Atlantic Ocean. In the Indian Ocean, however, the undercurrent is well developed only during the northeast monsoon season, when the

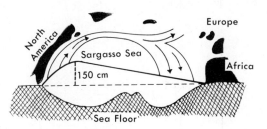

FIGURE 8.10 Surface circulation around the Sargasso Sea "hill." Sea surface topography is greatly exaggerated. (After Williams et al., 1968.)

North Equatorial Current is present. The equatorial undercurrents are thin ribbons of eastward-flowing water centered about the equator. They average 250 km in width and 250 meters in thickness. These currents increase in speed toward the core of the ribbon, where they may reach 150 cm/sec (about three knots). The cores are also characterized by high-salinity water and therefore can be traced easily.

The causes of undercurrents are not fully understood. The undercurrents probably represent a subsurface return flow of some of the water that has been piled up along the western margins of the oceans. Perhaps as a result of the westward-moving surface currents, the sea surface (and underwater pressure surfaces) slopes downward toward the east, resulting in a subsurface flow of water.

Climate and Surface Currents

The oceans play an important role in determining the climates of the earth (see also the discussion on Heat Budget in Chapter 5). Surface currents of the oceans determine, to a certain extent, the distribution of sea-surface temperature. If it were not for continental land masses and consequently the ocean currents, the surface temperatures would be nearly zonal in an east–west direction. The surface temperatures (see Fig. 5.10), however, deviate from a zonal pattern in a manner which can be easily accounted for by the current patterns shown in Figure 8.3. In the North Atlantic, for example, these deviations can be explained by the southward-flowing cold Labrador Current, and the northeastward-flowing Gulf Stream, North Atlantic Current, and Norway Current, which transport warm water poleward. The coast of Norway, therefore, is warmer than the coast of Labrador.

The oceans serve as a vast heat reservoir, contributing heat to a cold atmosphere and drawing it from a warm atmosphere. Meanwhile, the sea itself remains at a rather constant temperature. Thus, the sea serves as a moderating influence on climate, especially when the wind blows over the sea and onto the land. This is particularly true along the west coast of continents where winds generally blow onto the land from the west. Along the east coasts, where winds generally blow from the land,

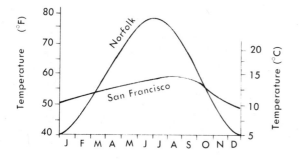

FIGURE 8.11 Seasonal temperature distribution for San Francisco, California, and Norfolk, Virginia, at about the same latitude. Both cities have an average temperature of approximately 12.8°C (55°F). (After Trewartha, 1954.)

the ocean's effect on climate is much less striking. Figure 8.11 shows the seasonal temperature distribution for San Francisco, California, and Norfolk, Virginia, which are at about the same latitude. The greater effect of the cool California Current produces a more moderate climate at San Francisco when compared to Norfolk, which is less affected by the ocean.

It is important to stress that wind patterns are not constant and that shifts in winds may cause relatively short-term weather conditions such as storms and fog. Although the development and movement of a storm depend on a number of factors, such as atmospheric pressure and wind, the strength of a storm may be modified by ocean currents. For example, the warm waters of the Kuroshio and the Gulf Stream evaporate rapidly when cold winter winds blow over them and are warmed. The warmed air can carry much more moisture than cold air. Therefore, a great amount of energy and water vapor is transferred to the developing storms in these regions, making the storms more severe than they would otherwise be. Conversely, when warm, moisture-laden winds blow over cold water, the air is cooled and its ability to hold moisture is decreased. Therefore, fog may form as the water vapor condenses. This situation frequently occurs when winds that have blown over the Gulf Stream meet the cold Labrador Current flowing over the Grand Banks area off Newfoundland. This poses a serious problem in the spring, when both fog and icebergs occur in this region. Fog may also develop when very cold air flows over warm water, resulting in rapid evaporation and subsequent condensation of water vapor. This phenomenon is known as *steam fog* and occurs in the fall over some lakes and semienclosed bodies of water such as Norwegian fiords.

Careful observation of sea-surface temperature and currents may eventually provide a base for effective long-term forcasts of weather and other phenomena such as El Niño. Furthermore, as part of the International Decade of Ocean Exploration (IDOE), an interdisciplinary study of *Climate: Long-range Investigation, Mapping, And Prediction* (CLIMAP) is gathering data that may allow predictions of long-term climate changes over the earth.

Current Measurements

There are numerous methods of measuring surface currents. Basically, they fall into two broad catagories; the float method and the flow method. In the *float method*, an object such as a piece of wood may be observed as it drifts with the current; the direction and distance traveled in some unit of time (and therefore the speed) may be determined. In the *flow method*, a current may be observed flowing past a fixed point such as a ship at anchor or a moored buoy. The current in this case may be measured with a specially designed current meter, which records the flow of water through the instrument.

Records of ships that have drifted off course, or have traveled faster or slower than expected, provide much general information concerning ocean currents. International cooperation in the compilation of ships' data has continued since 1853, and ships' records were analyzed even before this date. Based on this type of data, Benjamin Franklin published a chart of the Gulf Stream in 1786 (see Fig. 1.7).

In addition, sealed bottles known as *drift bottles* (Fig. 8.12) containing postcards have been used to study surface drift between the points of release and capture. The person who picks up the bottle is requested to record the location and time of discovery on the card and to mail it to the researcher (and to keep the bottle as a souvenir). This method has been simplified by replacing the bottle with a plastic covering for the post card. Of course, this method does not provide detailed information concerning the path but only the beginning and the end points of the drift, and it may have considerable error because of direct wind effects on the bottle. This technique is generally most useful in studying currents in coastal areas.

Although ship drift and drift bottle records have provided much of what is known about the general pattern of surface currents, other more sophisticated techniques are available for studying the details of ocean circulation. For example, a specially designed float called a *drogue* may be observed as it drifts with the current. A drogue (Fig. 8.13) is a weighted float with a subsurface vane or parachute that moves with the current at a desired depth, making the float less subject to drifting with the wind. It usually has a flag and a light attached to make it easier to follow and may also have a radio transmitter or a metallic radar reflector. Another float, the *Swallow float* (named after its developer, J. C. Swallow), may be adjusted to drift with the current at a particular density level (depth) in the sea. It produces a sound that may be picked up by a ship's sonar. It is particularly useful in the study of subsurface currents.

In addition to the above-mentioned devices, currents may be studied by marking the water with a dye that can be detected even when greatly diluted. This method is especially useful in aerial surveying of currents, as the dye may be dropped from the air and photographed as it drifts.

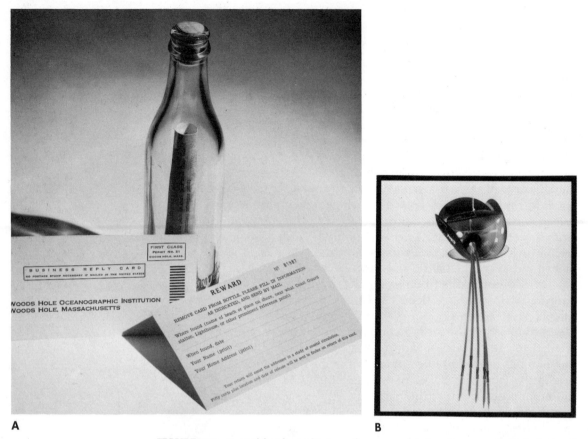

A   **B**

FIGURE 8.12   *A*, Drift bottle used to trace the movement of surface currents. *B*, A cluster of umbrella-like devices are dropped into the sea at a predetermined location. They sink, separate, and drift with bottom currents. Finders are requested to return them to the address shown on the label, and to state the locations at which they were found, providing the researcher with important data regarding current movements. (Courtesy of Woods Hole Oceanographic Institution.)

Current meters, or flow meters, measure the speed and frequently the direction of the current. In some instruments, the current flows past a propeller, the rotation of which indicates the speed of the current. A compass mounted on the instrument may provide information about the current's direction. A great variety of current meters are available (Fig. 8.14). Unfortunately, the use of current meters is an expensive and time-consuming operation compared to other methods and is therefore done only when the other methods are either too impractical or too imprecise. The flow of water past a point may also be estimated by measuring the deflection of a wire supporting a weight with a known drag (Fig. 8.15). This procedure may be done rather easily from even a small boat but is effective only near the surface, since many currents have quite complicated vertical structures.

Remote-sensing satellites are becoming increasingly important in the study of ocean currents. Infrared sensing equipment allows the tracing of currents by temperature. Radar altimeters accurately measure the distance from the satellite to the sea surface and hence can record tides and sea-surface topography. These techniques provide a great deal of information about the world's oceans in a short period of time.

DEEP OCEAN CURRENTS

The currents we have been discussing so far are produced primarily by wind. Most of them are restricted to the surface water, although the tremendous Gulf Stream carries water even at depths of 1500 meters or more. Water movement in surface currents is basically horizontal. Deep ocean currents involve horizontal and vertical flow extending into the deepest parts of the oceans. They are the result of extremely small horizontal and vertical density variations in the sea. The deep circulation, however, is inseparable from that at the surface. Even deep water is returned to the surface, becomes part of the surface currents, and eventually sinks to rejoin the deep circulation. Thus, a balance exists in the flow of water within each ocean as well as among oceans.

The deep ocean currents carry oxygen into the deepest parts of the seas, enabling marine animals to exist at all depths. Deep currents also return nutrients to the surface. If it were not for these deep currents, the seas would be stagnant and anoxic in the depths and surface productivity would be much less rich.

The average surface current flows at a speed on the order of 100 cm/sec. In contrast, deep ocean currents are very slow, frequently of the order of a few centimeters per second. Speeds of more than 40 cm/sec (about one knot), however, have been found at 4000-meter depths in the western Atlantic. Results of radioactive carbon dating of sea water show how sluggish deep ocean currents are. Such studies have revealed that it has taken about 750 years for water to travel from the

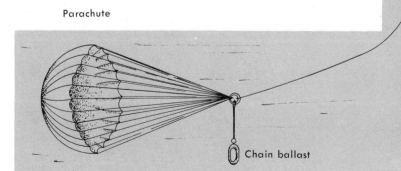

FIGURE 8.13 A parachute drogue, which drifts with the current. (After Von Arx, 1962.)

FIGURE 8.14 *A*, Typical current meter about to be lowered into the sea. *B*, "Pop-up" current meter. The two spherical floats return the instrument to the surface after bottom current measurements have been recorded. (*A* courtesy of WHOI; *B* courtesy of Lamont-Doherty Geological Observatory of Columbia University.)

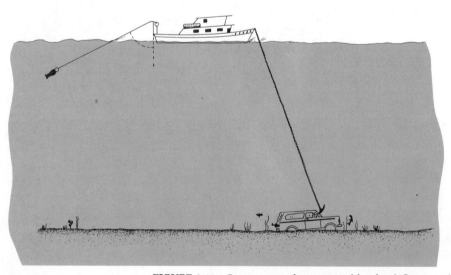

FIGURE 8.15 Currents may be measured by the deflection of a weighted line.

surface in the Antarctic to the bottom of the North Atlantic Ocean. It takes about 1500 years to travel from the Antarctic region to the bottom of the North Pacific Ocean. Because of their slow speeds, deep currents are much more difficult to study by direct means than are surface currents, although in recent times Swallow floats have been used to study them at depths of a few thousand meters. Much of our present knowledge of these currents is gained indirectly by studying density distributions in the sea. Since oxygen enters or is produced in the sea only near the surface, oxygen depletion during the flow of deep water can also be used to trace these currents.

Patterns

Figure 8.16 shows the pattern of deep circulation in the Atlantic Ocean. This circulation is predominantly in a north–south direction. The *Antarctic Bottom Current* (AAB) forms at the surface in the Weddell Sea near the Antarctic continent and flows along the ocean floor northward to about 40° N latitude. The *North Atlantic Deep Current* (NAD) forms at the sea surface east of Greenland and then sinks and flows southward over the Antarctic Bottom Current. Beyond 40° S latitude, this water, while mixing with surrounding waters, rises from about 3000 meters toward the surface until it joins the surface circulation in the West Wind Drift. Water from the North Atlantic is consequently carried into the Indian and Pacific oceans. This rising water, the *Antarctic Circumpolar Current* (AACP), is a global phenomenon, extending around the Antarctic continent. This "upwelling" provides nutrients to one of the world's most productive areas, the Antarctic Ocean. At about 50° S latitude in the Atlantic, water sinks to form the *Antarctic Intermediate Current* (AAI), which flows downward and northward to a depth of about 700 to 1000 meters and continues northward above the North Atlantic Deep Current to beyond 25°N latitude. In the North Atlantic, however, there is no comparable intermediate current flowing southward. All of these deep currents mix with surrounding waters and are thus weakened farther away from their sources.

In addition to the circulation shown in Figure 8.16, dense water from the Mediterranean flows into the Atlantic (Fig. 8.17). This water travels over the Gibraltar Sill and sinks to a depth of about 1000 meters as it fans out into the Atlantic.

FIGURE 8.16 Deep ocean circulation in the Atlantic Ocean. AAB, Antarctic Bottom Current; AAI, Antarctic Intermediate Current; AACP, Antarctic Circumpolar Current; NAD, North Atlantic Deep Current. (After Wüst, 1936.)

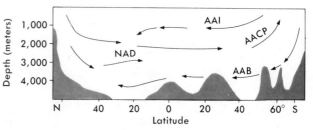

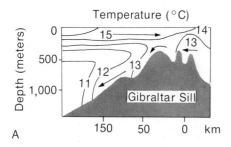

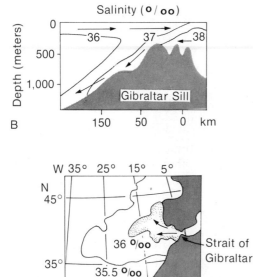

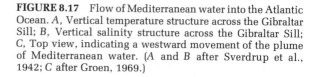

FIGURE 8.17 Flow of Mediterranean water into the Atlantic Ocean. *A*, Vertical temperature structure across the Gibraltar Sill; *B*, Vertical salinity structure across the Gibraltar Sill; *C*, Top view, indicating a westward movement of the plume of Mediterranean water. (*A* and *B* after Sverdrup et al., 1942; *C* after Groen, 1969.)

The deep water in the Pacific Ocean is much more homogeneous than that of the Atlantic Ocean, as the Antarctic Bottom Current occupies most of the bottom of the Pacific (Fig. 8.18). This current originates in the Weddell Sea. The water flows first into the Antarctic Ocean, then part of it flows into the Atlantic, part into the Indian Ocean, and finally the rest enters the Pacific, after mixing has modified its characteristics. This indirect route results in a sluggish current and greater depletion of oxygen in the Pacific compared with the bottom water in other oceans. A current similar to the North Atlantic Deep Current is absent in the Pacific; instead, two minor currents flow south toward the equator below the surface. These are the *Subarctic Bottom Current* (SAB), which flows out of the Okhotsk Sea near the Kuril Islands in the Northwest Pacific, and the *North Pacific Intermediate Current* (NPI),

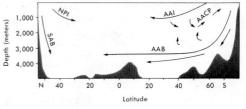

FIGURE 8.18 Deep circulation in the Pacific Ocean. AAB, Antarctic Bottom Current; AAI, Antarctic Intermediate Current; AACP, Antarctic Circumpolar Current; NPI, North Pacific Intermediate Current; SAB, Subarctic Bottom Current. (After Pickard, 1963, and others.)

which forms northeast of Japan in the region where the Oyashio Current converges with the North Pacific Current. The Antarctic Intermediate Current in the Pacific is weaker than the corresponding current in the Atlantic and extends northward only to the equator. Some of the water of the Antarctic Intermediate Current as well as of the Antarctic Bottom Current is returned to the surface in the Antarctic Circumpolar Current.

The deep circulation in the Indian Ocean is similar to that in the South Pacific. North of the equator, however, salty dense water from the Red Sea flows out and sinks to a depth of about 1000 to 2000 meters in the Northwest Indian Ocean. This source of water is similar to that found flowing out of the Mediterranean into the Atlantic Ocean.

Causes of Deep Ocean Currents

For water to sink, it must be denser than the surrounding water. Let us consider a hypothetical ocean (Fig. 8.19A), in which the density increase with depth is the same everywhere. The surfaces of equal

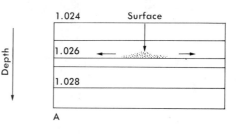

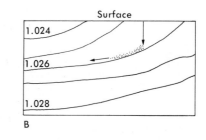

FIGURE 8.19 Currents (arrows) in hypothetical oceans with (A) horizontal and (B) sloping density surfaces; assuming that water with a density of 1.026 is introduced at the surface. Numbers indicate density in gm per cubic cm.

density would be horizontal, and there would be no deep ocean currents, because no water in this ocean could sink to a lower level unless denser water were introduced from outside. In contrast, real oceans have surfaces of equal density that are sloped (Fig. 8.19B); and it is along these slopes that currents flow. These currents are in geostrophic balance as discussed earlier; that is, the Coriolis force is balanced by the pressure-gradient force. If denser water is introduced at the surface, it will sink until it reaches a level at which its density equals that of the surrounding water and will continue to flow laterally along the sloping equal-density surface.

Maintenance of the deep ocean currents shown in Figures 8.16 and 8.18 requires a fairly constant production of dense waters at the surface, where they begin to sink. Evaporation (creating increased salinity) and cooling both result in dense water. Also, the freezing of sea water into relatively salt-free ice causes an increase in salinity (and density) of the adjacent residual water. In addition, wind-driven surface circulation leads to transport of water toward regions where mixing and sinking are possible. For example, in the southern oceans, warmer water from the north and colder water from the south converge at about 50° S latitude (*Antarctic Convergence*) and mix. The resulting water is of nearly the same density as the downward-sloping density surface at the convergence. Therefore, this water is able to sink below the lighter water to the north and becomes the Antarctic Intermediate Current.

The Antarctic Bottom Current, which flows northward along the bottom in all oceans, is probably produced by the formation of very dense water by cooling and by the formation of ice near the Antarctic continent, especially in the Weddell Sea. The southward-flowing North Atlantic Deep Current is formed by the sinking of water off Greenland. This water, although saltier, is warmer and has a lower density than that of the Antarctic Bottom Current and hence flows over it. The Antarctic Circumpolar Current is the result of water rising to compensate for the sinking of water associated with the Antarctic Intermediate Current to the north and the Antarctic Bottom Current to the south. In the Atlantic Ocean, the Antarctic Circumpolar Current is also fed by the North Atlantic Deep Current. Thus water from the North Atlantic eventually enters the other oceans by way of the West Wind Drift.

Indirect Study of Deep Ocean Currents

The slow-moving deep ocean currents are difficult to measure directly with the present technology. The deep ocean circulation patterns shown in Figures 8.16 and 8.18 are derived mostly from indirect studies which involve measurement of temperature and salinity of sea water at various depths at different locations. We will now attempt in brief to see how these data are obtained.

Starting with the premise that dense water tends to sink and lighter water tends to rise, and that density is a function of temperature and salinity, let us consider an idealized ocean. This ocean consists of two layers of water, each having a uniform temperature and uniform salinity (Fig. 8.20). Figure 8.20C shows a *T-S diagram,* with a plot of temperature versus salinity values of water samples taken simultaneously from the top to the bottom of this ocean. Since only two temperature and salinity values are involved, the T-S diagram contains only two points corresponding to the two layers of water. If these two layers of water are slowly allowed to mix vertically across the boundary, the vertical temperature and salinity distribution will change to that shown in Figure 8.20D. At the original boundary, for example, water from both layers will mix in equal proportions and will have a temperature and salinity intermediate to the initial values (point C on Fig. 8.20D) of the two layers. Farther away from the boundary, there will be less mixing between the two layers. Mixing of two *water types* (each having a single temperature and salinity value) results in a *water mass* characterized by a range of temperatures and salinities and represented by a straight line on a T-S diagram.

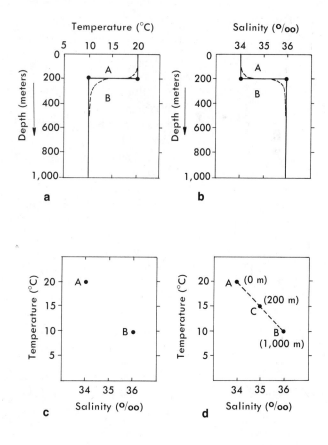

FIGURE 8.20 Temperature-salinity relationships in an idealized two-layer ocean (dashed lines represent results of mixing). a, Temperature; b, Salinity; c, T-S relationship before mixing; d, T-S relationship after mixing.

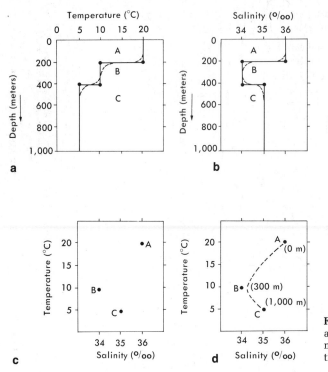

FIGURE 8.21 Temperature-salinity relationships in a three-layer ocean (dashed lines represent results of mixing). *a*, Temperature; *b*, Salinity; *c*, T-S distribution before mixing; *d*, T-S distribution after mixing.

Figure 8.21 shows the mixing of three layers of water (three water types) and the resulting *T-S curve.* Note that in this case a curve (Fig. 8.21*D*) consisting essentially of two straight lines (two water masses) is obtained.

Figure 8.22 shows typical T-S curves for various oceanic regions. These curves are the most useful tools developed to date for studying deep ocean currents, especially if they are used in conjunction with other techniques, such as Swallow floats. T-S curves are unique for individual oceanic areas; those of the South Atlantic, for example, are characteristically different from those of the North Atlantic. Specific water masses are found in the various oceanic regions, produced and maintained by the mixing of different water types. Most of the water types are formed near the surface of the oceans as a result of physical processes such as evaporation, precipitation, and freezing of sea water. The resulting water, after mixing with other water types, may be carried to varying depths by currents such as the Antarctic Intermediate Current.

T-S curves can be used in tracing deep ocean currents. Figure 8.23 shows T-S curves from four oceanographic stations located along a north-south line in the Atlantic. The prominent feature of these diagrams is a "core" of low-salinity water of the Antarctic Intermediate Current, although cores are not necessarily characterized by a salinity

FIGURE 8.22 Typical T-S curves for various oceanic regions. Note that the lines all converge toward a point near the bottom of the graph. This point represents the water of the Antarctic Bottom Current. ENPC, Eastern North Pacific Central; ESPC, Eastern South Pacific Central; IEW, Indian Ocean Equatorial; IOC, Indian Ocean Central; SAC, South Atlantic Central; NAC, North Atlantic Central; WM, West of the Mediterranean Sea. (After Sverdrup et al., 1942.)

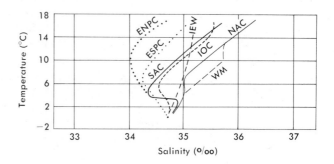

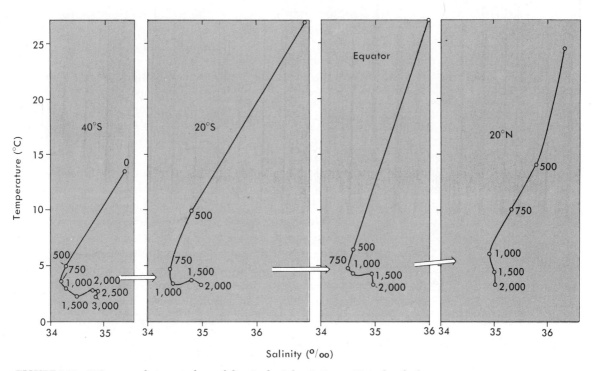

FIGURE 8.23 T-S curves along a north-south line in the Atlantic Ocean. Note that the low salinity core of the Antarctic Intermediate Current (indicated by arrows) diminishes northward. Numbers along the curves indicate depths in meters. (After data of Wüst, 1936.)

low. They may also consist of a salinity high, or a temperature low or high, or a combination of them. Note that the core gradually disappears at northern stations. Thus, we can trace the movement of deep ocean currents simply by the use of T-S curves and by following the cores, if there are any. This method is known as the *core method,* and was developed by George Wüst in 1935.

Mathematical analysis of T-S curves can be used to estimate the speed of the currents involved as well as to determine the degree of mixing of the water types in question. Most of our information about deep ocean currents is obtained in this manner. In addition, the core method may be a highly useful tool in tracing the movement of pollutants, especially in coastal areas.

SUMMARY

The surface circulation of the oceans is primarily caused by the force of wind, whereas the deep ocean circulation is caused by density differences of sea water. The major feature of surface currents is that they flow in closed loops or gyres, moving clockwise in the Northern Hemisphere and counterclockwise in the Southern Hemisphere.

The effect of wind, modified by Coriolis force, results in the transport of a surface layer of water known as Ekman transport. Ekman transport is 90° to the right of the wind in the Northern Hemisphere or 90° to the left of the wind in the Southern Hemisphere. It is also responsible for coastal upwelling. When winds blow parallel to the shore, warm nutrient-poor surface water is drawn offshore and cold nutrient-rich water rises to the surface near shore. This enhances biological productivity. Many of the world's great fisheries are in upwelling areas, such as along the coast of Peru.

The slow-moving deep ocean currents are very vital to life in the sea, since they carry oxygen to the depths of the oceans and bring the nutrients toward the surface. Deep ocean currents are produced when dense masses of water form and sink near polar and subpolar regions. Because of their slow speed, they are usually studied by indirect means from the measurement of temperature and salinity of water collected at different depths. The technique of tracing deep ocean currents by plotting these data on T-S diagrams is called the core method. Radiocarbon datings of deep ocean water indicate that these currents require 1500 years or more to move from one end of the ocean to the other.

IMPORTANT TERMS	Antarctic convergence	Doldrums
	Callao painter	Drift bottle
	Core method	Drogue
	Cromwell current	Ekman spiral

Ekman transport
Equatorial Counter Currents
Equatorial currents
Equatorial undercurrents
El Niño
Geostrophic balance
Geostrophic current
Gyres
Polar easterlies
Polar westerlies

Pressure-gradient force
Sea-surface topography
Swallow float
Trade winds
T-S curve
T-S diagram
Upwelling
Water mass
Water type
West wind drift

1. How is coastal upwelling produced?
2. How does wind produce surface currents?
3. How does the wind pattern shown in Figure 7.2A relate to the current pattern presented in Figure 7.3?
4. Discuss the relation of the El Niño phenomenon to the fisheries off South America.
5. Why do the major ocean gyres flow clockwise in the Northern Hemisphere and counterclockwise in the Southern Hemisphere?
6. How is the Equatorial Counter Current related to the sea-surface topography in the Atlantic Ocean?
7. Discuss the origin of the equatorial undercurrent.
8. Discuss the driving mechanism responsible for deep ocean circulation.
9. Describe the core method for the study of subsurface currents.
10. Discuss the possible relationship between ocean currents and climate. Compare the North American east coast and west coast climates in relation to ocean currents and wind.

Knauss, John A. 1961 (April). The Cromwell Current. *Scientific American.*
Lissaman, P. B. S. 1979. The Coriolis Program. *Oceanus.* 22:23–28.
McDonald, James E. 1952 (May). The Coriolis Effect. *Scientific American.*
Munk, Walter. 1955 (September). The Circulation of the Oceans. *Scientific American.*
Neumann, Gerhard. 1968. *Ocean Currents.* Amsterdam, Elsevier.
O'Brien, James J. 1978. El Niño: An example of ocean/atmosphere interactions. *Oceanus* 21(4):40–46.
Pickard, George L. 1963. *Descriptive Physical Oceanography.* New York, Pergamon Press.
Robinson, A. R. and William Simmons. 1980. A new dimension in physical oceanography. *Oceanus.* 23(1):40–51.
Stommel, Henry. 1958 (July). The Circulation of the Abyss. *Scientific American.*
———. 1965. *The Gulf Stream.* 2nd ed. University of California Press.
Wyrtki, Klaus. 1979. Sea Level Variations: Monitoring the Breath of the Pacific. *Trans. Am. Geophysical Union.* 60(3):25–27.

Coastal Oceanography

9

The parts of the ocean most familiar to us are the coastal regions. They are especially important because most of the fisheries of the world are located here. They are also valuable as sources of petroleum and minerals. In addition, coastal regions include sites for mariculture, harbors and ocean-oriented recreation. Although seemingly isolated, all of these uses interact and frequently conflict with one another. For example, port activities might result in coastal pollution, which in turn will have adverse effects on fisheries and recreation. The shore, which includes beaches and rocky coasts, undergoes natural changes owing to geological forces as well as to waves and coastal currents. When people interfere with these natural forces, detrimental effects may occur.

Neptune State Park is one of the many state parks along Oregon's 400-mile air conditioned seacoast. At sundown, breakers, weather beaten trees, and glistening Pacific waters combine to present this picture from Neptune Park's parking area. (Photo courtesy of the Oregon State Highway Department.)

BEACHES

The *beach* includes the entire region of unconsolidated materials extending from the low-tide line to the uppermost region of wave action (the coastline), represented by cliffs, sand dunes, or permanent vegetation. Beaches are dynamic. They can grow or erode naturally on a daily, weekly, or seasonal basis, depending on the heights of the waves and tidal fluctuations. Many of the changes taking place along or near beaches are relatively slow, requiring tens or hundreds of years to have any large-scale effects. However, during severe storms and tsunamis, major coastal changes may occur within a few hours. It is still uncertain whether long-term gradual change or short-term catastrophic change is the most effective coastal shaper. We may further affect the nature of beaches by building shore structures such as jetties and breakwaters, which may protect or destroy the beaches.

Figure 9.1 shows four kinds of beaches. Some beaches are composed of nothing but sand. Others may be composed of pebbles or cobbles, and some coastal areas of the world may not have any beach at all—just a solid rocky bottom.

Beach Materials

Beaches can be composed of different mineral grains of varying sizes, derived from the erosion of coastal rocks or brought to the sea by rivers, wind, and glaciers. Some beaches are composed largely of the shells and skeleta of marine organisms. Once on the beach, these materials may be transported considerable distances along the beach.

As granite and granite-derived rocks are among the most common components of the continents, many beaches are composed of quartz and feldspar, the main constituents of granite. Depending on the amount of sorting (separation by size or weight) and the distance traveled (greater distance tends to destroy softer or more soluble minerals), some beaches may be almost wholly composed of any one mineral such as quartz, one of the most resistant materials. Beaches located far from granite sources may be composed of bits of broken coral, as along the Florida coast. However, many Florida beaches are composed of about one half quartz material even though they are hundreds of kilometers from obvious sources. Beaches along volcanic coasts such as in Hawaii are generally composed of dark sands derived from the breakdown of volcanic rock. Some beaches in Brazil and India are composed of heavy black minerals that are a source of rare elements such as titanium and thorium. Waves, currents, and winds along these shores have removed the lighter materials and concentrated the heavy minerals along the beach. In fact, many economically important deposits of such minerals are found in areas that were once beaches.

FIGURE 9.1 Various kinds of beaches. *A*, A sandy beach in Portugal. *B*, A pebble beach on the Olympic Peninsula, Washington. *C*, A steep cobble beach at Yaquina Head, Oregon. *D*, A rocky coast, with no beach, in Brittany, France. (*A* courtesy of Hayward Associates, Inc.; *B* courtesy of National Park Service; *C* photo by Lynn O'Connor, courtesy of School of Oceanography, Oregon State University; *D* courtesy of French Embassy Press and Information Division.)

Beach Profile

Figure 9.2 shows a generalized beach profile. Not every feature shown in this figure may be developed on any given beach. The beach profile can be divided into three major zones: *offshore, foreshore,* and *backshore.*

The most seaward part of the profile, the *offshore* region, is the zone of breakers extending seaward from the low-tide line to the depth at which these waves first feel the bottom (see Chap. 6). In areas where the tidal range is relatively small, as well as in areas of gently sloping sea bottom, the offshore region is often characterized by several *longshore bars* and *longshore troughs* parallel to the coastline. A longshore bar is simply a submerged ridge produced by the deposition of sand seaward of the longshore trough. It may be exposed during low tides. Although the origin of longshore bars and troughs is not well understood, they are rarely developed in areas having large tidal ranges. This suggests that they develop only when surf and tidal conditions are fairly constant. Bars are often visible as light areas on aerial photographs owing to the shallowness of the water or waves breaking on the bars. Apparently, the size of bars is a function of wave height, because higher winter waves usually result in larger, longer bars. In many cases, bars may be almost absent during the summer, when the waves are much smaller. In addition, bars and troughs usually migrate landward during winter and seaward during summer in response to changing wave heights.

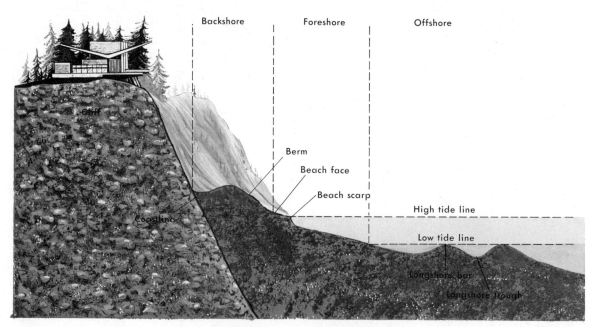

FIGURE 9.2 Typical beach profile.

The *foreshore* consists mostly of the zone that is regularly exposed between low and high tides (the *low-tide terrace*) and the *beach face* (the sloping region above high-tide line subject to the uprush of waves). The slope of the beach face depends on the grain size of the materials making up the beach face: the larger the grain size, the steeper the slope. For example, very fine sand (1/16 to 1/8 mm) produces a slope of approximately 1°, and very coarse sand (one to two mm) results in a slope of about 9°. Cobbles (64 to 256 mm) produce a slope of about 24°. The lower limit of the beach face is sometimes characterized by a nearly vertical *beach scarp* produced by large waves, and may have a relief of about one or two meters. The beach face is sometimes marked by a series of crescent-shaped depressions (small bays) separated by points called *cusps* (Fig. 9.3). Most are between one and 100 meters across. They tend to form under equilibrium conditions; that is, on beaches where the erosional and depositional aspects of the waves are balanced. Their characteristic shapes and patterns may be due to factors such as the size, shape, and period of the waves, as well as to the composition and topography of the beach.

The *backshore* is "the beach" to most people when they visit the beach during high tides, but in oceanographic terminology it is only a part of the beach. It extends landward from the beach face to the cliffs, sand dunes, or permanent vegetation of the coastline. It is the region above the normal high-tide line and is rarely covered by water except

FIGURE 9.3 Series of beach cusps, at El Segundo Beach, California. (From U.S. Army Coastal Engineering Research Center, 1973.)

during spring tides and severe storms. It is composed of one or more *berms.* A berm is a nearly horizontal surface produced by deposition of beach material by long-period (about 10 seconds) spilling breakers. Many berms are low ridges parallel to the coastline. Berm height increases with increasing wave height and increasing tidal range. The berm materials are composed of larger particles than those making up the beach face. Even the berm itself may be composed of materials of varying particle size, increasing in size as one goes up the beach. Berm widths also vary considerably.

Winter and Summer Beach Profiles

Beaches are constantly undergoing changes, which are often periodic (Fig. 9.4). During the winter (as well as during severe storms and tsunamis), wave heights are considerably greater than at other times. These high waves are capable of eroding the beach (mostly the berm) and transporting the materials offshore to help form longshore bars. The berm width is reduced considerably, but its height increases as the uprush of higher waves transports some of the materials up the beach. During the summer and non-storm conditions, when the waves are much smaller, the process is one of wave deposition rather than erosion. The berm height decreases, and the longshore troughs are filled in. Eventually, a wider berm is produced.

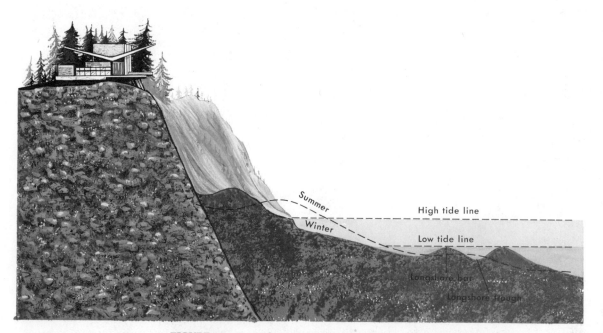

FIGURE 9.4 Typical summer and winter profiles of a sandy beach. (After Bascom, 1964.)

BEACH DRIFTING, LONGSHORE CURRENTS, AND RIP CURRENTS

Waves approaching the shore obliquely cause movement of the materials along the beach by a process called *beach drifting.* A sand grain, for example, is pushed up the beach face diagonally (parallel to the direction of wave advance) by the uprush (swash), but during the backrush (backwash) it may descend straight down the slope (Fig. 9.5). Beach materials can thus be transported hundreds of kilometers as the process is repeated.

When waves strike a coast obliquely, a current moving parallel to the shore is formed in the surf zone (Fig. 9.5). This flow of coastal waters is the *longshore current.* It is best developed along straight coasts but may also be formed along irregular coasts. Longshore currents, which may pose a threat to landing crafts, usually have speeds of about one to four km/hr. Like beach drifting, longshore currents are important in transporting sediments parallel to the coast. These sediments may be obtained locally by wave erosion or may be brought to the sea by rivers. Longshore currents are also responsible for the sediment deposition in many harbors and river mouths (Fig. 9.6), where the current is slowed because of greater water depth and the absence of an adjacent shoreline. Costly dredging operations are often required to counteract the effects of this type of deposition. The *longshore* (or *littoral*) *transport* of material may be quite substantial. For example, along parts of the southern Atlantic coast of the United States, the net southward transport past a given point may be as much as half a million cubic meters per year.

The water brought to the near-shore region by breakers must eventually return to the sea. Some of the returning water may be localized to form *rip currents.* Rip currents on a particular beach may vary depending on the nature of wave conditions present. Larger swells produce fewer but stronger rip currents than smaller swells. As seen in Figure 9.7A, a rip current consists of a *feeder,* a *neck,* and a *head.* The feeder is usually the longshore current. The neck, or rip current proper, may be up to 30 meters wide, and the head can have 10 times this width. The speed of rip currents may exceed seven km/hr, but they are usually much slower. They may extend across the surf zone for about one km.

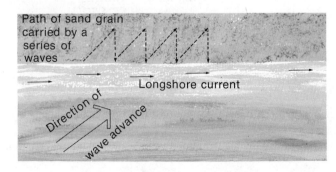

FIGURE 9.5 Beach drifting and longshore currents produced by waves striking the coast obliquely.

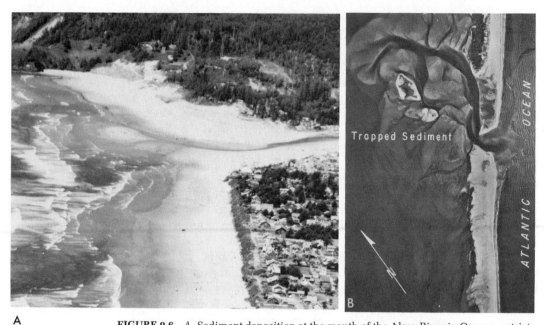

A

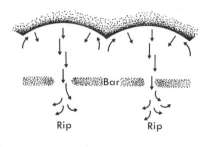

FIGURE 9.6 *A*, Sediment deposition at the mouth of the Alsea River in Oregon restricts its opening to the ocean. This in turn causes erosion on the southern bank of the river near its mouth. *B*, Sediment transported by longshore currents carried into Old Drum Inlet. North Carolina, by tides. (*A* courtesy of Oregon State Highway Department; *B* from U.S. Army Coastal Engineering Research Center, 1973.)

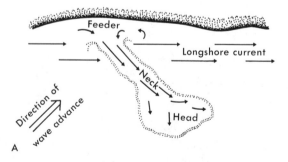

FIGURE 9.7 *A*, Rip current produced by waves striking the coast obliquely. Note the longshore current feeding the rip. *B*, Rip currents produced by waves entering small bays separated by cusps.

FIGURE 9.8   Rip currents along Ludlam Island, New Jersey. (From U.S. Army Coastal Engineering Research Center, 1973.)

Rip currents produced by waves reaching the shore perpendicular to the beach move out to sea straight across the surf zone, whereas those produced by waves striking obliquely usually move diagonally. Although rip currents can be found on straight coasts as well as on irregular coasts, they are usually more pronounced in small bays, such as on cusped beaches. The divergence of wave energy in bays and the resulting weak zone cause water to flow in from the convergent zones (points). Eventually, this water returns seaward, forming rip currents as shown in Figure 9.7B. Like the longshore currents, rip currents also transport sediments, but in this case sediments are moved to deeper water seaward of the surf zone. Rip currents are capable of cutting channels in the surf zone, even across the longshore bars.

Swimmers are sometimes caught in rip currents and swept out to sea. Even a powerful swimmer cannot swim against a strong rip. The best way to get back to land is to swim parallel to the shore until out of the rip and then swim back to shore. The presence of a rip current is usually indicated by the near-absence of breakers along the rip as well as by the higher turbidity of the water (Fig. 9.8).

OTHER COASTAL FEATURES

Barrier Islands and Barrier Spits

Along some coasts, one finds *barrier islands* and *barrier spits.* These a⁻₂ most common along the Gulf and Atlantic coasts of the United States. These features are usually formed nearly parallel to the shore by the

FIGURE 9.9 *A*, Apollo 9 photograph of the Cape Hatteras area, North Carolina, showing barrier islands. Cape Lookout is near the bottom of the photograph. Cape Hatteras juts farthest into the ocean. *B*, Aerial photograph showing Sandy Hook, New Jersey, a barrier spit. (Courtesy of U.S. Department of Agriculture.)

sediment-laden longshore currents. Barrier islands are essentially the same as longshore bars, except that they are permanently exposed (Fig. 9.9A). If one end is attached to the mainland, a barrier spit is formed. They are often curved toward the land like a *hook* (Figs. 9.6 and 9.9B). The curving of the barrier spits is believed to be caused by waves reaching the shore from two or more directions or by wave refraction taking place in these regions. *Lagoons* and *salt marshes* may develop behind barrier islands and spits, as the sediments brought from land accumulate and are prevented from reaching the open sea. Lagoons are coastal lakes or shallow bays with limited connection to the sea. The salinity of lagoons varies depending on the rates of evaporation, input of fresh water, and tidal exchange. They may be nearly fresh or saltier than the adjacent sea. Continued sand deposition may close the entrance from the sea, forming a saltwater lake. If there is sufficient river input of

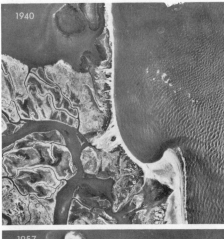

FIGURE 9.10 Changes in the coastline at Little Egg Inlet, north of Atlantic City, New Jersey. Note the growth of the spit in the lower two photos. The width of the area photographed is about four km. (Courtesy of U.S. Geological Survey.)

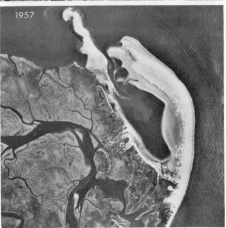

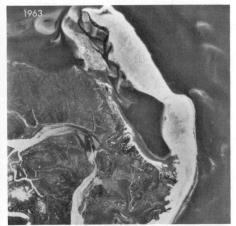

terrigenous sediment, a salt marsh may form in the lagoon or bay as sediments build up broad, flat, muddy beaches upon which marsh grasses grow. Salt marshes are subject to periodic submergence by tides. Eventually, salt marshes evolve into permanent land because of the continued supply of sediment. The circulation and life in brackish (lowered salinity) lagoons and salt marshes are discussed later in this chapter.

All coastal features undergo constant change (Fig. 9.10). Barrier spits and islands may grow or may undergo rapid erosion during severe storms, earthquakes, and tsunamis.

Sand Dunes

Sand dunes (Fig. 9.11) characterize many coasts, especially in tropical and temperate areas. They are located above the backshore. Sand dunes are mounds of various sizes and shapes produced by wind deposition. Wind is capable of transporting dried beach sand and then depositing it when an obstacle such as a plant is encountered. Once initiated, the dunes begin to grow in size by blocking and trapping sand carried by the wind. Eventually, plants and even trees may be buried by the sand. The orientation of these coastal dunes varies, depending on the direction of the wind.

Dunes provide natural protection for many coasts against beach erosion by large waves. They also prevent the loss of coastal sand to the ocean by acting as retaining walls. The growth of sand dunes can be accelerated, as well as stabilized, by planting various grasses, by construction of fences, and even by piling up discarded Christmas trees. In areas where there are no natural sand dunes present, artificial dunes may be produced in this manner.

FIGURE 9.11 Sand dunes on Cape Hatteras. (Photo courtesy of the U.S. Department of the Interior; National Park Service Photo, Cecil W. Stoughton, Photographer.)

Long-Term Coastal Changes

In addition to the action of waves and currents, geological forces can cause vertical movement of coasts. These are the same forces that are responsible for making mountains, volcanoes, and earthquakes. Except for volcanic activity and earthquakes, which cause rapid changes, their action is slow and their effects may require hundreds or thousands of years to be perceived. Examples of long-term changes of coasts have been provided by archaeologists. Some coastal areas around the Mediterranean have either been uplifted above or have sunk below the sea. Some ancient port cities built by the Romans, the Greeks, and the Phoenicians more than 2000 years ago serve to mark the ancient coastline. Today, remains of houses, storage tanks, quarries, and roads (Fig. 9.12) are underwater, indicating sinking of the land. Harbors are sometimes found several kilometers inland, indicating uplift of the land. Utica (Fig. 9.13), located about 30 km northwest of Carthage, was a seaport about

FIGURE 9.12 Partly submerged quarry and road surfaces along the Mediterranean coast. (From Fleming, 1969.)

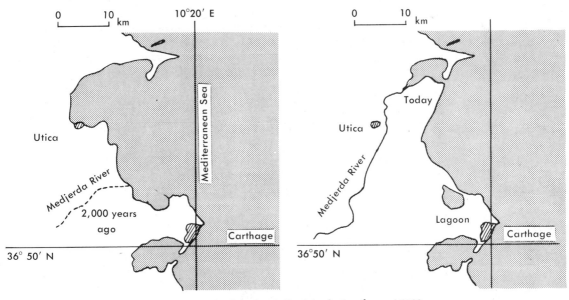

FIGURE 9.13 Changes in the coastline north of Carthage, Tunisia, during the past 2000 years. (After Fleming, 1969.)

2000 years ago in what is now Tunisia. Because of the uplifting of land and siltation of the bay, the town is now about 10 km inland. These forces also caused the Medjerda River to be deflected about 20 km to the north.

During the last ice age, portions of the continental shelves in many parts of the world were exposed. Georges Bank, east of Cape Cod, is believed to have been a forested island. Since that time broad expanses of the coastal plain have been submerged.

ARTIFICIAL COASTAL STRUCTURES

Waves are sometimes capable of causing tremendous damage to coastal areas such as headlands which they can inundate in relatively short periods of time. Valuable property may be lost forever. Large storm waves and tsunamis, in addition to causing numerous deaths, are capable of running up the shore and damaging coastal installations. Furthermore, wave-produced longshore currents may cause deposition or removal of sediments along the coast, which for some areas may not be desirable. In order to prevent such changes, people often intervene—actually, interfere—along the coast by building various coastal structures, which often result in serious unwanted side effects. Generally, it is best to avoid unnecessary building on low-lying areas, cliffs, and narrow barrier islands, all of which may be subject to wave erosion.

Sea Walls

Sea walls are built to protect the coastal areas behind them from large waves as well as to maintain a permanent coastline. They are built along the coastline using concrete, boulders, or other materials. Large waves break on them or are reflected back to sea. They may be built in a variety of shapes (Fig. 9.14). The seaward sides of some sea walls have a parabolic design, which reflects waves more effectively than walls with sloping sides. Others may be built like steps, so that higher waves will break first, farther offshore. Some sea walls may have a drain at the top, a design which may someday be utilized to obtain power from large waves as well as from tidal fluctuations (Fig. 9.14E).

Jetties

Jetties are long structures constructed perpendicular to the coast. They are usually built at the mouths of rivers and harbors. They may be single structures or may be in pairs, one on either side of the inlet. They are built to prevent sediment deposition in the inlet by longshore currents as well as to make stream and tidal flow favorable for navigation. In

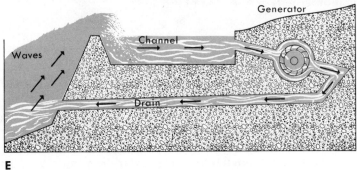

FIGURE 9.14 Different types of seawalls. *A*, Rocks, Fernandina, Florida; *B*, Parabolic, Galveston, Texas; *C*, Steps, Mississippi coast; *D*, Parabolic and steps, San Francisco, California; *E*, Channeled wall with possible application for power generation from waves and tides. (*A–D* from U.S. Army Coastal Engineering Research Center, 1973, *E*, after Smith, 1973.)

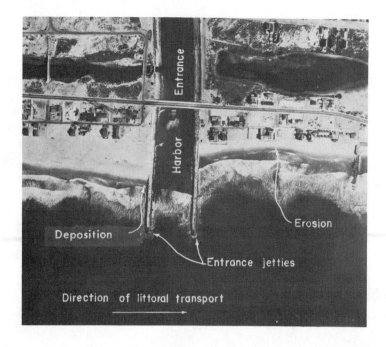

FIGURE 9.15 A pair of jetties protecting a harbor entrance at Ballona Creek, California. (From U.S. Army Coastal Engineering Research Center, 1966.)

addition, jetties are capable of shielding the inlet from large waves arriving at an angle.

Although jetties often perform their functions well, they may have unwanted side effects. Jetties sometimes cause deposition of sand on one side of the inlet and erosion on the other side (Fig. 9.15). In order to prevent erosion, sand must be continually supplied to this region—a costly and permanent operation. The erosion, however, may be retarded by building another jetty in the area of erosion.

Groins

Groins are similar to jetties but are built along the open coast. They are usually shorter than jetties. The purposes of groins are many. They include the prevention or slowing of beach erosion and the creation or widening of beaches by trapping sediments from longshore currents. Like a jetty, a groin will cause erosion and deposition along the coast (Fig. 9.16). A second groin is often built to prevent the erosion caused by the first groin. The result is a chain reaction of groin construction along the coast. Many coasts, therefore, are littered with numerous groins. Groins certainly do their job of trapping sand, but they do so at the expense of the next beach down the coast.

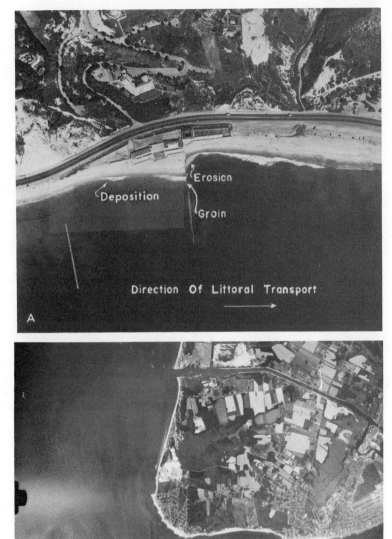

FIGURE 9.16 *A*, Groin at Santa Monica, California. *B*, Groins at Cape May, New Jersey. Note the erosion downstream from the groins. (*A* from U.S. Army Coastal Engineering Research Center, 1966; *B* courtesy of the U.S. Dept. of Agriculture.)

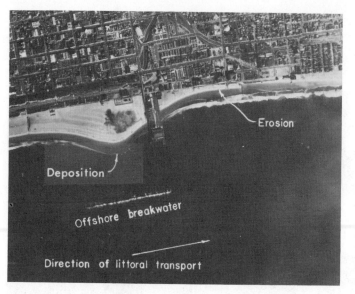

Erosion

Deposition

Offshore breakwater

Direction of littoral transport

FIGURE 9.17 A breakwater at Santa Monica, California. Note the buildup of sand in the quiet water behind the breakwater. (From U.S. Army Coastal Engineering Research Center, 1966.)

Breakwaters

Breakwaters are offshore structures built parallel to the shore (Fig. 9.17), mainly to protect beaches and harbors from high waves. Waves either break on them or are reflected back to sea. Because of the reduction in wave energy behind the breakwater, the sediments suspended in the longshore currents are deposited here, and may result in some erosion farther down the beach. Breakwaters may eventually be connected to the mainland by the accumulation of sediment. The connection is known as a *tombolo.* Tombolos may also connect islands to the mainland or to other islands, generally through growth of a sand spit.

ESTUARIES

An *estuary* may be defined as the region where river water meets and mixes with sea water, resulting in intermediate salinities. The influence of the river on the sea may extend several kilometers into the sea and may be marked by a plume of muddy water extending out from the river mouth.

Historically, estuaries have been attractive as nuclei for the development of cities. Navigable water and proximity to the ocean make them ideal for seaport activities such as shipping. In addition, the natural load of land-derived nutrients carried by the rivers results in highly productive water, suitable for the rapid growth of many fish, crabs, and clams. Consequently, important fisheries have developed on and around estuaries.

As areas surrounding estuaries have become used for agriculture and industry, the water has become increasingly polluted and the natural estuarine resources threatened. In addition, oil tankers pose a constant threat of oil spills. Many estuaries are too small for the new supertankers, but leaks from small tankers produce significant contamination of those estuaries used as oil ports. In many parts of the world, offshore floating oil ports appear to be the only solution to this problem.

Types of Estuaries

The nature of an estuary is determined not only by how it is used but also by its dimensions, by river flow, and by tidal flow. When river flow is large, tidal flow is small, and the estuary is rather deep, the fresh water tends to flow over the sea water with little mixing (Fig. 9.18A). Thus, a *stratified* estuary is maintained. The deeper salty water forms a sort of "salt wedge" under the less saline, predominantly river water. In this case the salinity difference between surface and bottom water may exceed 20°/oo (parts per thousand). The net flow of water is upstream near the bottom and seaward near the surface in a stratified estuary. The mouth of the Mississippi River is a good example of this type of estuary.

A *partly mixed* (or partly stratified) estuary occurs under conditions where depth and river flow are less and tidal flow is greater than in a stratified estuary. These factors all contribute to increased mixing of the fresh water with the salty water (Fig. 9.18B), and stratification is less. The salinity difference between surface and bottom may be from four to 20°/oo. Mixing causes some of the salty water to be brought into the surface water, where it flows back to the sea. Consequently, there may be a significant net flow upstream near the bottom as well as a seaward flow near the surface. The Columbia River estuary is partly mixed near its mouth when river flow is substantial.

If the river flow is small, the tidal flow is great, and the estuary is shallow and narrow, vertical mixing may be nearly complete, and the vertical difference in salinity generally will be less than 4°/oo. This is a *well-mixed* estuary (Fig. 9.18C), with a net outward flow of water at all depths. Many small estuaries—and some large estuaries during periods

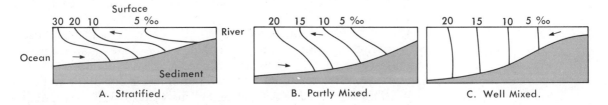

FIGURE 9.18 Types of estuaries, based on the distribution of salinity. Arrows indicate net flow of water.

of low river flow—are of this type. For example, the Columbia River may be well mixed during the dry seasons.

Some very wide estuaries, such as the lower Chesapeake and Delaware estuaries, exhibit a horizontal (transverse) gradient of salinity owing to the effects of Coriolis force (see Appendix C). In the Northern Hemisphere, this force causes salt water to be deflected to the right as it enters these estuaries (Fig. 9.19). The river flow is likewise deflected to the right. Consequently, one shore experiences higher salinities than the opposite shore.

The estuaries we have discussed so far were formed by the drowning of river mouths as sea level rose following the ice ages, or by the sinking of coastal areas. These are usually referred to as *coastal-plain estuaries.* The lagoons or bays that are formed behind barrier islands are also considered to be estuaries, as they may have many of the characteristics discussed above. Some lagoons, such as the Laguna Madre of Texas, become extremely saline when evaporation exceeds precipitation and runoff. Salinity in the Laguna Madre may exceed 60%/oo, decreasing toward the ocean entrances.

Another type of estuary is the *fiord* (Fig. 9.20). These have been carved by glacial erosion; they are quite deep (some are several hundred meters deep), and they usually have a shallow sill, or submerged ridge, across the mouth. The sill marks the limit of seaward advance of the glacier. Circulation is restricted in fiords, and thus they are quite distinct from the estuaries mentioned previously. Fiords tend to be highly stratified at times, and circulation is restricted by the sill, which reduces the flow of sea water in and out of the fiord. Consequently there is a tendency for the deep water to become stagnant and anoxic (deficient in oxygen). The anoxic condition causes an increase in the concentration of hydrogen sulfide because of its production by certain anaerobic bacteria.

During a cold winter, freshwater runoff is reduced and surface temperatures are lowered. Consequently the stratification breaks down and the fiord becomes mixed. This causes a rather sudden drop in the concentration of oxygen and an increase in hydrogen sulfide at the surface. This combined effect may cause fish kills, and the hydrogen sulfide which is released into the atmosphere as a gas may drift to shore and cause houses to become blackened. (Recall that the presence of hydrogen sulfide in sediment is indicated by black color and a rotten-egg smell.)

Although most estuaries have circulation patterns similar to those already discussed, special mention should be made of the Amazon River. A large area of the South American continent (Fig. 9.21) drains to the sea via the Amazon River, and the Amazon rain forest is one of the wettest regions of the world. About 20 per cent of the worldwide river input of fresh water into the oceans is contributed by this mighty stream. Al-

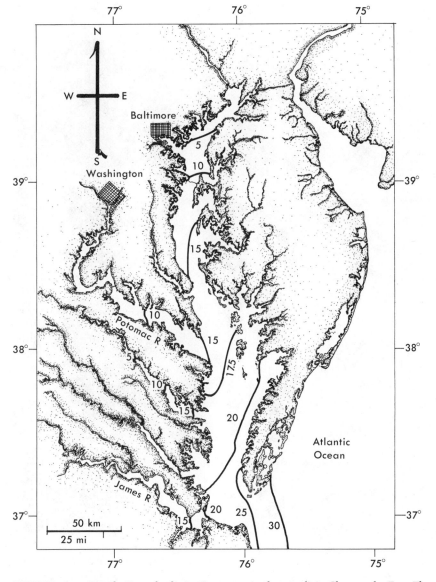

FIGURE 9.19 Distribution of salinity (in parts per thousand) in Chesapeake Bay. The shape of the surface salinity contours indicates the influence of Coriolis force deflecting the salty water to the right as it enters the estuary. It should be noted, however, that more fresh water enters the western side of the bay than the eastern side. (After Pritchard, 1952.)

FIGURE 9.20   A Norwegian fiord. (Courtesy of the Norwegian Information Service.)

though tides extend 725 km (450 miles) upstream, the mouth of the river is fresh. A freshwater plume extends far out to sea, and the influence of the Amazon can be followed in surface water from near the equator to 10° N latitude and beyond.

Life in Estuaries

Since most estuaries are well nourished by nutrients that are carried by the rivers, they are frequently the sites of highly productive fisheries. Clams, oysters, shrimp, and many species of food fish thrive in estuaries. Many animals breed there, in waters rich in the food necessary for their developing young, and others that breed in the sea migrate into estuaries as juveniles and thrive on the rich food supply.

Because of their high productivity, healthy estuaries are important nurseries for the juveniles of many migratory fish as well as some coastal shrimp and resident clams, oysters, crabs, and fish. The eggs or larvae of many species of crabs, shrimp, and fish, however, are intolerant of brackish estuarine water. In these species the adults migrate to the sea before reproducing, thus granting their offspring a start in higher salinity water. As soon as the maturing larvae become more tolerant of lowered salinities, they begin a migration up the estuary. The young of one species of Chinese crab, *Eriocheir sinensis,* may migrate hundreds of kilometers inland, even traveling on land briefly to get around dams.

The distribution of estuarine organisms depends especially on salinity and temperature fluctuations. Organisms attached to permanent structures such as pilings must tolerate changing salinities and temperatures as the tides move the salty water in and out. In addition, organisms in the intertidal (littoral) zone must tolerate periodic exposure to air. Barnacles and mussels can close their protective shells, trapping salty water within. Many attached seaweeds can tolerate periodic exposure and even drying of their surfaces. Many bottom-dwelling worms and

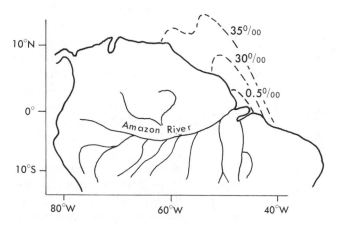

FIGURE 9.21 The fresh water plume of the Amazon River. (After Barnes, 1973.)

clams may burrow into the sediment in order to secure an environment that is very stable with regard to temperature and salinity. Plankton (drifting organisms) may be carried out to sea if they remain near the surface. Some plankton compensate for this danger by spending part of the time in deeper water, where the net water movement may be upstream. Estuarine fish can swim against currents as adults, and many have heavy eggs that sink to the bottom and thus avoid being washed out to sea. While most truly estuarine organisms can tolerate wide ranges in salinity, some are restricted to the ocean end of the estuary. Often these are also found along the open coast. On the other hand, some freshwater organisms can tolerate the low salinities found in the upper estuary.

With increasing concentrations of pollutants in an estuary, species that are less tolerant to pollution are eliminated. Very rugged species, on the other hand, survive and often thrive. Newark Bay is a well-mixed estuary in highly industrialized northern New Jersey. In the 1880's it was an important oyster breeding ground, and the shad that were caught in the bay were highly regarded in New York markets. Today, both fisheries are gone, having succumbed to the pressures of increased population and industrialization. However, some hardy organisms such as killifish and grass shrimp thrive.

Is it possible to clean up an estuary that has reached a highly polluted condition? Many estuaries are the sites of sewage and industrial waste disposal. If all pollution of water stopped today, it would still be many years before some of our estuaries returned to a condition suitable for fishing and recreation. Years of pollution have contaminated the sediments to the point that they would retain oil and grease for years. Heavy toxic metals such as mercury and lead may persist in the sediment for even longer periods, continuing to affect the water for many years.

Because many of the larger estuaries are important shipping ports, they are periodically dredged to maintain deep channels. This process re-introduces pollutants from the sediment back into the water, either during the dredging operation itself or from the dredge spoils (the dredged-up sediment that is deposited outside the channel, frequently on shore). Of course, dredging may also remove contaminated sediments from the estuary, making it less polluted, but then the area in which the contaminated sediments are deposited becomes more polluted.

When the shipping channels are dredged, the increased depth may affect the circulation of the estuary and may result in increased stratification of the water. In addition, construction of dams may reduce the flow of fresh water, increasing the mixing of the estuary. For example, a partly mixed estuary may become well mixed. Similarly, narrowing of estuaries by filling, or widening of estuaries may also affect their circulation. All of these changes may, in turn, have a significant effect on the distribution and survival of estuarine organisms. Finally, estuaries may be influenced by human activities such as farming and construction, which may introduce sediments into rivers. These sediments are carried down to the estuaries, where they are deposited.

Estuaries will continue to be under stress as long as urban centers and world trade are focused on them. However, the quality and utility of estuarine water may be improved considerably with care and patience.

BASINS

A *basin* is a large semienclosed depression filled with sea water and usually having a narrow, shallow opening to the ocean. The circulation in basins depends largely on the processes of freshwater input (precipitation and river runoff) and loss through evaporation.

If evaporation exceeds precipitation and river runoff, the basin becomes saltier than the ocean and a gradient of salinity occurs, with the less saline water located near the ocean. The Mediterranean Sea is an example of this type of basin. The sinking saline water causes the Mediterranean Sea to be well mixed rather than stratified. In this sea (Fig. 9.22), water flows in at the surface and saltier water flows out over the rim of the Gibraltar Sill and into the Atlantic Ocean. This salty warm Mediterranean water contributes to the circulation of the North and South Atlantic oceans at intermediate depths.

If precipitation and river runoff exceed evaporation, a typical estuarine circulation occurs. In this case the excess fresher water tends to flow out of the basin near the surface, diluting the adjacent sea. Smaller amounts of saltier water flow into the basin over the sill (Fig. 9.23). The Baltic Sea and the Black Sea are examples of this type of basin. The fiords, discussed earlier, may also be thought of as small basins of this nature. In silled basins of this type, there is a tendency toward isolation of stagnant salty water below the depth of the sill and, consequently, oxygen depletion and hydrogen sulfide buildup near the bottom of the basin. This effect is especially well illustrated in the Black Sea, where a sharp gradient of decreasing temperature coincides with the increase in salinity with depth. Here the plants and animals are restricted to the upper 200 meters or so. Below this depth only anaerobic microorganisms such as bacteria are found. The "rain" of dead organisms into the deep water provides energy for anaerobic bacteria, many of which release hydrogen sulfide into the water. The Baltic Sea is much shallower than the Black Sea, and consequently, anoxic conditions exist only in the deepest areas, where stagnation occurs.

COASTAL POLLUTION

Human activities may result in beneficial or detrimental modifications of the oceans, especially near shore. We have already discussed the building of shoreline structures and we will now discuss some of the effects of industrial and domestic pollution and the decision-making

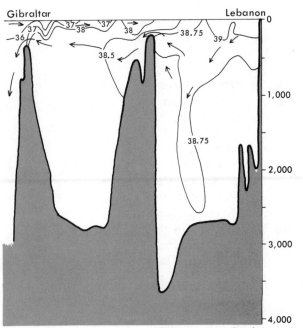

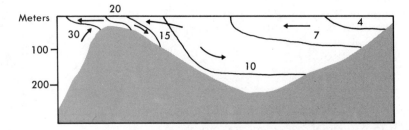

FIGURE 9.22 Circulation of water in the Mediterranean Sea. Salinity values are given in parts per thousand. (Modified after U.S. Naval Oceanographic Office, H.O. 700, 1967.)

FIGURE 9.23 Circulation of water in the Baltic Sea. Salinity values are given in parts per thousand. (After Segerstråle, 1957.)

processes that affect the use and misuse of the seas. These are largely political issues because they are controlled by local and national governments. For example, pollution of the coastal oceans may be prohibited by laws but permitted when the laws are not enforced, or may be permitted in the absence of appropriate laws.

The coastal waters are most susceptible to pollution, because of the concentration of urban activity near the coasts and near rivers that eventually enter the sea. Large cities and ports are located near the mouths of rivers. In fact, seven of the 10 largest metropolitan areas of the world are situated near estuaries. Other coastal areas are popular as resorts. Coastal environments are of special value to us as a source of fuel and food. Because of intense use, our coastal oceans are in danger of misuse and damage. Their protection is in the interest of all.

Pollutants may enter coastal waters by design, as when industrial or domestic wastes are dumped directly into the sea (Fig. 9.24), or by accident, as in an oil spill. Depending on the nature of the pollutant, whether toxic or nutrient, marine life may be either inhibited or stimulated. It is tempting to believe that "dilution is the solution to pollution." However, industrial wastes are frequently toxic to marine life near the source. Domestic sewage, whether treated or not, provides nutrients (such as phosphates) for marine plants, including the organisms that cause "red tides." (The role of red tides receives special consideration in Chapter 12.) In the case of oil on the sea, both effects may be seen. Oil kills some organisms, such as sea birds, but stimulates the growth of others, such as oil-digesting bacteria, which are slow-acting and which certainly do not fully counteract the damaging effects of oil pollution. In fact, some of these bacteria secrete substances that are more toxic than the oil itself.

The ocean is frequently regarded as a handy dumping ground. Usually the municipalities and industries that dump wastes into the sea do so as close to shore as they are allowed by law (or sometimes even closer). The result is that certain areas, especially near metropolitan centers, become more polluted than others. In 1974, more than 70 per cent of the municipal sludge dumping in the United States and about 60 per cent of the industrial dumping took place in the New York Bight, in 30 meters of water off the Hudson River. In this area, about 5 million cubic meters of sludge were being dumped 25 km from shore each year. (Dredge spoils are dumped less than 15 km from shore.) Heavier materials, such as sewage sludge, sink and pollute bottom sediments (Fig. 9.25), killing or displacing many of the natural animal species. The wastes are not restricted to the area where they are dumped but are carried by currents. Lighter materials such as oil and plastics (Fig. 9.26) may even reach the shore. Industrial chemicals such as acids, heavy metals, and other toxic materials are released into the sea. In the early 1950's a chemical plant in Japan was discarding waste mercury compounds into Minamata Bay. The mercury soon worked its way up the food web. The concentration of mercury was progressively increased as

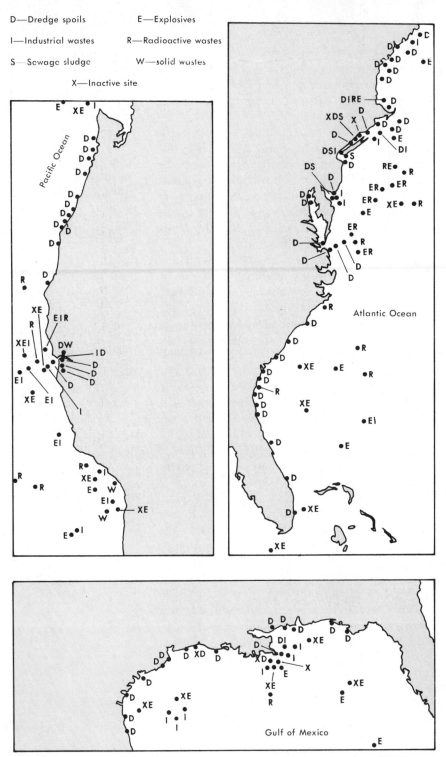

FIGURE 9.24 Known ocean dumping sites off U.S. coasts. (Modified after Council on Environmental Quality, 1970.)

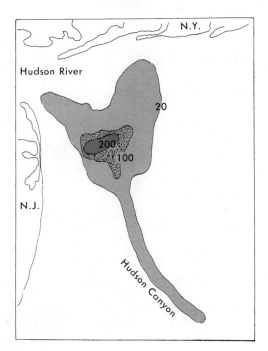

FIGURE 9.25 Distribution of lead in the sediments of the New York Bight in the vicinity of the sludge dumping grounds. Concentrations of lead (in parts per million) are indicated by curved lines. (After Carmody, Pearse, and Yasso, 1973.)

plants were eaten by herbivores which, in turn, were eaten by carnivores and ultimately by people. Eventually, people who ate contaminated fish suffered neurological damage (Minamata Disease), and at least 46 people died.

Oil pollution is a special problem in the coastal environment because oil floats on water and does not easily mix with it. Some oil seeps into the sea naturally from submarine oil fields such as in the Santa Barbara Channel off California. This oil is usually not a great problem since amounts are relatively small and some marine bacteria eventually digest the oil and render it harmless. On the other hand, drilling for oil or transportation of oil in tankers (Fig. 9.27) may occasionally result in large-scale oil spills. On March 18, 1967, the oil tanker *Torrey Canyon* ran aground off the British coast of Cornwall, releasing 118,000 tons of crude oil into the sea. Some of it was intentionally burned at sea, but most of it drifted along the coast and polluted beaches in Britain and France. In addition, many birds died as a result of the oil. In an effort to save the tourist trade, beaches were sprayed with detergents to disperse the oil. The detergent, however, was toxic. Many intertidal organisms, especially the snail-like limpets, were killed by the detergent. It is now known that limpets browse on the rocks, removing algae and oil deposits. The limpets eat the oil. It is unfortunate that many of these natural cleaners were killed in an attempt to disperse the oil. In fact, it is probable that more marine life was killed by the detergents than by the oil itself.

FIGURE 9.26 Plastics collected on a fine-mesh plankton net. Also shown, in the upper right, is a smudge of oil. (Courtesy of Harold Wes Pratt, National Marine Fisheries Service.)

Ultimately, the disappearance of oil from the sea is mostly due to the decomposition of oil by bacteria, a process that may take several years.

More recently, in 1974, the tanker *Metula* (Fig. 9.28) ran aground in the Strait of Magellan, spilling 54,000 of the 195,000 tons of oil that she carried. In 1975, the Liberian-flag tanker *Spartan Lady* broke in two in heavy seas off the New Jersey coast. The next year, another Liberian-flag tanker, *Argo Merchant,* ran aground off Nantucket. Before it could be freed, the weather worsened and this ship also broke in two. Each ship spilled about 30,000 tons of fuel oil into the sea. Fortunately, the oil drifted farther out to sea and apparently no coastal damage occurred.

Although large, the *Torrey Canyon, Metula, Spartan Lady,* and *Argo Merchant* are not the largest tankers of today. The supertanker *Universe Ireland* carries almost three times the oil carried by the *Torrey Canyon.* In 1973, there were 26 supertankers, each with a capacity of 400,000 tons or more. On March 17, 1978, the supertanker *Amoco Cadiz* ran aground in the English Channel off France. More than 400,000 tons of oil were spilled and caused considerable damage to the beaches of France, which could

FIGURE 9.27 The supertanker *Atlantic Sun* with a capacity of about 250,000 tons. (Courtesy of the Sun Oil Co.)

FIGURE 9.28 The tanker *Metula*, aground in the Strait of Magellan. (Courtesy of the U.S. Coast Guard.)

possibly affect the oyster fisheries along the Brittany coast for 15 years or more.

Not all oil spills come from ships. On January 28, 1969, an oil well being drilled off Santa Barbara, California, blew out, sending more than 3 million gallons of oil (about 20,000 tons) into the sea. This is about as much oil as reached the Cornwall coast after the *Torrey Canyon* went aground. Again, beaches were coated with oil, birds died, and marine life suffered for a time. Eventually most of the oil was removed or was destroyed by bacteria.

Pollution of the coastal environment will continue. Someday it may not be necessary to dump sewage and industrial wastes into the water, as they may be recycled. Sewage waste may even provide nutrients to grow food for people under controlled conditions (algae culture).

Oil will continue to be spilled accidentally, although one hopes that the chances of this disaster will be reduced. The value and scarcity of this fuel has resulted in special efforts to reduce the possibility of accidents and to facilitate the recovery of the valuable oil.

As our needs for power and development of coastal areas continue to grow, the coastal environment will be under increasing stress and perhaps in danger of even greater pollution despite stringent control measures. As the sea has a tremendous capacity to absorb heat, it is tempting to locate atomic power plants in or near the sea so that marine or brackish water can be used to cool the nuclear reactors. Of course, dangerous radioactive water pollution through accident or sabotage is a frightening possibility. But a more likely source of pollution lies in the discharge of hot cooling water. The water will warm the local environment, causing some organisms to die and new communities of marine life to develop. These effects may be beneficial if they provide a greater diversity of life in the area, or they may be harmful if local fisheries are disrupted. The heat may also deflect marine fish from their normal patterns of migration to breeding grounds. Some fish may be attracted to the warm water around the power plant and fail to swim upstream to spawn. In addition, massive deaths of fish and other life may occur if the power plant is shut down and the water temperature drops suddenly.

SUMMARY

Coastal oceanography includes the study of the processes that determine the characteristics of beaches and the near-shore waters. Beaches are dynamic and are constantly changing as a result of wave action and the transport of beach materials by longshore currents. The shape of beaches is also influenced by artificial structures such as jetties, sea walls, and breakwaters, all of which influence waves and currents and consequently the transport of sand. Beach sand may also be carried by wind and sometimes large sand dunes are formed which extend inland for several miles.

The coastline itself is dynamic and changing. Sea levels rise and fall as polar ice recedes and grows in relation to the ice ages. On the other hand, the land itself may rise or subside due to geologic processes. Some of these changes in sea level, although slow, have been recorded over the last several thousand years and the evidence of past shorelines is sometimes found through archeological study.

An important feature of coasts is the estuarine system. Estuaries, generally the sites of mixing of ocean water and fresh water, are usually well nourished by nutrient-rich runoff from land, causing them to be highly productive and important nursery grounds for clams, crabs, shrimp, and fish. Human activities have resulted in added nutrients and input of chemicals that may be harmful to the natural balances that exist among the various forms of estuarine life. This is a form of pollution and must be controlled in order to maintain the value of estuaries as providers of food for human consumption.

Basins are large coastal features that may have estuarine circulation. The Baltic Sea and Black Sea are examples of basins where precipitation and runoff of fresh water exceed evaporation. These basins are much like giant estuaries. The Mediterranean Sea is an example of a basin where evaporation exceeds precipitation and freshwater runoff. In the Mediterranean, the fresher water is found near the Atlantic Ocean.

Since coastal regions, especially estuaries, are most suitable for harboring large ships, these areas have included historically important centers of urbanization and concentration of population. Consequently they have also been subject to the detrimental effects of pollution. The dumping of domestic and industrial wastes and spilling of oil into harbors has lessened their value as fishing grounds. The estuary as provider and the estuary as urban cesspool are conflicting uses. Efforts to eliminate the latter have been only partially successful. Further progress toward cleaning up the estuaries may restore their value for fisheries.

Backshore	Beach	**IMPORTANT**
Barrier island	Beach drifting	**TERMS**
Barrier spit	Beach face	
Basin	Beach scarp	

Berm   Lagoon
Breakwater   Longshore bar
Coastal-plain estuary   Longshore current
Estuary   Longshore trough
Foreshore   Marsh
Fiord   Offshore
Groin   Sea wall
Jetty

STUDY QUESTIONS

1. What determines the steepness of a beach?
2. How are beach materials transported?
3. What is the mechanism responsible for rip currents?
4. What are the consequences of building groins along a coast?
5. How may a well-mixed estuary become a partly mixed estuary?
6. How might dredging of an estuary result in a change in its degree of stratification?
7. Discuss the relationships of estuarine circulation to the distribution of fish and other organisms.
8. What is the ultimate fate of oil that spills onto the sea and is not removed by cleanup operations? Approximately what period of time is involved?
9. Calculate the growth rate (in meters per year) of the spit shown in Fig. 9.10 from 1940 to 1957 and from 1957 to 1963. A severe storm hit the area in 1962. Could this have affected the shape of the spit? How?

SUGGESTED READINGS

Barnes, R. S. K. 1974. *Estuarine Biology.* London, Edward Arnold Publishers.

Bascom, Willard. 1964. *Waves and Beaches.* Garden City, N.Y., Doubleday.

Dorfman, Donald. 1980. Senegal River, Africa: fishes or dams? *Underwater Naturalist* 12(2):4–6.

Inman, D. L. and B. M. Brush. 1973. The coastal challenge. *Science, 181*:20–32.

King, Cuchlaine A. M. 1959. *Beaches and Coasts.* London, Edward Arnold.

Komar, P. D. 1976. *Beach Processes and Sedimentation.* Englewood Cliffs, N.J., Prentice-Hall.

Lauff, George H. (ed.). 1967. *Estuaries.* Washington, D.C., American Association for the Advancement of Science (pub. 83).

Marx, Wesley. 1967. *The Frail Ocean.* New York, Ballantine.

Officer, C. B. 1976. *Physical Oceanography of Estuaries (and Associated Coastal Waters).* New York, John Wiley.

U.S. Army, Coastal Engineering Research Center. 1966. *Shore Protection, Planning and Design,* 3rd ed. Washington, D.C., Government Printing Office.

———. 1973. *Shore Protection Manual.* Washington, D.C., Government Printing Office.

Williams, Jerome. 1979. *Introduction to Marine Pollution Control.* New York, John Wiley.

Marine Environments

10

Life in the oceans is unevenly distributed, both vertically and horizontally. Some areas are rich in life; others are nearly desertlike. Some of the important factors that limit an organism's distribution in the sea include its mobility and its relation to light, temperature, pressure, and nutrients (or food supply).

MOBILITY IN MARINE ENVIRONMENTS

Marine organisms may be classified with respect to their mobility in the sea. Thus we speak of *plankton* (drifters), *nekton* (swimmers), and *benthic* organisms (bottom dwellers).

Plankton (Fig. 10.1) drift at the mercy of ocean currents. They drift passively or swim very weakly in the sea. They include plants *(phytoplankton)* and animals *(zooplankton)*. Many of the zooplankton feed on phytoplankton. Some plankton, called *pleuston,* actually float on the surface of the water, being held up by gas floats, as is *Physalia,* the Portuguese man o' war, or by the surface-tension membrane of the

A slug-like nudibranch crawling over encrusting sponges, which are also inhabited by sea anemones. (Photo by Harold Wes Pratt.)

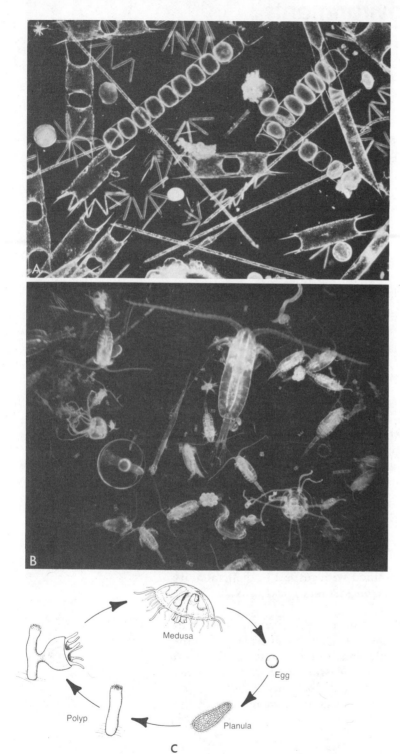

FIGURE 10.1 *A*, A variety of marine phytoplankton, mostly diatoms (60×). *B*, Marine zooplankton. The largest animal is a copepod, as are the smaller similar ones. There are also two small jellyfish with tentacles, a straight arrow worm, two wormlike tunicates, and a spherical fish egg (12×). *C*, The life cycle of a small jellyfish, indicating both planktonic (medusa) and benthic (polyp) stages. This type of life cycle is found in the fresh-water jellyfish *Craspedacusta*, as well as in certain marine forms. (*A* and *B* photos © Douglas P. Wilson; *C* from Barnes, 1974.)

water as is *Halobates,* the oceanic water strider. Most plankton are very small, less than one mm in length, but a few are quite large, such as jellyfish and *Sargassum* weed, the latter of which is actually more like its attached benthic relatives than most other phytoplankton.

Many organisms, called holoplankton, are planktonic during their whole life. These include many species of diatoms and copepods, which are the most abundant marine plants and animals, respectively.

Some organisms are planktonic during part of their life cycle and either nektonic or benthic at another stage. These plankton are called *meroplankton.* For example, some jellyfish are planktonic during the sexually reproducing part of their life cycle and attached (sessile) during a stage that reproduces asexually by budding. Figure 10.1C illustrates the life cycle of a small jellyfish.

The larvae of most fish and invertebrates, including clams and crabs, are too small to swim against ocean currents and must be considered meroplankton. They mature into nektonic or benthic adults. In some cases the food of the larva differs considerably from the food of the adult.

The larvae of most fish that live near the surface are nearly transparent and thus are harder for predators to see. As the larvae mature, they develop the pigments characteristic of the adult. Larvae may also develop the vertical migration pattern characteristic of the adult of that species. Many fish migrate into deeper water during the day, thus avoiding predators that might see them. It is interesting to note that the larvae of deep-water fish, which live in the dark, usually have black pigmentation.

Animals that can swim rapidly enough to counteract ocean currents are called nekton. They include many fish, squid, porpoises, and whales. They may carry out extensive vertical migrations, as in the case of lantern fish, or much longer horizontal migrations as in the case of some eels and salmon. Some nekton, such as tuna, can swim for days at speeds about 10 km/hr. Nekton are adapted for swimming by the presence of fins, a generally flat tail, and some degree of streamlining.

Many fish have gas bladders that help maintain buoyancy, enabling the fish to remain above the bottom with little effort. Sharks, however, have no gas bladder and tend to sink when they are not swimming. In compensation for this tendency they have an upturned tail (Fig. 10.2). When the tail beats, it gives a downward and forward thrust to the rear and consequently an upward and forward thrust to the head, lifting the shark from the bottom.

The shape of nektonic animals has much to do with their habitat and way of life (Fig. 10.3). Most fish that live near the bottom are slow-moving and awkward-looking but have great maneuverability at slow speeds. They usually feed on clams, crabs, and other sessile or slow-moving benthic species. Fish that live higher in the water tend to be much more streamlined and fast but are not very maneuverable.

FIGURE 10.2 Sharks have upturned tails, which give them an uplift when they swim. They have no buoyant gas bladder as do most fish, so they tend to sink if they become motionless. (Photo courtesy of NOAA.)

Benthic animals may live on the bottom as *epifauna,* or within sediment as *infauna.* The epifauna may be attached (sessile), such as sponges and sea anemones, or they may move around (vagrant), as do crabs and snails. The shallow-water epifauna are subjected to fluctuating temperatures and salinity. The infauna, such as various worms and some clams, are more protected from fluctuations in the environment. Even in shallow-water sediment, temperature and salinity vary less than in the water above it. There is little circulation of water in the sediment, so oxygen tends to be almost lacking. Many sediment-dwelling worms are adapted to this habitat, but clams must receive oxygenated water from above, via an *incurrent siphon,* a fleshy tube which also draws in small particles of food. Waste water is then pumped out through a separate *excurrent siphon.*

CLASSIFICATION OF MARINE ENVIRONMENTS

Pelagic Environments

The *pelagic* environment includes the entire ocean except the sea floor. The plants and animals that live in the open sea but which are not closely associated with the shore or sea floor are known as pelagic organ-

A

B

FIGURE 10.3 *A*, A maneuverable but slow-moving fish, the butterfly-cod; *B*, A pelagic, rapid-swimming, stream-lined fish, the tuna. (*A* courtesy of Australian Tourist Commission; *B* courtesy of NOAA.)

isms. The pelagic environment is further subdivided into zones on the basis of water depth, light distribution, and temperature (See Figure 10.4).

Because light that is sufficiently intense for plant growth penetrates only a short distance into the sea (generally less than 200 meters in the clearest open ocean water and only a few meters in cloudier coastal water), plants—which require light for photosynthesis—are found only in the upper layer of the sea. Essentially all of the conversion of light energy into the stored energy of organic matter occurs in the well-lighted water, the **euphotic zone.** The production of plant life in this zone provides food for various plant-eating animals (herbivores). Some of this stored organic matter may eventually settle to the sea floor, become

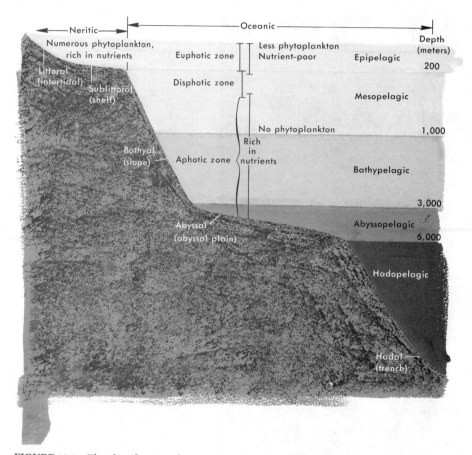

FIGURE 10.4 The classification of marine environments. Distribution of phytoplankton, light, and nutrients in neritic and oceanic water. Penetration of light is much less in the turbid neritic water than in the clearer oceanic water.

buried, and be converted into oil and gas. More oil and gas was formed on continental shelves, over which production tends to be greater, than in areas farther from shore. The deeper, dark water, the *aphotic zone,* is cold; the only light found there is produced by some animals. Between the euphotic and aphotic it is convenient to include a *disphotic zone,* where light exists but is not intense enough for effective production of plants. This intermediate zone may extend as deep as 1000 meters or so in the clearest ocean water. Figure 10.4 illustrates the classification of marine environments.

NERITIC PELAGIC

The *neritic* pelagic region comprises all the water that lies over the continental shelf, generally extending downward to a depth of about

200 meters. This area includes the most productive parts of the oceans, because it is close to rich supplies of nutrients from land as well as from upwelling of deeper nutrient-rich water. It is continuous with the surface water farther from shore, and the processes that apply here apply also in the water beyond the shelf. Note that the depth of light penetration (Fig. 10.4) is less in the neritic water than in the open ocean, farther from shore. This effect is due to the greater amounts of suspended sediment and marine life in the coastal water.

OCEANIC PELAGIC

The *oceanic* pelagic region includes all the water beyond the continental shelf, from the surface down to the greatest depths found in the trenches. These waters are generally much poorer in nutrients than the neritic water; hence the production of life is much less.

Pressure in the sea increases at a rate of about one atmosphere for each 10 meters of depth. Thus, at a depth of 2000 meters the pressure is about 200 times that at the surface. In spite of the tremendous pressures in the deep sea, animals exist at all depths. Apparently the distribution of marine life is less affected by pressure than by other factors such as nutrient availability, food supply, light, and temperature.

Photosynthesis is restricted to the euphotic upper layer of the sea, the *epipelagic* zone. The animals which eat the plants, however, are not so restricted and may take part in daily vertical migrations upward as the sun sets and down again in the morning.

The migrant animals may be eaten by carnivores at any of the depths at which they are found. This may be near the surface, or at depths of 400, 500, or even 700 meters or more in clear water. Organic energy-rich matter in living organisms is thus carried down into the *mesopelagic zone,* which extends from about 200 to 1000 meters and is sometimes called the "twilight zone" (disphotic) because of the reduced light in this zone.

Another way that energy may be transferred to deeper zones is by the sinking of *detritus* (dead organic matter). This may be in the form of dead plants or animals or fecal material. Some detritus even reaches the bottom in the trenches, below 6000 meters, and may be the main source of food for deep-sea organisms.

The mesopelagic and the deeper *bathypelagic zone* (1000 to 3000 meters) are isolated from the surface waters and the benthic environments except where they touch the continental slope. Thus, the organisms in this zone depend on migration of other animals and on descending detritus for food.

The animals of the *abyssopelagic* (3000 to 6000 meters) and *hadopelagic* (deeper than 6000 meters) zones are closely associated with the animals of the deep-sea floor. These deep-sea pelagic animals, along with the ever-present detritus, provide food for the deep-sea benthic animals. The most striking features of these zones are the monotonously constant cold temperature (below 4° C) and darkness.

Benthic Environments

The *benthic* environment is the sea floor, and the plants and animals that live there are called benthic organisms. Benthic organisms require a proper substrate such as rock, sand, wood, or any other material upon or within which they may survive. The great variety of substrate materials results in a great variety of niches. A *niche* consists of all the requirements (physical and biological) of an organism. Most clams, for example, require a sandy or muddy bottom. They also require water from which they can filter phytoplankton food. Different species tolerate different temperature ranges, salinity ranges, oxygen concentrations, and so forth. As a rule *(Gause's Law)*, no two species may have exactly the same environmental requirements, and thus no two species occupy the same niche. If two species with similar environmental needs compete for the same place on the substrate, only the one best suited to the total environment will survive. A rocky coast generally has a greater number of niches than a sandy coast or the deep ocean floor. Thus the diversity of benthic species tends to be greater on the rocky coast than in the more uniform environments.

LITTORAL ZONE

The part of the sea floor most readily seen by people is the intertidal or *littoral* zone. Organisms at the edge of the sea are subject to periodic wetting by waves and tides. The pattern of waves and tides varies from place to place, depending on the shape of the ocean basin, weather, and other factors discussed in the chapters on waves (Chapter 6) and on tides (Chapter 7). Most coasts have semidaily tides, with two high tides and two low tides each day. Some areas, for example along the Gulf of Mexico, experience daily tides with one high and one low tide per day. Some examples of tidal patterns are shown in Figure 10.5.

Casual observation of a piling or rock jetty shows a vertical distribution of attached organisms, in which discrete zones or stripes of color and pattern are seen (Fig. 10.6). The organisms found above the high-tide levels are only occasionally covered by the splashing of waves *(splash* or *supratidal zone)* and are adapted to greater exposure and less water coverage. Those found between the high and low tides *(littoral)* are immersed in water at least once a day, but they must also be able to survive exposure to warm and cold air temperatures. They may accomplish this by closing their shells, as do barnacles and mussels, or they may live near the substrate, covered by a layer of seaweed. Organisms living at levels below tide *(sublittoral)* are always covered.

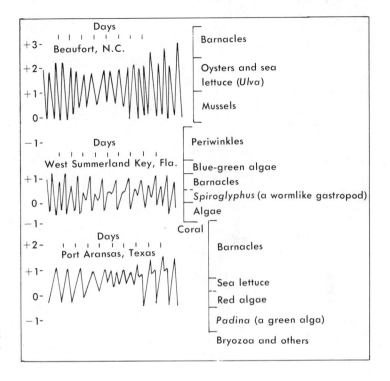

FIGURE 10.5 Tidal patterns in relation to the distribution of attached organisms. The scales shown on the left are in feet (3.28 feet = 1 meter). (After Hedgpeth, 1953.)

There are three general ways in which organisms adapt to life in the intertidal zone: (1) some organisms can adapt to alternate periods of submersion and exposure to air; (2) others migrate with the tides, so that their environments remain fairly stable; and (3) still others live either buried in constantly moist sand or in tide pools (depressions where water remains even when the tide is out). Organisms in sand or in tide pools must still contend with fluctuating temperatures and salinity, especially during the summer, when the water that surrounds them may be warmer and saltier during low tide than at high tide. These graded conditions result in zones characterized by different fauna and flora, and each zone has a community of specially adapted organisms.

SUBLITTORAL ZONE

The sublittoral zone, which extends from the low tide levels down to the edge of the continental shelf (about 200 meters) is generally well nourished by nutrient-laden rivers and upwelling of deeper nutrient-rich water from below, as well as by the rain of detritus from the pelagic organisms above. Consequently, a rich and often diverse collection of

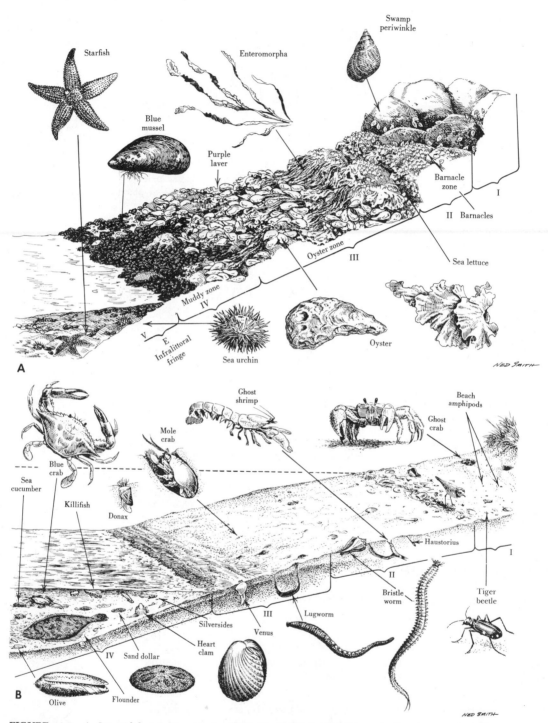

FIGURE 10.6 *A,* Intertidal zonation on a rocky surface at low tide. *B,* Sandy beach zonation. Dashed line indicates high tide. (From Smith, R. L.: Ecology and Field Biology. New York, Harper and Row, Publishers, 1966.)

fauna and flora abounds in this benthic zone. The flora, of course, are limited to the shallower areas, where the euphotic zone reaches the bottom.

Because organisms in this zone are never exposed to the drying influence of the atmosphere, they need not have the same kinds of adaptations as the intertidal life. Their main needs are adequate food and protection from predators.

Some of the sublittoral (and littoral) animals feed on benthic plants. Some snails and limpets, for example, are herbivores and browse on the algae that are attached to rocks. They glide over the surface and scrape off the plants with a filelike plate (radula) near the mouth. Many clams and oysters, on the other hand, feed on the phytoplankton in the water above them. These animals are restricted to water containing sufficient plant life for growth. Not so restricted are the scavenging crabs. They crawl around, feeding on dead organisms and invertebrates on the sea floor. They, in turn, may be food for bottom-dwelling fish, including some sharks. Figure 10.7 illustrates some views of the sublittoral zone.

Another kind of environment that is always submerged and must also be considered sublittoral is that inhabited by fouling organisms. Fouling organisms include plants and animals that invade pilings, floating docks, buoys, and the hulls of ships. One of these is the boring clam *Teredo,* or "ship worm," which causes millions of dollars' worth of damage yearly to docks and other wooden structures in the sea. Also included are certain types of barnacles, which attach to the hulls of ships and slow them down owing to increased frictional drag. Antifouling paint, however, can be used to prevent the young larval forms from attaching and becoming established.

DEEPER BENTHIC ZONES

The deeper benthic zones comprise the entire ocean floor beyond the continental shelves, where light is insufficient for plant growth. The amount of organic matter is much reduced in these zones compared with that in shallower environments. The principal sources of food for the animals here are sinking detritus from above, migrating pelagic animals, and large food-falls such as dead whales.

The *bathyal zone* is the benthic environment of the continental slope, about 200 to 3000 meters in depth. Temperatures in this zone are usually less than 10° C and are nearly unchanging, resulting in a very monotonous environment. One view of the bathyal environment is shown in Figure 10.8. A wide variety of animal life exists in this zone, including many species of sponges, soft corals, sea lilies, sea stars, sea cucumbers, crabs, shrimp, and fish.

The *abyssal zone,* which comprises the abyssal plains and hills, represents more than 80 per cent of the sea floor. Less food is transported onto this zone than onto the shallower benthic environments;

FIGURE 10.7 *A,* Sublittoral environment dominated by sea stars, brittle stars, and feather stars, at 92 meters on the continental shelf off Antarctica's Palmer Peninsula. *B,* An Australian coral reef. *C,* Sublittoral environment on the New England shelf, dominated by sea urchins. Note also the sea star and hermit crab. (*A* courtesy of Smithsonian Insitution, Oceanographic Sorting Center; *B* courtesy of Australian Tourist Commission and Qantas Airlines; *C* courtesy of Bruce Reynolds, National Marine Water Quality Laboratory, Narragansett, Rhode Island.)

FIGURE 10.8 Sea cucumbers (the three larger organisms) and feather stars at about 600 meters on the continental slope off Antarctica. (Courtesy of Smithsonian Institution, Oceanographic Sorting Center.)

hence the abundance of animal life is greatly reduced. Contrary to earlier belief, however, a great diversity of life is found in the abyss due to the stable conditions found there. Abyssal diversity is comparable to that found in shallow tropical seas. Most of the abyssal animals (Fig. 10.9) are either mud-eaters, such as sea cucumbers and brittle stars, or predators, such as sea stars and certain grotesque fish with large mouths. The distribution of life in the abyssal zone is not uniform. It may be abundant in one area and very sparse in others, depending largely on the productivity in the epipelagic zone above, or the introduction of large organisms, such as whales, which have sunk to the sea floor after dying. In addition to darkness, this zone is characterized by temperatures of less than 4° C.

FIGURE 10.9 Life in the abyss. *A*, Sea cucumber (right) and trail of feces left by a worm (left) at about 4000 meters in the southern Indian Ocean. *B*, Sea urchin and brittle star at about 4000 meters between South America and Antarctica. (Photos courtesy of Smithsonian Institution, Oceanographic Sorting Center.)

The *hadal zone* consists of long, narrow oceanic trenches located close to many island chains or continents. With temperatures generally ranging from 1.2 to 3.6° C and pressures in excess of 600 atmospheres (depths greater than 6000 meters), this area rates as the most extreme of ocean habitats. Larger predators such as fish, sea stars, and crabs are generally absent in the deep trenches, apparently because of the low concentration of food organisms. The concentration of life may be more than 1000 times as great on the shelf as it is in the hadal benthic environment. The dominant species tend to be slow-moving mud-eaters such as sea cucumbers and other animals that are able to make the most of the sparse food available. These include sea anemones and strange worms with no mouth or stomach (the pogonophorans). These worms can absorb dissolved food through their skin.

SAMPLING THE MARINE LIFE

Sampling of ocean life may be simple, as in scraping organisms from a measured area of a rock or piling; or it may be very difficult and time consuming, as in sampling the deep-sea benthic life. In a day's time, many samples may be collected in the littoral and sublittoral zones, but it may take months to collect an equivalent number of samples from the abyss. The equipment used in deep water is frequently very heavy and must be lowered slowly. For example, if a deep-sea grab-sampler (Fig. 10.10) is lowered to 4000 meters at 50 meters per minute and retrieved at 30 meters per minute, it will take about four hours to get a single bottom

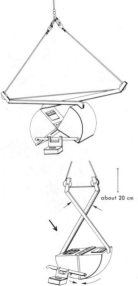

about 20 cm

FIGURE 10.10 A van Veen grab, open on the way down (top), and closed on the way up, with a sample of the mud (bottom). (After Hedgpeth, 1957.)

sample. This figure is increased to about 10 to 12 hours if a dredge (Fig. 10.11) is used to sample the sediment-dwelling life, because much more line must be let (at an angle), for the sampler may be dragged nearly horizontally. Use of a plankton net (Fig. 10.12) or a mid-water trawl (Fig. 10.13) to sample the deep pelagic zones involves similar complications.

Thus, the time and expense of obtaining ecological data increase dramatically as one works in deeper and deeper water. Neritic waters have been studied most extensively because of this factor and also because of the presence of economically important food resources in this zone.

An economical way around the need for specialized expeditions to sample the epipelagic plankton was devised by Sir Alister Hardy in

FIGURE 10.11 A large dredge, used to collect large samples of ocean sediment and infauna. (Courtesy of Woods Hole Oceanographic Institution.)

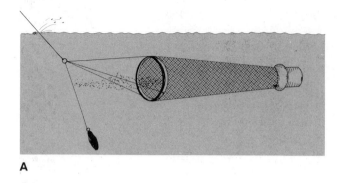

A

FIGURE 10.12 *A*, Plankton net in tow. *B*, Plankton net being lowered for collection run. (Photo courtesy NOAA.)

B

1925. His *continuous plankton recorder* (Fig. 10.14) may be towed behind a commercial vessel traveling across the ocean. The plankton are caught, preserved, and continuously wound between two gauze bands and stored until they are "unwound" and identified. This method provides continuous sampling of the plankton between two points. Of course, many of the organisms, especially the zooplankton, become crushed, making identification more difficult. Consequently, experts in identifying crushed plankton have evolved out of these studies.

BIOLUMINESCENCE

Although sunlight is the most obvious source of light in the sea, it is not the only one. At night and in disphotic and aphotic regions, the phenomenon known as bioluminescence is a significant source of light. This light source can have a profound influence on many aspects of life from vertical migration to reproductive behavior to predation.

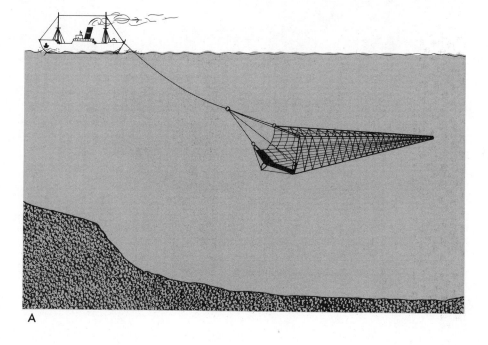

A

B

FIGURE 10.13 The midwater trawl being towed *(A)*. The Isaacs-Kidd midwater trawl *(B)* collects marine creatures living in waters a mile or more deep. The wide metal vane keeps the trawl level and holds it at the required depth during towing. *(B* photo courtesy of Scripps Institution of Oceanography, University of California, San Diego.)

Actually, even bioluminescence depends indirectly on the sun, because sunlight is the source of energy for the plant life which make organic matter out of inorganic nutrients. Bioluminescent animals may eat plants or other animals that had eaten plants and thereby attain energy to live. Some of this energy is released in certain organisms as a cool blue-white light. Little heat is given off. Bioluminescence also provides light in surface waters at night, when it frequently lights up the surf or the wake of a ship.

Pliny and other ancients had observed luminescence in some of the larger jellyfish. But the glow sometimes seen in the surf and in the

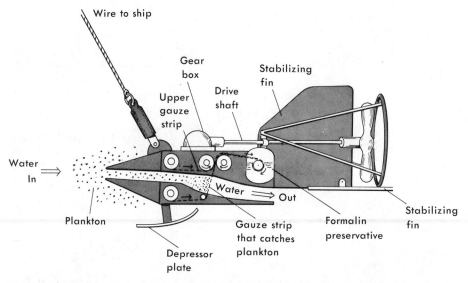

FIGURE 10.14 The Hardy continuous plankton recorder, which collects the plankton on a gauze band that is continuously wound into a tank of preservative. (After Hardy, 1956.)

wakes of ships was still a mystery in the seventeenth century. It was thought by some that the sea absorbed the sun's energy by day and gave it back at night. The British chemist Robert Boyle, in the seventeenth century, thought that it was caused by friction of waves against the air or the side of a ship. This light was observed only at night and frequently on the tips of waves or in the wake of a ship or along its hull. In the eighteenth century, it was discovered that the glowing of waves and surf is often due to a tiny one-celled dinoflagellate, *Noctiluca*, which means "night light." Although an individual is less than one mm in diameter, the presence of great numbers of these organisms causes the water to glow. Other dinoflagellates are also responsible for bioluminescent displays in various parts of the world; for example, *Gymnodinium* in the Caribbean and *Gonyaulax* off California.

The reaction involved in bioluminescence is oxidation of a substance called *luciferin* in the presence of an enzyme called *luciferase.* The exact composition of the raw materials and the nature of the reactions involved vary with the kind of organism. A typical reaction might be something like this:

$$2LH_2 + O_2 \text{ (in the presence of luciferase)} \rightarrow L + H_2O + light$$

where L = oxidized luciferin. This reaction is much like the burning of fuel (carbohydrate + $O_2 \rightarrow CO_2 + H_2O$ + light + heat), except that bioluminescence releases very little heat, whereas burning results in a great deal of heat as well as light.

A wide variety of organisms exhibit bioluminescence, including some bacteria, protozoans, jellyfish, polychaete worms, shrimp, squid,

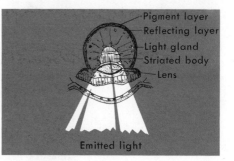

FIGURE 10.15 Photophore (light-producing organ) of an euphausid shrimp. (After Hardy, 1956.)

and fish. Few bioluminescent species, however, exist on land compared with the sea. Only a few terrestrial bacteria, fungi, and insects (such as the well-known firefly) light up during the night. The nature and distribution of bioluminescence in the sea are so diverse that one wonders about its significance. Of what value is it to organisms that produce it?

The structures causing the light range from the single cell of a bacterium or dinoflagellate to the complex light organs that very much resemble an eye found in some fish. In fact, it was once thought that these complex light organs, or *photophores,* such as those found on the shrimplike euphausid (krill), some squid, and lantern fish, actually represented eyes that were adapted to giving off light rather than receiving it (Fig. 10.15). The nerve pattern and development of eyes, however, differ greatly from photophores, so this theory has been abandoned.

An interesting relationship exists between the flashlight fish, *Photoblepharon* (Fig. 10.16) and the luminescent bacteria that it maintains in pouches on its cheeks. The fish provides food and lodging for

FIGURE 10.16 The petit Peugot or flashlight fish (*Photoblepharon palpebratus*) of the Indian Ocean, Red Sea, and western Pacific Ocean. The light organ beneath the eye emits light continually, but blinking is effected by raising the black lid below, and may aid in mating or attracting prey (and predators). (Photograph by David Powell, Steinhart Aquarium, California Academy of Sciences.)

the bacteria. The bacteria produce light for the fish. The light may help the fish by illuminating the nearby surroundings, by attracting a mate, or by attracting small fish which may be eaten by *Photoblepharon*. It may even attract a larger fish which might eat *Photoblepharon*.

Some angler fish (Fig. 10.17) have a similar relationship with luminescent bacteria which they carry in a light organ (photophore) on the end of a spine extending over the mouth. The photophore resembles a large, luminescent copepod, a favorite crustacean food for many small ocean fishes. Small fish may be lured by the light to within grabbing distance for the angler fish.

Some animals produce colors other than blue-white by covering the light source with a colored filter. For example, the deep-sea squid *Thaumatolampas diadema* (Fig. 10.18) produces lights of red, white, and blue (leading some to suggest that it should be our "National Squid"). Structurally, the light organs closely resemble those of some fish and euphausids. This similarity is an example of convergent evolution as these three groups of animals are not otherwise closely related. Most of the lights of *Thaumatolampas*, euphausids (the shrimplike krill), and lantern fish are on the undersides. The light may provide camouflage through countershading, a distinct benefit to vertical migrants, as is discussed later in this chapter.

MARINE MIGRATIONS

Many marine animals move at one time or another from one environment to another, quite often either to breed or to feed. These migrations may seem mysterious at first because they can involve considerable distances and great navigational precision. Human beings would need quite sophisticated navigational instruments to duplicate some of these remarkable journeys. Marine animals can migrate either horizontally or vertically, and we will discuss each type of migration in turn.

FIGURE 10.17 Some angler fish have luminescent lures that resemble copepods, which are food for fish.

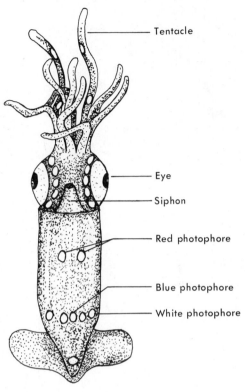

FIGURE 10.18 *Thaumatolampas diadema*, the amazing squid with red, white, and blue lights. Its total length, including tentacles, is about 8 cm. (After Hardy, 1956.)

Horizontal Migration

Horizontal migrations cover distances of many miles. The direction is often well defined and accurately aimed at a target destination.

Some migrations of fish may be related to chemicals in the water. The chemicals washed from the land by a particular stream may be characteristic and may "label" that stream as unique. Anadromous fish, that is, those which return from sea water to fresh water in order to spawn, can probably recognize their home stream (the place of their birth) by a chemical sense similar to that of smell. In the case of anadromous salmon, migration involves journeys into fresh water and up rivers to specific breeding grounds (Fig. 10.19). Most of the feeding, however, takes place in the sea, perhaps thousands of miles from the home stream. This migration is made possible by physiological adaptations that permit the organism to be able to tolerate contact with waters having a salinity difference of about 35‰.

Chemical pollution may someday pose a problem to organisms that rely on their sense of "smell" for migrational orientation. Thermal pollution might misguide some fish into a detour on their journey to breed,

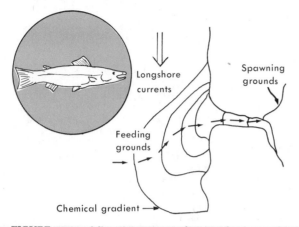

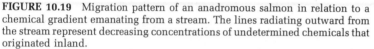

FIGURE 10.19 Migration pattern of an anadromous salmon in relation to a chemical gradient emanating from a stream. The lines radiating outward from the stream represent decreasing concentrations of undetermined chemicals that originated inland.

causing a decrease in the production of a species. Offshore power plants may heat water locally, causing not only the formation of a new community around the plant but also the deflection of some marine migrants from their course.

Perhaps some migration patterns, even though they are no longer really needed by a species, are imprinted from past times when they were necessary. For example, the American and European eels migrate from the east coast of North America and from Europe, respectively, and meet in the Sargasso Sea (in the west central Atlantic), an area of weak currents and low productivity, to breed separately (Fig. 10.20).

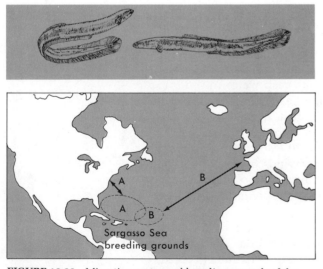

FIGURE 10.20 Migration routes and breeding grounds of the American eel (A) and the European eel (B).

The larvae return to the freshwater home of their parents without ever having made the trip before. What mechanism could these eels, especially the larvae, use to navigate such distances? Is it the earth's magnetic field, sun arcs, chemicals, or some other agent? Whatever the factors, they are definitely part of the hereditary makeup of the eels.

One theory postulates that when the continents were joined together, the Sargasso Sea was a freshwater lake and that the two types of eels were a single species that fed in tributaries that emptied into the lake. When the continents of Europe and North America drifted apart, the tributaries were separated, and the Sargasso Sea became part of the Atlantic Ocean. Through evolutionary adaptation, the eels became two species, reproductively isolated and adapted to life in both fresh and salt water. The larvae retained an instinct to migrate into the tributaries of either Europe or North America, but the adults still migrate back to the Sargasso Sea to breed. The process of continental drift is very slow, requiring millions of years, which is plenty of time for a gradual adaptation, by many steps, to the new migratory pattern. The migration itself takes about three years, during which the larvae mature. The eels remain in the fresh water for seven to 15 years. The reproductive organs do not mature until the return trip, when the adults are 10 or more years old.

Some sea turtles, seals, and whales also migrate great distances. The navigational skills of these animals are no less amazing than those of birds and fish. Green turtles (*Chelonia mydas*) are known to migrate about 2000 km from their feeding grounds along the coast of Brazil to their breeding grounds on Ascension Island in the South Atlantic (Fig. 10.21). This journey is all the more amazing because Ascension Island is only about eight km wide and the adult turtles must swim against the South Equatorial Current. The young turtles, however, are aided by the current during their westward migration back to Brazil.

The baby turtles head for the sea as soon as they hatch. Somehow they know where the water is, even when sight of it is obscured by barriers such as sand dunes. Once they find the sea they swim away from land until they are picked up by the South Equatorial Current and carried to Brazil, where they feed and mature. As adults, they migrate back to Ascension Island every two to three years to breed. Their primary navigational aid appears to be the sun. The height of the sun at noon, the arc made by the sun as the earth turns, plus a seasonal sense cued by temperature changes, could provide sufficient information to determine latitude. Although such capabilities have not been proven for the green turtle, if the turtle can find the latitude of Ascension Island (about 8° S latitude), it can migrate due east, using the arc of the sun as an east–west guide. Once the turtle is within range of a chemical gradient emanating from Ascension, it can swim toward the island, using a sense of smell or taste until the island can be seen. After mating in coastal waters, visual and chemical orientation guide the female turtle to a specific beach that is suitable for egg laying.

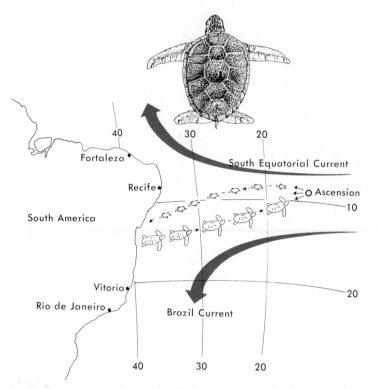

FIGURE 10.21 Migration route of the green turtle in the South Atlantic. (After Carr, 1965.)

Vertical Migration of the Zooplankton

Almost as soon as biologists began the systematic sampling of plankton, in the nineteenth century, they noticed that more zooplankton were collected at the surface during the night than during the day. It was suggested by some that the plankton avoided the nets during the day because they could see the nets (this is frequently a problem when sampling for larger animals such as fish). It was subsequently found, however, that deep samples contained more animals during the day than during the night (Fig. 10.22), suggesting that the zooplankton were migrating downward at dawn and upward at dusk. The vertical migration of animals has now been proven experimentally.

In the 1930's Sir Alister Hardy and his co-workers invented a "plankton wheel" (Fig. 10.23), which they used to measure the speed of migration of various small plankton animals in response to various light conditions. Most of the zooplankton studied swam actively toward a dim light and away from a bright light. In other words, they tended to migrate toward an area of optimum light (a movable "twilight zone"). Each species was found to have its own light optimum and con-

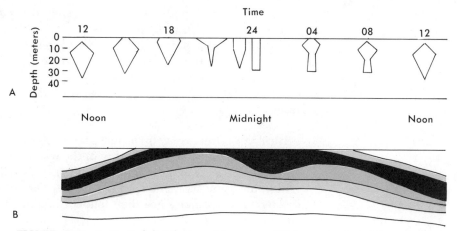

FIGURE 10.22 *A,* Vertical distribution of the copepod *Calanus* at selected times during a 24-hour period. The width of the patterns shown indicates abundance at a particular depth. (After Russell, 1927.) *B,* Continuous representation of (*A*), where the darkest shading indicates the greatest abundance of *Calanus*.

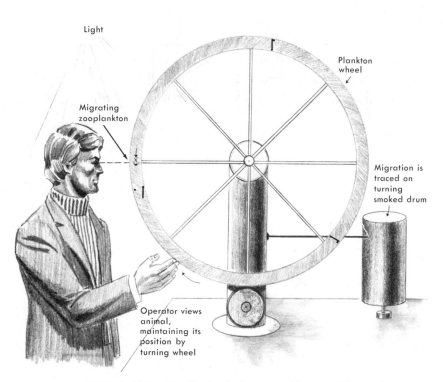

FIGURE 10.23 The Hardy plankton wheel in use. (After Hardy, 1956.)

sequently its own path of migration. They discovered that copepods could migrate as fast as 30 meters per hour (about 0.8 cm/sec). The shrimplike euphausid, which is much larger than a copepod, can migrate at a speed of 135 meters per hour, and the polychaete worm *Tomopteris* swam in the wheel at a rate of more than 200 meters per hour. These are all upward speeds; downward rates were frequently even faster.

Some crustaceans may migrate 800 meters or more during their vertical trek. Others may travel just a few meters. The range of migration generally depends on the turbidity of the water, the brightness of light, and the sensitivity of the animal to light, or its own light optimum. Even within the same area, one species may descend to depths several times greater than another since they follow different light intensities (Fig. 10.24).

During maximum darkness, some migrants stay near the surface, others tend to become scattered by random swimming, and some even sink to slightly lower depths during what Hardy refers to as "midnight sinking." As dawn approaches and a light optimum is re-established, the animals re-aggregate in the "twilight" of the sea. Then they follow this light into the depths. This is probably a mechanism of vertical migration for many, though perhaps not all, day-night migrants.

Many zooplankton require phytoplankton as food. The phytoplankton are restricted to the upper waters, the epipelagic zone. Therefore the zooplankton that migrate between the epipelagic and mesopelagic zones are in water that is rich in food during the night and in water that is poor in food during the day. They could not survive if they remained in the deeper water; they must come up to the surface water to feed.

It has been observed that some zooplankton will not migrate into extremely dense phytoplankton blooms but will engage in a shorter migration route until the bloom has diminished or drifted away (Fig. 10.25). The phytoplankton may secrete chemicals that are toxic or perhaps merely repulsive to the zooplankton. In support of this hypothesis, it has been shown that "old" phytoplankton raised in the

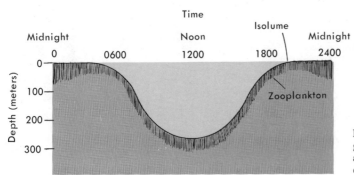

FIGURE 10.24 Path of vertical migration in relation to light intensity, as indicated by the isolume, or line of equal light intensity.

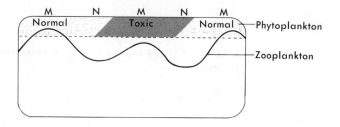

FIGURE 10.25 The effect of a toxic phyto-plankton bloom on vertical migration. Note that the zooplankton do not rise to the surface layer when it is occupied by poisonous phytoplankton. (M = midnight; N = noon.)

laboratory can cause the water flea *Daphnia* to die, apparently because of a toxic effect of the plants. Vertical migrations can be advantageous in that different water is sampled by the zooplankton on succeeding days, if the currents are different at the surface than in deeper water. It is also possible that some phytoplankton secrete more toxins during the day, when they are photosynthesizing, than during the night.

It has also been suggested that by migrating to deeper, darker water during daylight hours, the zooplankton avoid being eaten by predators that could see them if they were in bright light. Interestingly enough, however, some zooplankton are followed by their predators, usually small fish, during their migrations. These fish may be following a light optimum or the prey itself.

The Deep Scattering Layer

Layers of migrating fish are among those animals that may be responsible for "false bottoms" recorded on echo sounders. When echo sounding replaced sounding lines as a means of determining water depth, mariners were puzzled by the appearance of reflections from something in the water above the bottom. Occasionally these echoes were thought to be the bottom itself and were reported as isolated reefs. Later, when continuous echograms were made over long distances and over many hours, the "false bottom" or *deep scattering layer* (DSL) was observed to rise in the evening and descend during the morning. The DSL could be differentiated from the true bottom as a soft trace rather than the hard trace produced by the sea floor.

Now it is known that the DSL's are actually layers of vertically migrating animals that swim toward the surface as night approaches and descend in the morning. In fact, several DSL's may be seen, one above the other (Fig. 10.26). Many marine animals, including copepods, shrimp, krill, squid, and lantern fish engage in vertical migrations. Copepods are probably too small to produce distinct echoes, but the larger shrimp, krill, and fish probably cause DSL traces or echoes. Different species may migrate to different depths during the day, producing multiple DSL traces.

Fish with gas bladders (swim bladders) are presumably the most

Depth
(meters)

(fathoms)

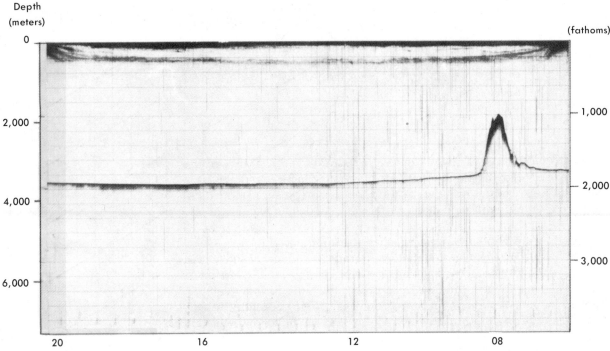

Time of day (hours)

16 August 1968

FIGURE 10.26 Multiple Deep Scattering Layers, descending at dawn (0700 hours) and rising at dusk (2000 hours) in the open sea west of Africa. The lower trace indicates the ocean bottom. (From Lowrie and Escowitz, 1969.)

effective sound reflectors, owing to the difference in density between the gas and the sea water. Fish that are mesopelagic and have swim bladders, such as lantern fish, probably are among the most important animals in causing DSL traces.

Lantern fish carry their own light source as they trek to the epipelagic regions. They cover a migratory pattern similar to that of the copepods and other small crustaceans that they eat. At first it might be thought that the lights would attract predators that eat the lantern fish, but on a closer examination of the fish, it is noticed that most of the lights are directed downward (Fig. 10.27). Larger fish looking down

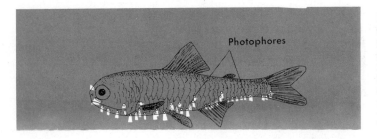

Photophores

FIGURE 10.27 A lantern fish (*Diaphus theta*) countershaded (camouflaged) in the "twilight zone" during its vertical migration. White areas indicate biolumi-nescence. The fish is less readily seen from above or below. Countershading also makes the fish less conspicuous from the side, as a shadow along the belly vanishes when the ventral lights are on. The fish is about 10 cm long.

FIGURE 10.28   The Palolo worm *(Eunice viridis)*, whose reproductive tail sections (lower part) migrate vertically in response to the light of the moon. (From Barnes, 1974.)

from above would see nothing, since the dark upper surface blends in with the dark, deeper water. The same predator viewing the lantern fish from below would probably still see nothing. The glow produced by the lantern fish tends to blend in with the "twilight" glow from above. The light, therefore, provides the lantern fish with a type of camouflage called *countershading*.

Lunar Migration

Some vertical migrations are dependent not on the sun but on the moon. An example of this type is the migration of the Palolo worm of the South Pacific (Fig. 10.28). It lives most of its life among rocks in clear tropical water, emerging at night to feed on zooplankton. During the eighth to tenth day after the full moon in October or November, the Palolo worms, male and female, lose their reproductive hind ends, which contain sperm and eggs. These tail sections, which can swim on their own, migrate to the surface of the water, where they burst, allowing the sperm to fertilize the eggs. These develop into new Palolo worms. The reproductive stages of the worm are regarded as a delicacy by the natives of Samoa and other islands of the area. Once a year they have quite a feast of Palolo worm tails, which they catch in nets at night. The most remarkable aspect of this process is the timing of the migration. Why is it eight to 10 days after the full moon in October or November every year? The most likely stimulus is the lunar light itself. A plausible suggestion is that the worms, which are exposed to moonlight at night, mature to a certain point, then stop until the bright light of the full moon triggers the final stage of maturation, which lasts for eight to 10 days and culminates in the sexual migration.

SUMMARY

The distribution and diversity of life in the sea is determined by the biology of the marine organisms and their relation to the physical, chemical, and geological factors that determine the character of the marine environment.

Drifting phytoplankton are mostly restricted to the upper 200 meters of the sea, called the epipelagic zone, where light sufficient for their growth exists. Pleuston, such as the Portuguese man o'war and the marine water strider, float on the surface of the sea, supported by gas-filled bags or the surface tension of the sea surface. Zooplankton, which frequently feed on phytoplankton, are more mobile than their food and often migrate into deeper water during the day, returning to the food-rich surface waters at night. Nekton (the swimmers), such as fish and squid, may migrate vertically or horizontally.

Bottom-dwelling organisms, the benthic life, are restricted to the sea floor, where they feed mostly on plankton, dead organic matter, or other benthic organisms. They include the crabs, lobsters and clams.

Life in the deeper zones of the sea, at bathyal, abyssal, and hadal depths, are subject to near-total darkness. Only the light from occasional bioluminescent organisms reaches their eyes. Since there are no plants at these depths, food input comes largely in the form of detritus—dead organic remains that fall from above.

Bioluminescence is a fairly common and widespread phenomenon in the sea. Some fish that carry out daily vertical migrations may achieve a form of camouflage through countershading. Others may use the light to lure prey, to scare away predators, or to recognize others of its species. For some luminescent marine life, the light has no known function.

The marine environment is extremely varied and consequently its inhabitants are also varied. Knowledge of marine ecology is important in fisheries management, since the production, distribution, and migrations of economically important fish and shellfish are influenced by the ever-changing character of the marine environment.

IMPORTANT TERMS

abyssal
abyssopelagic
aphotic
bathyal
bathypelagic
benthic
continuous plankton recorder
deep scattering layer
disphotic
epifauna
epipelagic
euphotic
Gause's Law
hadal
hadopelagic
infauna
littoral

luciferase
luciferin
meroplankton
mesopelagic
nekton
neritic
niche
oceanic
pelagic.
photophore
phytoplankton
plankton
pleuston
splash zone
sublittoral
zooplankton

1. What causes deep scattering layers and why do they migrate?
2. Why do vertically migrating fish descend to greater depths in oceanic than in neritic regions?
3. How might the migration of American and European eels be related to continental drift?
4. How might the operation of an offshore nuclear power plant affect the migration of fish?
5. Why is bioluminescence common among mesopelagic fish?
6. Describe the abyssal environment with respect to temperature, salinity, light, and food.
7. Discuss possible roles for photophores of the flashlight fish (Photoblepharon).
8. How are organisms adapted to a pelagic or benthic way of life?
9. Describe the classification of marine environment based on light and temperature.

Briggs, J.C. 1974. *Marine Zoogeography.* New York, McGraw-Hill.
Carefoot, Thomas. 1977. *Pacific Seashores : A Guide to Intertidal Ecology.* Seattle, University of Washington Press.
Carr, Archie. 1965 (May). The navigation of the Green Turtle. *Scientific American.*
Costlow, J. D., and Richard Barber. 1980. IDOE biology programs. *Oceanus* 23 (1): 52–61.
Dietz, Robert S. 1962 (August). Deep scattering layers. *Scientific American.*
Ekman, Sven. 1953. *Zoogeography of the Sea.* London, Sidgewick and Jackson.
Hardy, Alister C. 1956. *The Open Sea, the World of Plankton.* Boston, Houghton Mifflin.
———. 1959. *The Open Sea, Fish and Fisheries.* Boston, Houghton Mifflin.
Harvey, E. N. 1952. *Bioluminescence.* New York, Academic Press.
Hedgpeth, Joel W. (ed.). 1957. *Treatise on Marine Ecology and Paleoecology, I: Ecology.* New York, Geological Society of America (memoir 67).
Limbaugh, Conrad. 1961 (August). Cleaning symbiosis. *Scientific American.*
Southward, A.J. 1965. *Life on the Seashore.* Cambridge, Mass., Harvard University Press.
Sumich, J.L. 1976. *Biology of Marine Life.* Dubuque, Wm. C. Brown.
Tait, R. V., and R. S. De Santo. 1972. *Elements of Marine Ecology.* New York, Springer-Verlag.
Yonge, C. M. 1949. *The Seashore.* London, Collins.

Life in the Sea

11

Anchovy school in the North Atlantic. (Photo courtesy of NOAA.)

The oceans contain a bewildering assortment of living things—bewildering because of the sheer variety and also because we simply do not see most marine creatures in our everyday lives. How can we study these organisms? One way to tackle the problem is to look at the roles played by various organisms in marine ecosystems.

Perhaps the most convenient approach is to classify organisms according to the way in which they acquire or manufacture food. Thus, they may be *producers, consumers,* or *decomposers.* Plants produce organic matter using the sun's energy and inorganic nutrients, whereas animals consume organic matter, deriving their energy from the organisms that they eat. Bacteria and fungi function as a part of the decomposition process, whereby dead organic matter (detritus) is broken down into organic nutrients that may be used by plants in producing new organic matter.

The role or function of an organism in the sea (its niche) depends on its relationship to the physical environment and to other organisms through the *food web* (feeding relationships, discussed in the next chapter) and *symbiotic* relationships (close interactions among two or more species).

Symbiotic relationships may take several forms. If one species benefits and the other is harmed, the relationship is called *parasitism,* as in the case of a roundworm living in the intestine of a fish. In *mutualism,* both species benefit. An example of this is the relationship of certain single-celled algae to some sea anemones and corals. The algae live within the tissues of the host animal and supply it with organic substances that stimulate its growth. The anemone, in turn, gives the algae a home and inorganic nutrients. In *commensalism,* one species benefits and the other is neither harmed nor helped. For example, some marine worms form a burrow which is shared by other species such as shrimp or crabs. The worm is not harmed, but the others have a free home and may even share the worm's food.

All of the interactions among organisms constitute a closely interrelated community of producers, consumers, and decomposers, the composition of which is largely determined by the physical environment.

In this chapter, we will discuss some of the more familiar or important kinds of marine organisms. Familiar species are not necessarily the most important to the ecosystem, because they may have low productivity or they may not be available to marine food webs. The tall plume grass *Phragmites,* a common plant of disturbed salt marshes, is of questionable value. Some researchers regard it as worthless except as a source of detritus; others point out its value as a nesting ground for birds. On the other hand, certain species may be unknown to most people, but they are of tremendous importance to life in the sea. For example, the bristlemouth *(Cyclothone),* a small oceanic fish, is probably the most abundant vertebrate in the world, but few people have seen one as they are not found near shore. Some organisms, such as bacteria and diatoms, are minute and hence rarely noticed except when present in numbers great enough to discolor the water. Bacteria are essential to the recycling of nutrients used by plants. Diatoms, which are tiny single-celled plants, are probably the most important producers of organic matter in the sea.

PRODUCERS

Plants are frequently called the *primary producers,* since they incorporate light energy with carbon dioxide, water, and inorganic nutrients in the presence of chlorophyll to make organic matter, which directly or indirectly becomes the food for the rest of the organisms. This is true in the oceans as well as in fresh water and on land. Estimates vary as to the relative amounts of photosynthesis in the sea versus photosynthesis on land and in fresh water. One of the most acceptable is about 50 per cent for the sea and 50 per cent for land and fresh water.

In littoral environments, simple marine plants (seaweeds or algae) often flourish on rocks, pilings, and other solid objects. In the open waters of the euphotic zone, many kinds of drifting plants (phytoplankton), mostly microscopic in size, fill the role of primary producer. In addition, flowering plants such as salt-tolerant grasses are found growing in shallow bays, estuaries, and salt marshes. They also provide food and shelter for a wide variety of animal life. The major groups of plants are categorized into *divisions* (equivalent to the term *phyla,* which is applied to animals) based on their structure, mode of reproduction, and the nature of their pigments. The marine plant divisions may be roughly classified into the seaweed divisions, the phytoplankton divisions, and the flowering plants.

Seaweed

Seaweeds are marine plants familiar to most who visit a rocky coast. These are generally simple attached plants. The attachment region is called the holdfast and differs from a root in not absorbing nutrients from the substrate. The colors of the seaweeds are quite varied, suggesting terms such as "brown algae" (division Phaeophyta), and "red algae" (division Rhodophyta). All seaweeds contain the green pigment chlorophyll, required in photosynthesis, although other pigments may be present in large enough amounts to mask the green chlorophyll and give the seaweed a brownish, reddish, or blue-green color. Although they are usually attached to solid objects and to the shallower parts of the sea floor, some, such as the brown alga *Sargassum,* drift on the surface of the open sea and hence may be regarded as phytoplankton. The Sargasso Sea is a region of the central Atlantic Ocean famed for the large masses of *Sargassum* that drift there.

BLUE-GREEN ALGAE (DIVISION CYANOPHYTA)

Blue-green algae are commonly seen as very dark encrusting masses of tiny filaments, a few millimeters in length, in the splash zone of rocks or wood pilings. Some species, however, may grow as large erect seaweeds. Most appear black to the naked eye, but closer examination of a thinned-out sample under an ordinary light microscope frequently reveals a blue-green color. No internal structure is seen, but sometimes a protective gelatinous sheath surrounding the chains of cells can be observed. In order to see the internal structure of the cells (Fig. 11.1), it is necessary to use a powerful electron microscope. Because of their simplicity, they are thought to be similar to the first plants to inhabit the earth.

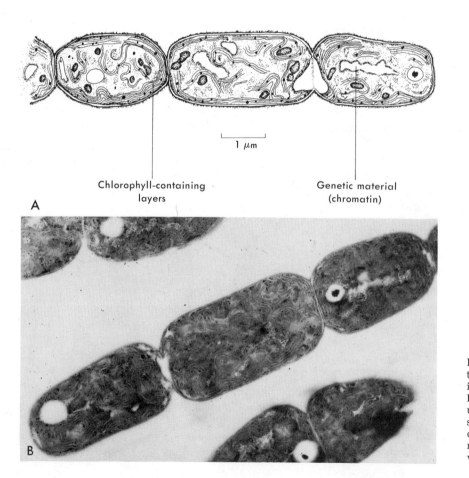

1 μm

Chlorophyll-containing
layers

Genetic material
(chromatin)

A

B

FIGURE 11.1 *A*, Structure of *Anabaena*, a typical blue-green alga. *B*, Electron micrograph of ultrathin longitudinal section of *Anabaena*. (*B* courtesy of G. B. Chapman, Georgetown University.)

Green Algae (Division Chlorophyta)

The members of this division (Fig. 11.2) that are perhaps the most obvious appear as green sheets (sea lettuce, *Ulva*) or tubes (sea intestines or *Enteromorpha intestinales*) or clumps of thin filaments (for example, *Cladophora*). *Ulva* may grow in sheets up to a meter or more in length and may be an important food for intertidal animals such as amphipods and sea urchins.

Brown Algae (Division Phaeophyta)

"Brown algae" (Fig. 11.3) are more brownish-green than brown. The largest of the algae belong to this division, including the kelp *Macrocystis* (up to 60 meters in length), which is important as a source

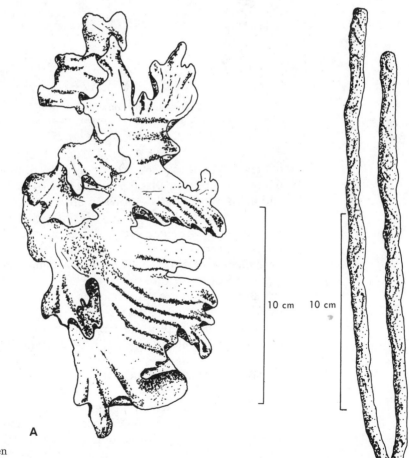

10 cm   10 cm

FIGURE 11.2 *A*, The green alga *Ulva* (sea lettuce). *B*, *Enteromorpha*, a tubular-shaped green alga.

A

B

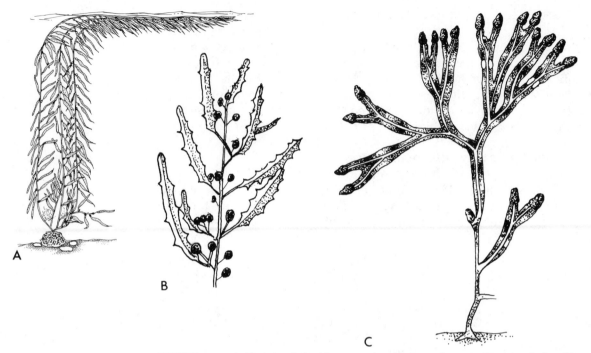

FIGURE 11.3 *A*, The giant kelp, *Marcrocystis*, which may be up to 60 meters in length; *B*, *Sargassum*; *C*, The common rock-weed *Fucus*. (*B* and *C* are shown approximately natural size.)

of "algin," used as a smoothing agent in commercial ice cream and for many other things. Bladder wrack *(Fucus)* is common on rocky coasts and in harbors, where it grows in great masses on pilings. Some species of *Sargassum* are the dominant algae floating in the Sargasso Sea.

Large brown algae such as those mentioned above have gas-filled floats built into their blades, keeping the plants well exposed to light near the water's surface. Thus they become the dominant seaweeds in water too deep for other algae.

Brown algae are important as a source of food for certain snails living on their blades. Also, a wide variety of other algae and animals live among the kelp or are attached to them. The food and protection offered by kelp attracts a tight little community of great diversity.

RED ALGAE (DIVISION RHODOPHYTA)

The red algae, which are generally quite sensitive to exposure, are usually found only at lower levels on intertidal rocks or in tide pools. These algae vary from delicate lacy forms to flat sheets, as in *Porphyra*. Some species that are common in tide pools and on coral reefs have a tough calcareous covering.

Several species of red algae are of special economic importance.

Gelidium, for example, is the source of agar, a non-nutritious medium used in culturing bacteria. *Porphyra* is cultured in the Orient as a source of food. It resembles the sea lettuce *Ulva* and is used in "seaweed soups." Red algae and other seaweeds frequently provide a home for small marine animals (Fig. 11.4).

Phytoplankton

Even though the most easily observed marine plants are the seaweeds, the most abundant with regard both to number and biomass are the phytoplankton, minute in size (usually less than 0.1 mm), widely distributed, and tremendous in numbers. The phytoplankton discussed in the following sections belong to two divisions, the Chrysophyta (diatoms, silicoflagellates, and coccolithophores) and the Pyrrophyta (dinoflagellates).

CHRYSOPHYTA

Diatoms.   The most prominent marine plants are generally the diatoms (Fig. 11.5), sometimes called the "grass of the sea" because of their importance as a food for some small shrimplike herbivores (copepods), which are called the "cattle of the sea." (Perhaps fish should be called the "wolves of the sea" since many feed on herbivores.)

FIGURE 11.4 The red alga *Ceramium*, harboring two unusual attached jellyfish. The animal on the left is about two cm in diameter. (Photo © Douglas P. Wilson.)

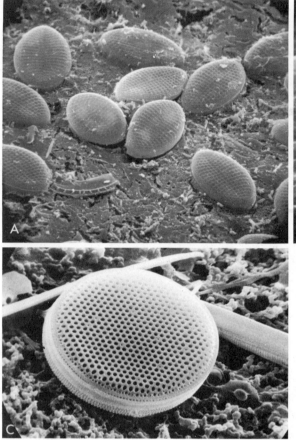

FIGURE 11.5 A variety of microscopic diatoms as revealed by the scanning electron microscope (magnified about 1000×). (*A* and *B* courtesy of Paul E. Hargraves, University of Rhode Island; *C* courtesy of Susumu Honjo, Woods Hole Oceanographic Institution.)

Diatoms are brownish in color and are surrounded by a glassy "shell" made of silicon dioxide. The shell remains after death and contributes to the composition of the sediment. If diatoms are the predominant constituent, the sediment is called *diatomaceous ooze.* These deposits may be 1000 meters thick and may contain more than 5 million shells per cubic millimeter.

The "shell" of a diatom is built much like two Petri dishes (Fig. 11.6). One half is small enough to fit inside the larger half. The walls are porous, allowing dissolved gases and nutrients to pass through. When the cell divides asexually (the usual means of reproduction), each original shell becomes the outer half of the new shell. A new smaller half-shell is secreted. The average size of the cells that are produced decreases until they can get no smaller and still live. These small cells form a "spore" and start growing, shedding the old shell. Then they form a new, larger shell and continue as a renewed "large" cell. Note that there

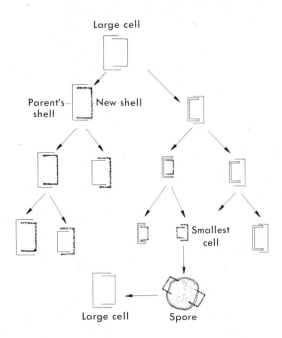

Large cell

Parent's shell   New shell

Smallest cell

Large cell   Spore

FIGURE 11.6 Asexual reproduction in a diatom. Note the overlapping half of the shell and decreasing size of successive "offspring" owing to the production of new shells inside the margins of the parents' shells. When a cell is produced that is too small to divide, it forms a spore that grows and produces a new large cell.

is no natural death in this type of reproduction. These organisms are, in a sense, immortal. Death may come when they are eaten, are poisoned by pollutants, or sink below the photic zone. This feature is characteristic of most phytoplankton that reproduce principally by asexual means.

Long ornamented spines help to retard the sinking of some diatoms, especially in warm water, which is less viscous than cold water. Some diatoms attach to other organisms. Others form a slimy coating on rocks and wood pilings.

When nutrients become sparse or conditions become unfavorable for the support of great diatom populations, other phytoplankton become dominant. These are frequently small cells with whiplike flagellae that beat and move them through water. Three major groups of flagellates are discussed below.

Silicoflagellates. These organisms are widely distributed as a minor constituent of plankton and may be quite abundant locally (Fig. 11.7). They contribute their glassy SiO_2 skeletons to the sediment when they die. In the North Sea, these chrysophytes may be the dominant primary producers among the plankton.

Coccolithophores. These tiny round flagellates are usually less than 10 μm (0.01 mm) in diameter with minute, buttonlike plates made of calcium carbonate (Figure 11.8). These "buttons" may be abundant in the sediments of warm shallow water (calcium carbonate tends to dissolve in cold or deep water), reflecting their sometimes great abundance in the water above.

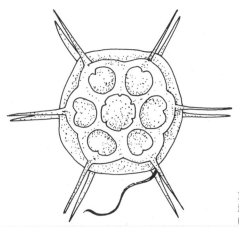

FIGURE 11.7 The silicoflagellate *Distephanus* is about 0.1 mm in diameter, including spines. (After Hardy, 1956.)

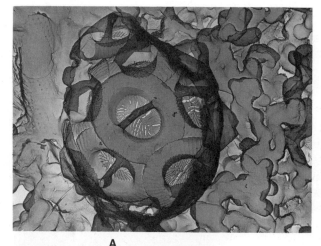

A

B

C

FIGURE 11.8 Coccolithophores with many calcareous plates, or coccoliths, sometimes referred to as "collar buttons." (*A*, transmission electron micrograph; *B* and *C*, scanning electron micrographs, courtesy of Susumo Honjo, Woods Hole Oceanographic Institution.)

THE ELECTRON MICROSCOPE

Several of the photographs in this chapter were produced with the aid of electron microscopes. Most of these organisms are less than a millimeter in diameter, and some, such as bacteria, may be less than 0.001 mm (1 micrometer) in diameter. Most conventional light microscopes effectively magnify only as much as 1000 times (1000×). Greater magnification is needed to show the internal structure of simple organisms such as blue-green algae and the surface features of the minute plankton. Electron microscopes have the advantage of using a beam of electrons, instead of light, to magnify objects as much as tens of thousands of times. The transmission electron microscope passes the electrons through a thin piece or slice of the object and focuses on a fluorescent screen or a photographic plate. This is an excellent technique for showing internal structure. The scanning electron microscope scans a beam of electrons over the surface of the object and causes other electrons to be thrown off the surface and produce an image of the organism on a television screen. Then a photograph is taken of the television image. This is ideal for showing great surface detail of small opaque organisms, such as plankton, and the skeletons of ooze-forming organisms, such as diatoms, foraminifera, and radiolaria.

DINOFLAGELLATES

Among the most abundant of the phytoplankton are the dinoflagellates (Fig. 11.9). They may be naked or armored with heavy, often ornamented cellulose plates. Most are tiny, generally less than 0.1 mm, and move about by means of a pair of flagellae. They are more abundant in warm water but may grow into blooms in temperate water when conditions are just right and may cause "red tides," some of which are poisonous. Red tides are discussed in detail in Chapter 12.

Many dinoflagellates are bioluminescent (see Chapter 10), producing their own cool blue-white light. This light is frequently observed illuminating the surf or the wake of ships on moonless nights. The division name for dinoflagellates, Pyrrophyta ("fire plants"), is especially appropriate for these light-giving phytoplankton.

Flowering Plants (Division Anthophyta)

Flowering plants are absent in the open ocean, as they require a substrate for attachment and light for photosynthesis. However, some flowering plants may be found in the marine environment. Eelgrass is frequently seen in rather protected shallow coves with sandy bottoms. Mangroves frequently grow in shallow water along subtropical shores. Salt marshes are usually covered with grasses that are partly submerged by the high tides. These plants contribute much detritus to the food webs of shallow estuarine bays.

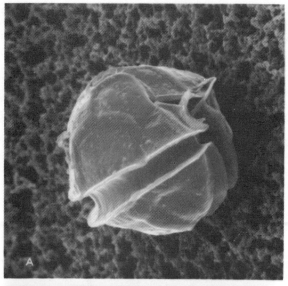

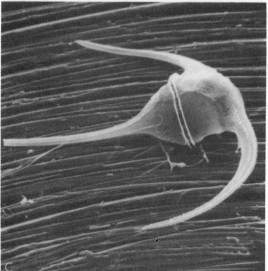

FIGURE 11.9 *A*, Red tide–producing dinoflagellate, *Goniaulax tamarensis*. *B*, The spindle-shaped dinoflagellate *Oxytoxum*, along with two diatoms and several coccoliths. *C*, *Ceratium*, a common dinoflagellate of coastal waters. (Scanning electron micrographs, magnified about 1000×: *A* courtesy of SIELAB, University of Rhode Island; *B* courtesy of Susumu Honjo, Woods Hole Oceanographic Institution; *C* courtesy of Paul E. Hargraves, University of Rhode Island.)

CONSUMERS

Animals are consumers; that is, they require a source of organic food that they themselves cannot make. This is in contrast with the plants, which can utilize the sun's energy and inorganic nutrients for growth. Animals that feed on plants are termed herbivores *(secondary producers).* Those that eat other animals are termed carnivores.

For classification purposes, the animals are divided into *phyla* (similar to the divisions of plants) and the phyla into *classes,* based mainly on structure and reproduction and in some cases on the method of feeding. Consumers vary considerably in their mode of feeding. Some of the animals in the sea are *filter-feeders:* they strain the food (phytoplankton, zooplankton, and detritus) out of the water with feather-like appendages, gills, or tentacles around or in the mouth. These animals may be tiny, such as the copepods, or huge, as are the baleen whales. The *scavengers,* including many crabs, feed on the dead remains of other organisms. *Deposit-feeders* eat the sediment and digest the organic food, living and dead, contained in the bottom materials, Many of the so-called worms survive in this manner. *Browsers,* such as many snails, cover the surface of rocks and pilings, scraping off and eating the algae that grow thereon. *Predators* consume other living animals. They may actively pursue their prey, as many fish do, or they may engulf prey that come their way by chance. For example, jellyfish paralyze the unwary organism with stinging cells located on the tentacles and then eat it.

The Simple Life

These organisms, including protozoans, sponges, and the so-called jellyfish, are among the simplest animals. They have no brain and therefore their behavior is relatively simple; feeding is semiautomatic, responding to simple stimuli such as food, touch, and light.

SINGLE-CELLED ANIMALS (PHYLUM PROTOZOA)

These little one-celled animals (Fig. 11.10) are usually less than a millimeter in size and are abundant in the sea, feeding on detritus, bacteria, and probably even dissolved organic matter. Ciliates are fairly common among grains of sand and among detritus debris and masses of hydroids and other benthic organisms. They move by waving tiny whiplike threads called cilia on the surface of the cell. Radiolarians and foraminiferans are amoebalike protozoans that are common in the plankton. Foraminiferans are also found in the sediment. Radiolarians have SiO_2 skeletons that resemble cages and intricate baskets, and foraminiferans have porous $CaCO_3$ shells. Both of these kinds of skeletons or shells contribute to the formation of the sediment in the areas where they abound. (See also Chapter 3.)

FIGURE 11.10 Single-celled marine animals. *A*, Stalked, ciliated protozoa around the openings of "moss animals" (bryozoa), 120×. *B*, Planktonic spiny foraminiferan. *C* and *D*, Foraminifera with calcareous skeleta. *E* and *F*, Radiolaria with siliceous (opal) skeleta. (*A* courtesy of J. McN. Sieburth, from Sea Microbes. Williams and Wilkins Co., Baltimore, Md., 1976. P. W. Johnson and J. McN. Sieburth, scanning electron micrograph; *B*, scanning electron micrograph courtesy of Allan W. H. Bé, Lamont-Dougherty Geological Observatory of Columbia University; *C–D*, scanning electron micrographs courtesy of James Kennett, University of Rhode Island; *E–F*, scanning electron micrographs courtesy of ETEC. Corp.)

SPONGES (PHYLUM PORIFERA)

Sponges were familiar to past generations in the form of bath sponges; now, natural sponges have been replaced almost totally by synthetic types. Only the flexible but tough spongin skeleton was used as a bath sponge after the living parts were removed by drying and washing. The cells, called amebocytes and collar cells, are similar to amoebas and flagellates. The collar cells have whiplike flagellae (similar to, but much longer than, cilia) that beat, driving water through canals in the sponge. As the water flows through, food is filtered and transferred to the amebocytes, which engulf and digest it.

Sponges are quite variable in shape and size, ranging from thin encrusting forms, which may be yellow, orange, red, or bluish green, to large grayish masses one meter in diameter. Some may be quite prickly owing to tiny, sharp, calcareous or glassy spicules, which, along with spongin, make up the structural support for sponges. In some glass sponges, the spicules form an elaborate network (Fig. 11.11).

FIGURE 11.11 The skeleton of the glass sponge *Euplectella*, about one-half natural size. (Courtesy of the American Museum of Natural History.)

JELLYFISH AND THEIR KIN (PHYLUM COELENTERATA)

Coelenterates are a diverse group of simple animals with radial symmetry, tentacles, and soft, often jellylike bodies. They are carnivorous, stinging their prey with poison darts called nematocysts (Fig. 11.12) and then manipulating the food into their blind digestive tract with their flexible tentacles. They have two distinct body plans: medusae (jellyfish) and polyps (hydroids, sea anemones, and corals), as shown in Figure 11.13. A polyp is like an upside-down medusa. Many small jellyfish have a sessile polyp stage. They have an alternation of asexual (polyp) and sexual (medusa) generations. In other words, the jellyfish that we see drifting in the water is the sexual generation. On the other hand, sea anemones and corals have only the polyp stage, which may reproduce asexually or sexually.

Coelenterates are found in all oceans and at all depths. They feed on small crustaceans, fish, or any other animals that they can ingest. Because they feed on the same food as many fish, they may be serious competitors and may deplete the supply of the fish's food. Some fish eat coral polyps, and sea slugs, as well as a few arthropods such as "sea spiders," eat hydroid polyps. Otherwise coelenterates generally are not eaten.

Some drifting or floating jellyfish and Portuguese men-of-war produce such a strong sting that they may cause painful welts on human swimmers. The cubomedusa jellyfish (especially *Chironex,* the sea wasp), found off the coast of Australia, may even cause death.

Corals are close relatives of the sea anemones. They differ, however, in that they form a protective calcareous exoskeleton around their polyps. Many corals grow in tremendous colonies that form reefs (Fig. 11.14). Others may grow as isolated polyps.

Reef-building coral generally requires warm (greater than 18° C) clear water such as is found in the upper 20 meters of the tropics. These coral polyps contain symbiotic algae, which apparently supply dissolved organic matter that stimulates the growth of the coral. The algae benefit in having a home and nutrients excreted by the coral. The algae require light, and therefore so do the corals.

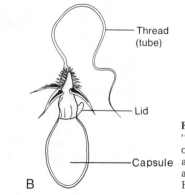

Thread (tube)

Lid

Capsule

Cnidocyte
Capsule
Thread
Nucleus

A   B

FIGURE 11.12 A coelenterate nematocyst, its "poison dart" undischarged (*A*), and discharged (*B*). The nematocysts are microscopic and are most abundant on the tentacles and around the mouth of the coelenterate. (After Hyman, 1940.)

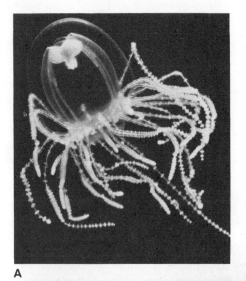

A

B

C

D

FIGURE 11.13 A variety of coelenterates. *A*, Hydromedusa, *Gonionemus*. *B*, A common sea anemone. *C*, The Portuguese man-of-war, *Physalia*, eating a fish. *D*, Polyps of the star coral. *E*, A stalked sea pen at 5066 meters on the abyssal plain west of Africa. (*A* photo © Douglas P. Wilson; *B* photo by Harold Wes Pratt; *C* courtesy of American Museum of Natural History; *D* courtesy of New York Zoological Society; *E* courtesy of Naval Oceanographic Office.)

E

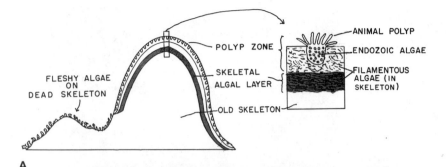

The living, reef-building coral is found only at the surface outer edge of the mass of coral rock; the interior is made up of dead coral skeletons. On or near the surface of dead coral rock grow algae, some filamentous, some fleshy, and others with calcareous "skeletons" of their own. This concentration of epibenthic life results in a characteristic environment dominated by the reef life but including a wide variety of fish that are adapted to feeding on the reef plants and animals. The coral polyps provide food for some fish and seastars.

COMB JELLIES (PHYLUM CTENOPHORA)

Comb jellies are frequently seen as small globs of jelly, a few centimeters across, washed ashore by the surf. They are small, round or oval jellyfish-like animals with eight vertical rows of ciliated comb-plates, their locomotive organs (Fig. 11.15).

Ecologically, comb jellies are similar to medusae. They are voracious carnivores and are rarely eaten by other animals. They differ from the true jellyfish in that they tend to be roundish or elongated, rather than bell-shaped. They have no sessile polyp stage and no stinging cells but only sticky-cells, with which they ensnare their prey (mostly small crustaceans and fish larvae). Their distribution is very patchy, and hence they appear to be absent one day and abundant the next. When they are abundant, they may foul fishing nets and decimate the food of fish.

Worms

What we commonly call "worms" really comprise a number of phyla that share similar shape and habitats. Most worms are long and slender and live in the mud or sand. Some live in tubes that they construct out of sand or other materials. Several of these phyla include worms that are parasitic or have commensal relationships with other animals. The shape of worms is well adapted for crawling in and out of small spaces. A few, including some flat worms, segmented worms, and arrow worms, no longer are benthic and have assumed a pelagic swimming life.

FIGURE 11.15 A comb jelly, *Pleurobrachia*, about three cm in diameter. This one is also known as the sea-gooseberry. (Photo by Harold Wes Pratt.)

FLAT WORMS (PHYLUM PLATYHELMINTHES)

Flat worms (Fig. 11.16) are fairly common on or under rocks, in shallow water, and on the sediment. They also crawl around benthic growths of hydroids and other small attached organisms. Most flat worms are carnivorous and may eat dead fish, clams, and other soft-bodied sessile animals. One genus, *Bdelloura*, lives on the gills of the horseshoe crab as a commensal. Many flat worms are small, but some species may reach five cm in length. Some may swim through the water by undulating their leaf-shaped bodies. Although many flat worms live in fresh water, more are marine. Most of the marine species are free-living; however, some live as parasites in fishes and other marine animals.

ROUND WORMS (PHYLUM NEMATHELMINTHES)

Round worms are small, round, unsegmented worms with ends that generally are pointed. They survive well under conditions that would be lethal to most animals. Round worms are important in the decomposition of detritus and are common in sediments that are rich in organic matter but which contain little oxygen. Many species of round worms are parasites of other organisms.

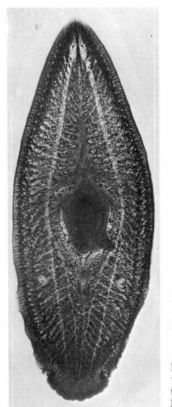

FIGURE 11.16 The flatworm *Bdelloura* (about 1.5 cm long) found as a commensal on the gills of the horseshoe crab *Limulus*. (Courtesy of Ward's Natural Science Establishment.)

SEGMENTED WORMS (PHYLUM ANNELIDA)

Most of the segmented worms (Fig. 11.17) in the sea fall into the class Polychaeta, having many spines that emerge from lateral projections on each segment. They may be predators, filter-feeders, or deposit-feeders. The predaceous worms have strong jaws on the head. Filter-feeders may use nets composed of a ring of featherlike tentacles to capture their prey. Hence they have popular names such as "plume worms," "fan worms," or "feather-duster worms." The plume worms live in mucus-lined burrows or in tubes that the worm makes out of calcium carbonate, sand, or debris cemented by a mucus that it secretes. Some tube worms form sandy tubes that accumulate in cemented masses and form distinctive reefs which provide homes for many other organisms.

Deposit-feeders (Fig. 11.18) consume detritus that may be found on or in the sediment. They may use tentacles to pick up the food or may eat the sand or mud and then digest the food that it contains. Some, such as the lugworm, *Arenicola,* leave coils of sandy feces at the rear end of the burrow. It pumps water in at the tail end of the burrow and forces it through the sand at the head end. Food is filtered out by the sand which is then eaten by the worm. Then the worm backs up to the burrow opening to defecate.

Some polychaetes share their burrows in a commensal relationship with other organisms. One such relationship exists among the tube-dwelling, filter-feeding polychaete *Chaetopterus* and small crabs that share the worm's burrow and food. The worm apparently is neither benefited nor harmed by the crab. The movements of the worm (Fig. 11.19), however, bring in plenty of food for both the worm and the crab.

ARROW WORMS (PHYLUM CHAETOGNATHA)

Arrow worms are among the most voracious predators for their size (about 1 to 2 cm). They may compete with small fish for their copepod food or may even eat fish larvae that are as long as the arrow worms themselves (Fig. 11.20). They behave like arrows but look more like torpedoes. They can dart forward and grasp their prey with strong spines attached to their head. They then manipulate the prey into the stomach, sometimes bending a fish larva in half. They are common in the epipelagic zone of most ocean water. In terms of numbers of individuals, arrow worms are thought to be the most abundant carnivores in the sea.

SAUSAGE WORMS (PHYLUM ECHIUROIDEA)

Most of these saclike worms are found in shallow water, but some may inhabit depths greater than 8000 meters. The majority feed on

Text continued on page 341.

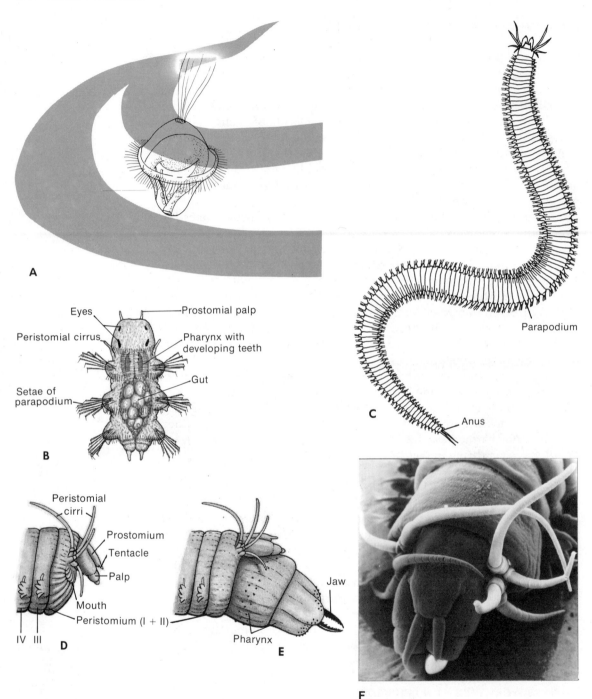

FIGURE 11.17 Larvae and adult of a common predaceous polychaete worm, *Nereis*. The trochophore larva *(A)* develops into a nektochaete larva *(B)*, which matures into an adult polychaete *(C)*. *D*, A closer view of the head. *E*, Same, with jaws extended. *F*, An electron-micrographic view. *(A–E* from Villee et al., 1973; *F* courtesy of J. M. Sieburth, *Microbial Seascapes*, 1975, University Park Press, Baltimore.

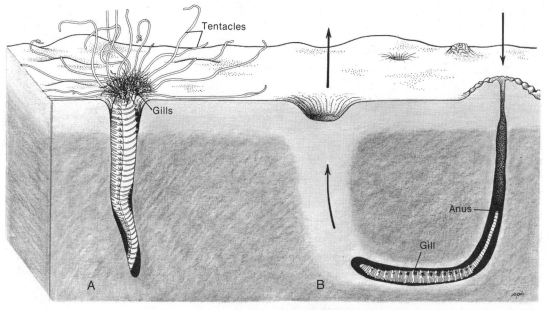

FIGURE 11.18 Two deposit-feeding poly-chaetes. *A, Amphitrite; B*, The lugworm. *Aren-icola*. Arrows indicate the direction of water flow (*A* modified from Wells; *B* after Villee et al., 1973.)

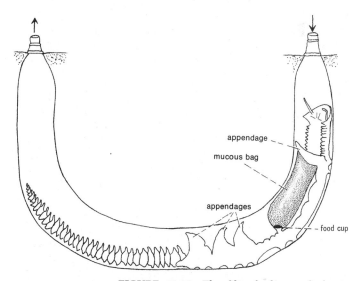

FIGURE 11.19 The filter-feeding polychaete *Chaetopterus* in its U-shaped burrow. (After MacGinitie. From Barnes, 1974.)

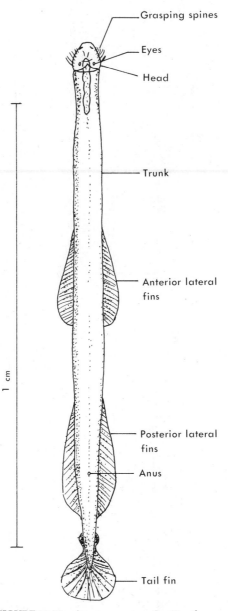

FIGURE 11.20   An arrow worm, *Sagitta elegans.*
(After Hardy, 1956.)

PLATE 17 Mont Saint Michel, France, connected to the mainland by intertidal sand flats, is an island at high tide. (Photo courtesy French Government Tourist Office)

PLATE 18 A *Spartina* marsh along Dennis Creek, a small estuary in southern New Jersey. (Photo by J.M. McCormick)

PLATE 19 A rocky coast at Acadia National Park, Maine. (Photo courtesy National Park Service)

PLATE 20 Waves pounding the Acadia shore. The dark patches in the foreground are composed of marine life adapted to the intertidal, wave-swept rocks. (Photo courtesy National Park Service)

PLATE 21 Egrets flying over a mangrove swamp in the Florida Everglades. (Photo courtesy National Park Service)

PLATE 22 The fiddler crab, *Uca minax,* inhabits muddy burrows in *Spartina* marshes. (Photo by J.M. McCormick)

PLATE 23 A clam bed in the Galapagos thermal vent area. (Photo courtesy WHOI)

PLATE 24 Crabs near the Galapagos thermal vents. (Photo by Robert Hessler, Scripps Institution of Oceanography, courtesy WHOI)

PLATE 25 The larva of a polychaete worm. (Photo by Tom Pendlebury)

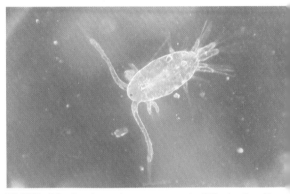

PLATE 26 An immature copepod. (Photo by Tom Pendlebury)

PLATE 27 A colonial coelenterate, related to the Portuguese man of war, in the Galapagos Rift area. (Photo by Al Giddings, Sea Films, Inc.© National Geographic Society)

PLATE 28 A gorgonian or "soft" coral in the Caribbean (Photo by Gray Multer)

PLATE 29   Reef life in Biscayne Bay, Florida. (Photo courtesy National Park Service)

PLATE 30   A nudibranch from Fiji. (Photo by Betty Barnes)

PLATE 31   A filter-feeding serpulid worm on a coral head. (Photo by Gray Multer)

PLATE 32   The spiny, coral-eating sea star *Acanthaster* from Fiji. (Photo by Betty Barnes)

PLATE 33   Two sea stars from Fiji (red and white —*Fromia monilis*; brown and white—*Fromia indica*). (Photo by Betty Barnes)

PLATE 34   Fish swimming around the legs of a drilling rig. (Photo courtesy American Petroleum Institute)

PLATE 35 A lizard fish at the base of a coral reef. (Photo by Gray Multer)

PLATE 36 Two blue-head wrasse approaching the spines of a sea urchin. (Photo by Gray Multer)

PLATE 37 Grunts and a queen angelfish near an offshore drilling rig. (Photo courtesy American Petroleum Institute)

PLATE 38 A school of moonfish. (Photo courtesy American Petroleum Institute)

PLATE 39 Purse seining in Java. (Photo courtesy FAO)

PLATE 40 A trawler in Vietnam. (Photo courtesy FAO)

detritus trapped by a mucus secretion. They have no eyes or sense of hearing. Even though they live at a wide range of depths, including the deep aphotic zones, they have sexual reproduction. Some species have apparently evolved a unique way of getting the male and female together (Fig. 11.21). If a larva settles in an area of the bottom that has a moderate to low carbon dioxide (CO_2) concentration, it develops into a female. If it settles in an area of high CO_2 concentration, it develops into a male. An actively respiring female would have a "cloud" of CO_2 produced around it. This causes a settling larva to develop into a male. The male grows and becomes parasitic on the female sausage worm, ensuring a high probability of reproductive success.

PEANUT WORMS (PHYLUM SIPUNCULIDA)

Peanut worms are fairly common, burrowing in mud and sand or under rocks, from the intertidal zone to depths greater than 4500 meters. Most feed on detritus trapped in mucus on their tentacles, but a few species are predators on other worms. Most peanut worms are less than 30 cm in length (Fig. 11.22).

ACORN WORMS (PHYLUM HEMICHORDATA)

Acorn worms are mostly wormlike animals that burrow in mud and sand or hide under rocks. They are thought by some to be closely related to the chordate phylum, which includes fish, because of the

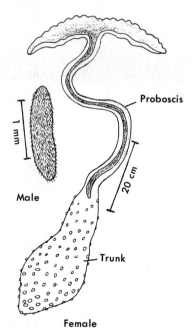

FIGURE 11.21 An echiuroid, the "sausage worm" *Bonellia viridis*. Note the small "pygmy" male, about 1–3 mm long, which is actually a parasite in the larger (about one meter) female. (After MacGinitie and MacGinitie, 1968.)

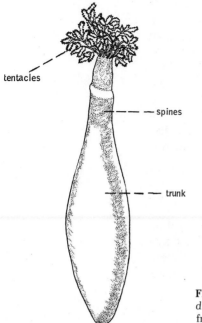

tentacles

spines

trunk

FIGURE 11.22 A typical peanut worm, *Dendrostomum*, about 12 cm long. (After Fisher, from Tétry, 1959.)

presence of gill slits. They are usually between 10 and 50 cm in length, with a stubby proboscis followed by an overlapping collar, giving the forward end the appearance of an acorn (Fig. 11.23). The mouth is located between the proboscis and the collar. Plankton, detritus, and sediment are trapped by mucus on the proboscis and carried to the mouth by the beating of cilia.

Mollusks

One of the characteristics shared by the phylum Mollusca is a soft body frequently covered by a protective shell of $CaCO_3$. A tissue called the *mantle* enshrouds the internal organs of the animal and produces any shell it may have. Most mollusks have a muscular "foot" that is used in locomotion. Most are adapted to life in the benthic environments, where they burrow into sediment, crawl over the bottom, or are attached to rocks and other substrata (Fig. 11.24). In the squid and octopi, however, the foot has been modified into arms and a jet-propulsion siphon, which enables them to swim through the water.

Table 11.1 presents characteristics of the most common classes of mollusks.

Snails, limpets, and chitons are generally browsers, crawling over the substrate, scraping food into their mouths with a filelike radula (Fig. 11.25). Some snails, such as the oyster drill, use their radula to bore

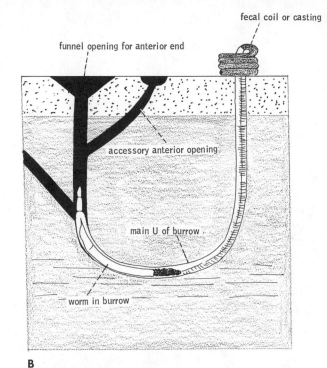

funnel opening for anterior end

fecal coil or casting

accessory anterior opening

main U of burrow

worm in burrow

B

A

C

FIGURE 11.23 *A,* A shallow-water acorn worm, *Saccoglossus; B,* the burrow system of an acorn worm; *C,* An abyssal acorn worm with its trail of feces. (*B* after Stiasny from Hyman, 1959; *C* courtesy of Lamont-Doherty Geological Observatory of Columbia University.)

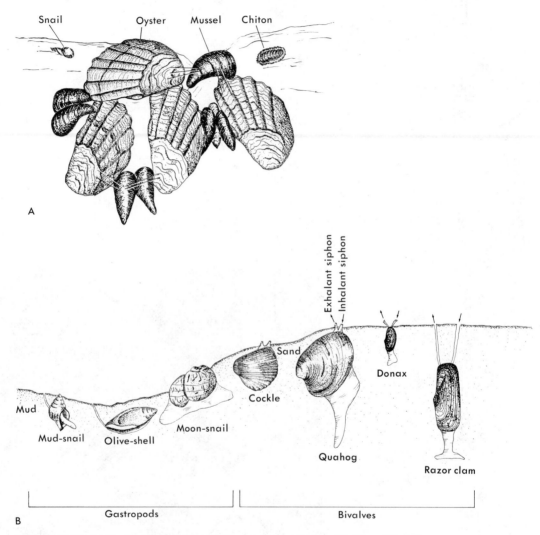

Snail   Oyster   Mussel   Chiton

A

Exhalant siphon   Inhalant siphon

Sand

Donax

Mud

Cockle

Mud-snail   Olive-shell   Moon-snail   Quahog   Razor clam

B

Gastropods   Bivalves

FIGURE 11.24 Mollusks on rocks (A) and in the sea-floor sediments (B). (A modified after Villee et al., 1973; B after Pearse, Humm and Wharton, 1942.)

Table 11.1 CHARACTERISTICS OF COMMON CLASSES OF MOLLUSCA

CLASS	COMMON NAMES	CHARACTERISTICS
Polyplacophora "many plates"	chiton	Row of plates along back, sometimes covered by mantle; gills along side of foot
Gastropoda "stomach foot"	snail, limpet, sea slug, pteropod	Single shell, usually a coiled spire (snail), or uncoiled (as in the limpets and pteropods), or missing (sea slugs); digestive tract twisted in most types
Bivalvia "double shell"	clam, oyster	Two shells, joined by a hinge; gills and foot enclosed by shells when closed
Cephalopoda "head foot"	squid, octopus, nautilus	Shell coiled (nautilus), internal (squid), or missing (octopus); eight or ten tentacles with suction cups; generally well-developed eyes

through the shells of oysters and clams and then devour their flesh. Others, such as the mud snail (*Nassarius obsoletus*), are scavengers and deposit-feeders. They thrive in great numbers in shallow mud flats.

Some gastropods, the sea butterflies, or pteropods, have winglike ciliated appendages that are evolved from the foot. These "wings" are used to propel them through the water and to aid in feeding. Pteropods are quite common in the oceanic plankton the world over. Upon death, their shells contribute to the formation of calcareous ooze on the sea floor.

Most clams are benthic filter-feeders, drawing in water through an inhalant siphon and expelling it through an exhalant siphon (Fig. 11.24). As the water passes over the gills, food particles such as phyto-

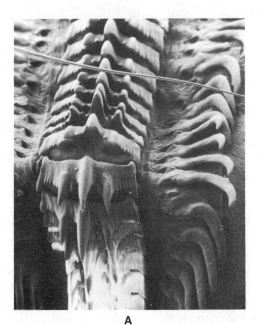

A

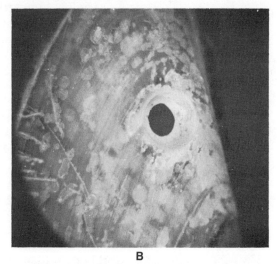

B

FIGURE 11.25 *A*, The file-like radula of a boring snail. *B*, A hole in a bivalve shell, drilled by a boring snail. (*A*, scanning electron micrograph from Carriker, 1969; *B*, photo by Betty M. Barnes.)

plankton are filtered out. Most of the economically important clams live on sandy or muddy bottoms and are thus easy to collect in large numbers. Many others, such as mussels, live on rocky substrates. Yet others have evolved the ability to burrow into rock or wood, thus achieving a great deal of protection.

The wood-boring shipworm (Fig. 11.26) has lost its ability to filter-feed and now eats the sawdust that it produces as it bores into wood. Shipworms may become abundant in untreated wood, causing millions of dollars' worth of damage.

In the cephalopods, which include the octopus, squid, cuttlefish, and nautilus, the foot has evolved into tentacles or arms with suction cups (Fig. 11.27). Of these, only the nautilus has an external shell. In the cuttlefish, the shell has been reduced to a calcareous internal "cuttle bone." Cuttle bone is often hung in the cages of pet birds for the birds to peck and sharpen their beaks. The squid has a thin flexible "pen" resembling the point of an old quill pen. It adds some support to the squid's body. Many cephalopods have an ink sac. If disturbed they can expel an inky cloud, which can confuse a predator and allow the cephalopod to escape. The octopus has lost all traces of a shell. The cephalopods are thought to be the most advanced of the mollusks. They have well-developed brains and eyes and are generally predaceous. Squid and even the clumsy-looking octopus (Fig. 11.28) may become quite active in pursuit of their prey.

Arthropods

The phylum Arthropoda is ecologically one of the most important groups. The greatest diversity is found in the class Crustacea, which includes the copepods, shrimp, lobsters, crabs, and sand fleas. They inhabit all regions of the seas and live as herbivores, carnivores, scavengers, and parasites. Along with the mollusks and fish, they are of great importance to us as a source of food.

Arthropods are characterized by jointed appendages and a hard protective exoskeleton covering their bodies. The life cycle of many arthropods involves several distinct stages (Fig. 11.29). When crabs molt, or shed their old skeleton, they lose their protection and are especially vulnerable and become ready prey for other animals. Many compensate in part for this vulnerability by living in the sand or among rocks.

Copepods, the "cattle of the sea" (Fig. 11.30A and B) are mostly less than 2 mm in length, but some are as long as a large grain of rice, about 5 mm. There are probably far more individual copepods in the world ocean than all the free-living animals on land and fresh water—including the insects—put together. In spite of their small size, they are among the most important herbivores in the sea, providing a key link between the phytoplankton which they consume and the fish that eat them.

Text continued on page 350

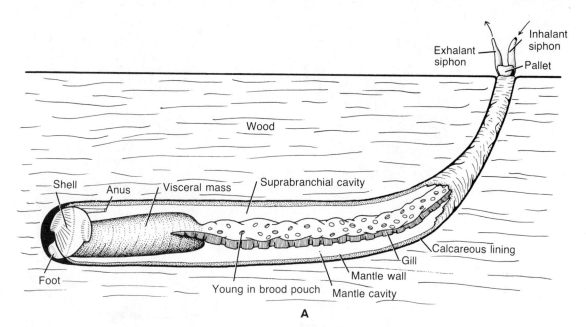

Inhalant
siphon

Exhalant
siphon

Pallet

Wood

Shell

Anus

Visceral mass

Suprabranchial cavity

Foot

Young in brood pouch

Mantle cavity

Mantle wall

Gill

Calcareous lining

A

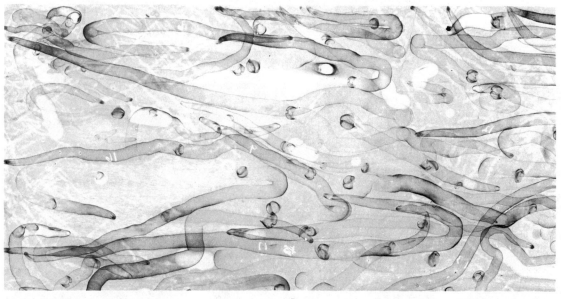

B

FIGURE 11.26   The work of a wood-boring shipworm. *A*, Detail of the bivalve mollusk *Teredo*. *B*,
X-rays of the shipworms in wood. (*A* from Villee,
et al., 1973; *B* photo by C. E. Lane. Copyright
© 1961 by Scientific American, Inc. All rights
reserved.)

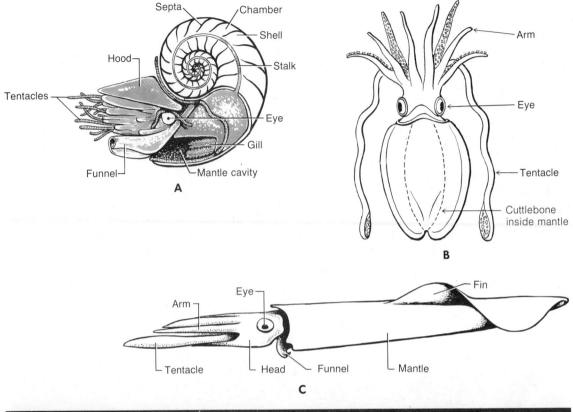

FIGURE 11.27 *A*, Side view of the chambered nautilus, showing internal anatomy. *B*, Top (*A–C* From Villee et al., 1973. *D* photo by Harold Wes Pratt.)

FIGURE 11.28   An octopus in pursuit of a crab. (Photo by Fritz Goro, courtesy of *Life* Magazine, © 1955 Time, Inc.)

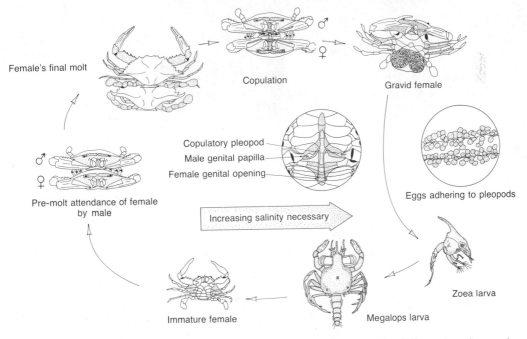

Female's final molt

Copulation

Gravid female

Copulatory pleopod
Male genital papilla
Female genital opening

Eggs adhering to pleopods

Pre-molt attendance of female by male

Increasing salinity necessary

Immature female

Megalops larva

Zoea larva

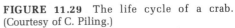

FIGURE 11.29   The life cycle of a crab. (Courtesy of C. Piling.)

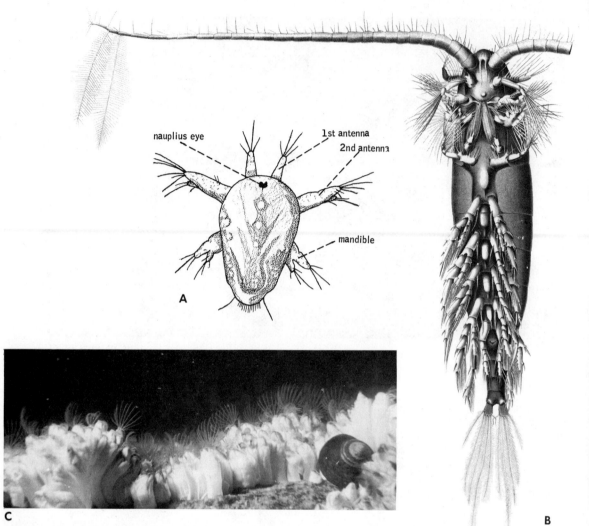

nauplius eye

1st antenna

2nd antenna

mandible

A

C

B

FIGURE 11.30 *A*, Typical copepod larva. *B*, Adult copepod *Calanus*. *C*, Barnacles feeding. (*A* after Green, 1961; *B* after Giesbrecht, 1892; *C* photo by Harold Wes Pratt.)

Copepods swim through the water with jerky movements caused by their thoracic appendages (those on the midsection, directed downward in Fig. 11.30*B*). When they stop swimming, the antennae project outward and away from the body, thus retarding sinkage. Feeding is usually accomplished by filtering the water with the smaller hairy appendages around the mouth. These legs set up water currents that draw plankton, including diatoms, dinoflagellates, and the tiny larvae of other animals, toward the mouth. The food is caught in the hairs of some of the legs and transferred by other appendages into the mouth. Although most copepods are herbivores, some may be carnivores when phytoplankton is scarce, and some are always carnivorous. Other copepods that live in the less productive oceanic water or in the deep sea rely on zooplankton and detritus as sources of food.

As with the copepods, barnacles begin their pelagic life as tiny larvae. Barnacles, however, are destined to mature into attached adults cemented to a substrate (Fig. 11.30C). This may be a wooden piling, intertidal rocks, or even another organism such as a whale. The typical barnacle has a hard calcareous shell around it with a hinged trap door consisting of several calcareous plates. The barnacle is really a rather ordinary crustacean except that it is positioned in its shell lying on its back, and it kicks food into its mouth with its feet. It functions as a filter-feeder, commonly living in the littoral zone.

Shrimp, lobsters, and crabs play an important role as scavengers, helping to prevent an accumulation of detritus on the ocean floor. They also feed on worms, clams, and fish and are, in turn, eaten by other fish and people. The shrimplike krill (Fig. 11.31A) are the principal food of the plankton-feeding baleen whales.

The isopods and amphipods (Fig. 11.31B and C) are common small crustaceans, usually less than 2 cm in length. They occupy similar niches in the benthic environments, especially in littoral communities. They differ in that the isopods have legs that are similar in size and shape (like the terrestrial pill bug), whereas the amphipods have two kinds of legs, dissimilar in size and shape. Some of the amphipod legs are aimed forward and some are aimed backward, enabling amphipods such as sand fleas to hop about on the beach. The food of many isopods and amphipods consists of algae and soft-bodied invertebrates such as hydroid polyps, among which they crawl. Some isopods, such as the wood-boring gribble (Limnoria), may cause damage to marine pilings. This is a role similar to that of the molluscan shipworm.

Sea spiders (pycnogonids) are not true spiders but resemble them in a superficial way, having long legs that enable them to crawl over and prey upon colonies of hydroids and bryozoans that abound on littoral rocks (Fig. 11.31D). The littoral species are usually less than a centimeter in leg span. Those found in the abyssal zone, however, may have a leg span of more than 60 mm (two ft).

The common Atlantic Coast horseshoe crab (Limulus) is not a true crab (Fig. 11.31E). It is believed to be more closely related to spiders, because of similarities in their feeding appendages. It is another scavenger in the littoral zone, also feeding on worms and small clams. Horseshoe crabs bear a striking resemblance to the early arthropods such as the extinct trilobite, suggesting that they have evolved little over a period of millions of years. Because they are not economically important and have few natural enemies, they may survive for millions of years more if their home is not destroyed by human activities.

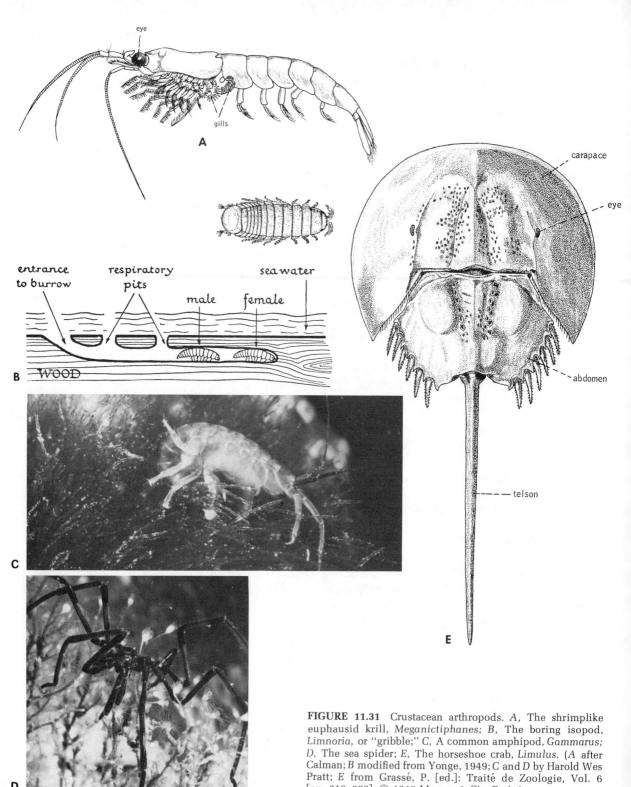

FIGURE 11.31 Crustacean arthropods. *A*, The shrimplike euphausid krill, *Meganictiphanes; B*, The boring isopod, *Limnoria*, or "gribble;" *C*, A common amphipod, *Gammarus; D*, The sea spider; *E*, The horseshoe crab, *Limulus*. (*A* after Calman; *B* modified from Yonge, 1949; *C* and *D* by Harold Wes Pratt; *E* from Grassé, P. [ed.]: Traité de Zoologie, Vol. 6 [pp. 219–262]. © 1949 Masson & Cie, Paris.)

Echinoderms

Members of the phylum Echinodermata generally are found in shallow water, but there are echinoderms at all depths, including the trenches. They usually have spiny skins, biradial symmetry, and five arms. However, the body shape and location of arms vary considerably from one class to another (see Table 11.2 and Fig. 11.32).

Sea stars are especially familiar to vistors of intertidal rocky areas. They are basically carnivorous. Most feed on living or dead animals such as crabs, clams, oysters, snails, or dead fish. Some can even remove hermit crabs from their shells and eat them. Some, however, feed on organic matter in the sediment or even on suspended detritus and plankton. They are found over a wide range of habitats, especially along rocky coasts, where they may feed on clams and mussels.

Sea urchins and sand dollars have no arms as sea stars do. Their calcareous shell is covered by many spines, which are used in locomotion over the sea floor. They have five spinelike jaws surrounding the mouth. Sea urchins in shallow water usually feed on dead or living algae and animals living on rocky substrates. Sand dollars and deep-sea urchins usually eat detritus from the sea floor.

Brittle stars have long, slender arms used in locomotion and feeding, and small mouths surrounded by five jaws. They usually feed on small benthic animals and detritus, although some with branched arms can capture relatively large and active crustaceans. A few even catch suspended organic matter by waving their arms about and then eat the food that adheres. Some of the largest forms are found in the deep sea, where they browse on the detritus on the sediment surface.

Sea cucumbers lack arms. They may be found at any depth, from the littoral zone to the deep sea. They have been found on the floor of the Peru–Chile Trench, at a depth of more than 6000 meters. At depths of more than 8000 meters in the Kurile Trench, sea cucumbers may constitute more than 80 per cent of the fauna by weight. Sea cucumbers

Table 11.2   CHARACTERISTICS OF COMMON CLASSES OF ECHINODERMATA

CLASS	COMMON NAMES	CHARACTERISTICS
Crinoidea "lily-shaped"	sea lily, feather star	Branched, jointed arms, some of which are stalked and look somewhat like flowers
Asteroidea "star-shaped"	sea star	Star-shaped; arms with ventral groove; arms taper into central disc of body
Ophiuroidea "shaped like a snake's tail"	serpent star, brittle star, basket star	Star-shaped; some with branched tentacles; arms without ventral groove; arms joined sharply to central disc
Echinoidea "spiny-form"	sea urchin, sand dollar	Most are biscuit-shaped and have many spines and no arms
Holothuroidea "sea-cucumber form"	sea cucumber	Most are long and tubular, with tube feet arranged in five longitudinal bands; many resemble cucumbers

FIGURE 11.32 Diversity among the echinoderms. *A*, Hawaiian sea urchin, *Echinothrix*; *B*, the arrowhead sand dollar, *Encope*; *C*, (left) larva of the brittle star *Ophiothrix fragilis* and (right) the adult form; *D*, sea cucumbers; *E*, crinoid or feather star; *F*, *Asterius vulgaris* on kelp. (*A* and *B* from Barnes, 1974, photo by Betty M. Barnes; *C* (left) photo © Douglas P. Wilson. *C* (right) and *F* photos by Harold Wes Pratt; *D* and *E* photos courtesy U.S. Navy.)

may be scavengers, mud-eaters, filter-feeders, or browsers. Many browsers of the deep sea crawl over the sediment and pick up detritus in a manner similar to vacuum cleaning. They are frequently found browsing among deep-sea brittle stars (Fig. 11.33).

Crinoids may be common in sublittoral and deeper zones. Most crinoids lack stalks and are called feather stars; a few in deeper water are attached to the substrate by a stalk and are called sea lilies. They are apparently all suspension-feeders, eating detritus and plankton that get stuck in mucus on their branched arms. The food is carried toward the mouth in slender grooves that run along the arms. Most species of crinoids are extinct.

Lophophorate Animals

The phyla discussed here include rather loosely related animals that are dissimilar in looks but that share a common method of feeding, by

FIGURE 11.33 Sea cucumbers browsing among brittle stars on the continental slope off New England. (From Heezen, B. C., and Hollister, C. D.: The Face of the Deep, Copyright © 1971 by Oxford University Press, Inc. Reprinted by permission.)

the use of a *lophophore,* which is a loop, spiral, or ring of tentacles bearing tiny whiplike cilia. The beating of the cilia causes water currents to form, which draw food toward the mouth. They are benthic animals, attached to the bottom, and are unable to go after their food. Consequently they have become adapted to a filter-feeding habit, drawing their food toward them.

PHYLUM ENTOPROCTA

The small (less than 5 mm) sessile animals belonging to this phylum superficially resemble hydroids because of their tentacles. However, they are filter-feeders and they have no stinging cells (Fig. 11.34A). They have a U-shaped gut with a mouth and anus, both opening within the ring of ciliated tentacles. The entoprocts are not generally considered to be truly lophophorate animals because their development differs from that of the lophophorates. However, the ring of ciliated tentacles of the entoprocts is similar in structure and function to the true lophophore, borne by the moss animals discussed below. Entoprocts' food consists of detritus and minute plankton such as diatoms and protozoa. They may be preyed upon by small crustaceans such as sea spiders.

MOSS ANIMALS (PHYLUM BRYOZOA)

The moss animals closely resemble the entoprocts except that the anus opens outside the whorl of tentacles, providing a more efficient and cleaner means of waste disposal. They occupy the same kinds of niches as entoprocts and compete for the same kind of food. Some moss animals have a hard protective shell (Fig. 11.34B). Most are colonial, growing on rocks and shells. They may encrust the substrate or they may grow in a branchlike fashion, giving the colony a mosslike appearance (Fig. 11.34C).

PHYLUM PHORONIDA

This phylum consists of small wormlike animals that are found mostly in shallow sediments, where they lie partly buried. They secrete a chitinous tube in which they live, filtering the water with their spiral-shaped lophophore. Like the moss animals, they have a U-shaped gut with the mouth within the spiral of ciliated tentacles and the anus just outside the spiral (Fig. 11.34D).

LAMP SHELLS (PHYLUM BRACHIOPODA)

Brachiopods may be mistaken for clams. They are benthic and have two shells (bivalved) and a shape similar to clams. However, they differ

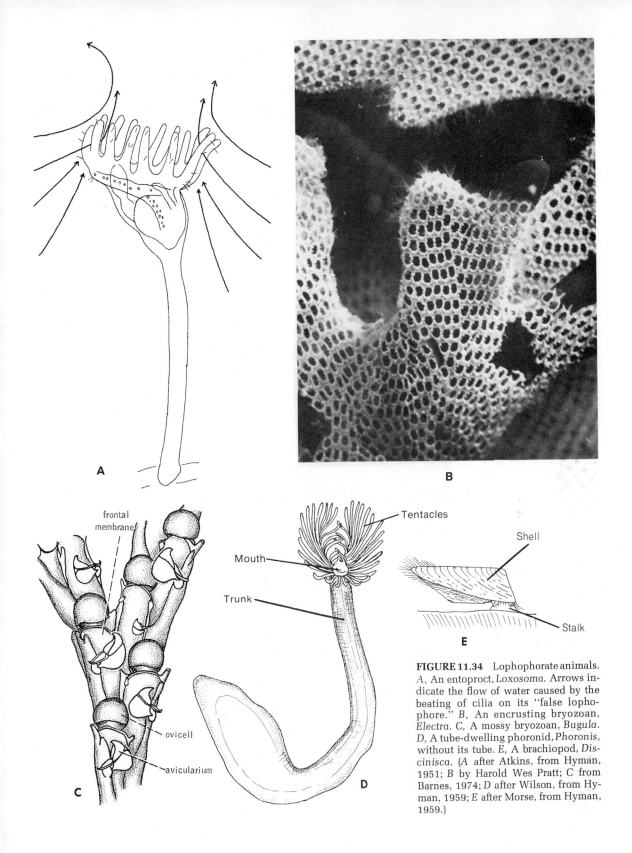

A

B

frontal
membrane

ovicell

avicularium

C

Tentacles

Mouth

Trunk

D

Shell

Stalk

E

FIGURE 11.34 Lophophorate animals.
A, An entoproct, *Loxosoma*. Arrows indicate the flow of water caused by the
beating of cilia on its "false lophophore." *B*, An encrusting bryozoan,
Electra. *C*, A mossy bryozoan, *Bugula*.
D, A tube-dwelling phoronid, *Phoronis*,
without its tube. *E*, A brachiopod, *Dis-
cinisca*. (*A* after Atkins, from Hyman,
1951; *B* by Harold Wes Pratt; *C* from
Barnes, 1974; *D* after Wilson, from Hyman, 1959; *E* after Morse, from Hyman,
1959.)

from clams in that they have a stalk and possess a lophophore that is used to filter food from the water. The shells are unequal and frequently give the animal the appearance of an old-fashioned oil lamp (Fig. 11.34E). About 99 per cent of the approximately 25,000 known species are extinct and are commonly found as fossils.

Chordates

This phylum includes the most advanced animals. They share the common characteristic of a dorsal strengthening rod along the back, the *notochord.* This rod becomes the core of the developing backbone in the vertebrates. These groups are quite diverse, but only one of them includes attached animals: the sea squirts in the subphylum Urochordata. Some members of the subphylum Vertebrata (including fish, reptiles, birds, and mammals) can swim. Some fish, however, may burrow for a time in the sediment.

TUNICATES (SUBPHYLUM UROCHORDATA)

Tunicates are small chordates that have a notochord when they are larvae. They have no skeleton or shell and the body has a gelatinous consistency. Some are attached to rocks, as are the sea squirts (Fig. 11.35); others, such as salps, are free-swimming. They usually collect food in a basketlike filter which strains the water that is drawn through their bodies by ciliary action. The food thus obtained is similar to the plankton food of clams and small crustaceans, and thus tunicates may compete with them for food.

FIGURE 11.35 A sea squirt. *Ciona intestinales,* with some sea anemones, *Edwardsia,* in the background. The sea anemones are about one cm in diameter. (Photo by Harold Wes Pratt.)

Subphylum Vertebrata

Fish (Class Pisces).   Many species of fish (Fig. 11.36) exist in the sea, occupying niches from tide pools to deep-sea and oceanic-surface waters. Some fish carry out amazing horizontal and vertical migrations related to breeding and feeding patterns. These migrations were discussed in Chapter 10. Many fish are important as food for people; others, even more abundant, play a vital role in the marine food webs.

Some of the most abundant fish in the sea are only a few centimeters in length. These include the anchovy *(Anchoa)*, lantern fish (generally known as myctophids), and the bristlemouth *(Cyclothone)*. The bristlemouth may be the most abundant fish in the sea, but it is rarely seen because it is usually found in the oceanic realm and is seldom caught by fishing fleets.

Most fish, although nektonic, have tiny planktonic larvae (Fig. 11.37). Epipelagic larvae are frequently transparent. The larvae generally look far different from the adult, and in the case of larger fish that eat other fish, the larva's diet is usually also different from that of the adult. The small larvae of these fish often feed on other zooplankton such as copepods.

Many fish that live in the surface waters, such as tuna, mackerel, and many sharks, have strong, sharp, conical or triangular teeth and are voracious predators on other fish and squid. However, many other fish, including the small anchovies and even the large filter-feeding, basking, and whale sharks, which may reach lengths in excess of 15 meters, feed on tiny zooplankton such as copepods. Some anchovies even feed on chains of phytoplankton diatoms, using tiny projections from the gill supports *(gill rakers)* as filtering nets. Fish, such as skates and rays, that feed on clams and crabs usually have flat-topped, crushing teeth.

The neritic and epipelagic fish that live in highly productive surface waters and are of economic importance—for example, herring, tuna, mackerel, salmon, and sharks—are familiar to most of us who eat them. Actually, about 50 per cent of the fish caught for human consumption are in the herring family (including anchovies and sardines).

There are many deep-sea fish, however, that we never see in the markets that are well adapted to seemingly inhospitable environments. Many of these bizarre-looking fish, such as the angler fish and the gulper eel *Saccopharynx* (Fig. 11.38), have mouths that are huge relative to the size of the fish. Some can eat prey as large as themselves. This is because their jaws open wide and their gut and body walls can expand to allow room for a large meal. Deep-sea fish do not always have a ready supply of food; therefore, they cannot afford to be particular. They must eat whatever they can, whenever they can. Some have luminous lures that can attract prey in the otherwise total darkness.

Some angler fish have evolved an unusual method of mating in the deep, dark sea. The young male bites the female, usually on her underside and becomes a permanently attached parasite on the

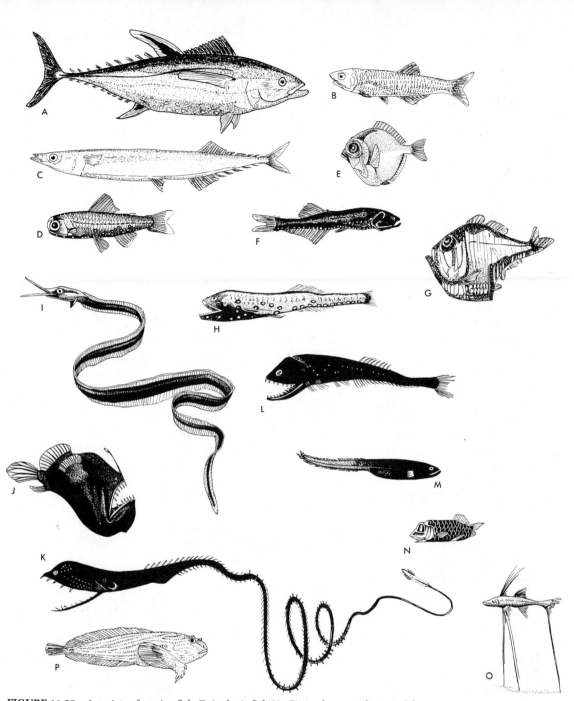

FIGURE 11.36 A variety of marine fish. Epipelagic fish (*A–C*) are shown at the top of the
figure, mesopelagic fish (*D–G*) in the middle, and deeper forms below. *A*, Yellowfin tuna
(*Thunnus*), 1 meter. *B*, Anchovy (*Anchoa*), 6 cm. *C*, Pacific saury (*Cololabis*), 40 cm. *D*,
Lantern fish (*Diaphus*), 5 cm. *E*, *Diretmus*, 5 cm. *F*, Bristlemouth (*Cyclothone*), 6 cm. *G*,
Hatchet fish (*Argyropelecus*), 2 cm. *H*, Deep-sea lantern fish (*Lampanyctus*), 3 cm. *I*,
Snipe-eel (*Nemichthys*), 35 cm. *J*, Angler fish (*Melanocetus*), 5 cm. *K*, Gulper-eel
(*Saccopharynx*), 20 cm. *L*, Swallower (*Gonostoma*), 5 cm. *M*, *Parabrotula*, 4 cm. *N*,
Opisthoproctus, 3 cm. *O*, Tripod fish (*Benthosaurus*), 20 cm. *P*, Snail-fish (*Liparid*), 20 cm.

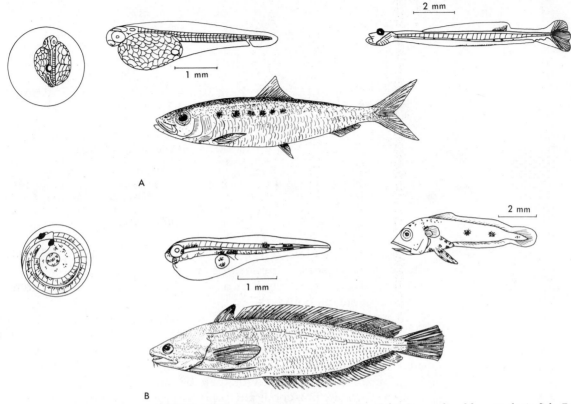

FIGURE 11.37 *A*, The larva and the adult pilchard, a sardine-like epipelagic fish. *B*, The larva and the adult hake. (Larvae after Newell and Newell, 1966; adults after Bigelow and Schroeder, 1953.)

larger female (Fig. 11.38*C*). The tissues of the male's head fuse with the female's body and she does the eating for both. As a result, the two sexes are always available to each other for reproduction.

The color of a fish is frequently related to its habitat. Most fish are somewhat camouflaged by color or pattern. Surface-dwelling fish, such as herring and tuna, are usually dark-colored above and light-colored or silvery below, resulting in countershading. When viewed from above, the dark brownish, greenish, or blue back blends in with brownish estuarine water, greenish coastal water, or blue oceanic water, respectively. Predators or prey in the water looking upward cannot easily see the fish, since the lighter sides and belly reflect light from the sky or light that has been scattered by particles in the sea. At greater depth, where the light is dim, many fish achieve countershading through the use of bioluminescence, as discussed in the previous chapter.

Fish that live on the bottom, such as flounder, usually are mottled, at least on the side that faces upward. Some even have *chromatophores,* variously pigmented patches that can be expanded or contracted, resulting in an appearance that matches the background.

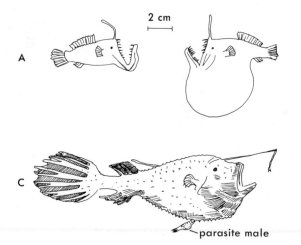
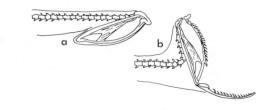

FIGURE 11.38 *A,* The angler fish *Melanocetus* before and after eating a large meal. *B,* The jaws of *Saccopharynx,* the gulper-eel, are hinged in a way that allows a large food organism to enter the gut, which expands to hold the food (a, mouth closed; b, mouth open). *C,* The angler fish *Ceratias* carrying a parasitic male on her underside. (After Hardy, 1956.)

Some fish that live in dim light are red. Since red light penetrates only a short distance in the sea, the fish appears black to another fish. Fish that live in the totally dark abyss are usually black, though some have luminous photophores.

Coral reefs contain a myriad of habitats, and the plant and invertebrate life there is often brightly colored. It is not surprising that fish living in coral reefs are also among the most brightly colored. There is a great diversity of reef fish, with widely varied shape and coloration. Many achieve a blending with their background through color and pattern. Some are poisonous and color could be a warning to predators. Coloration may also be of value for recognition of a mate. Some fish eat parasites off the surfaces of other fish and may use color to attract "customers" for their "cleaning service." Though brightly colored, none of these fish is at any apparent disadvantage.

Reptiles (Class Reptilia). Not many reptiles have adapted to life in the sea. Sea turtles, iguanas, and sea snakes are exceptions. However, they still depend on air for breathing, and most still breed on land. Because they may derive much of their food from the sea, they are considered part of the marine food webs.

The Galapagos iguana is the only marine lizard. Although it feeds completely immersed in the subtidal zone, it buries its eggs above the high-tide line.

Sea turtles are widespread in tropical and subtropical seas and frequently carry out extensive migrations between feeding and breeding grounds. Most species are carnivorous, feeding on large invertebrates such as conchs, crabs, and echinoderms. Only the green turtle is a herbivore, feeding on submerged eelgrass.

The poisonous sea snakes of tropical waters are related to cobras. Most are adapted to a strictly marine life, feeding on fish. Only one species returns to land to lay eggs; the others give birth to actively swimming young.

Birds (Class Aves). As with most reptiles, birds are not, strictly speaking, aquatic animals. However, many birds feed on marine mollusks, worms, and especially fish. Figure 11.39 illustrates some of the ecological relationships of marine birds. The fish-eating guano birds drop their feces (guano) on rocks and on land. The nutrients thus generated may find their way back into the sea, enhancing the growth of phytoplankton. The phytoplankton are eaten by copepods and other herbivores, which are in turn eaten by fish.

Over-fishing may deplete the stock of fish available to birds. The birds then go elsewhere or the bird populations decline. Less guano is deposited and phytoplankton numbers decline. In some areas, such as Peru, much guano is collected and carried away to be used on land as fertilizer, leaving less food for the remaining fish. Thus problems that affect one part of the ecosystem tend to affect the other parts as well.

Mammals (Class Mammalia). The fur-bearing animals are air-breathing, so they are not strictly marine. Some depend on the sea for a supply of food. Sea lions, seals, and walruses move between land and sea (Fig. 11.40). Whales (including porpoises) have become adapted to life in the surface waters of the oceans. Whales and porpoises lack hind legs necessary for movement on land, although their bone structure indicates that a four-legged land mammal was their ancestor. Mammals are the most intelligent of sea creatures, and many have been trained to perform tricks in circus and aquatic shows. Whales and porpoises emit a wide variety of sounds. Efforts are being made to understand their

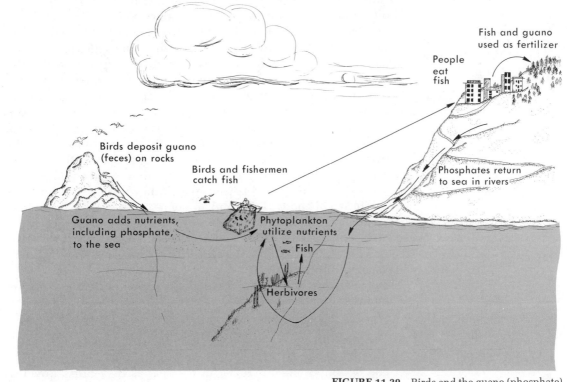

FIGURE 11.39 Birds and the guano (phosphate) cycle.

FIGURE 11.40 Sea lions along the Oregon coast. (Courtesy of Oregon State Highway Department.)

meaning, if any. Because of their economic importance as a source of food and oil, some whales are approaching extinction (see Chapter 12).

Whales (order Cetacea) fall into two major subcategories, the toothed whales (Odontoceti) and baleen whales (Mysticeti). The toothed whales include the porpoises, dolphins, sperm whales, and killer whales. They feed primarily on fish and squid. Killer whales may eat large numbers of smaller whales, seals, and sharks. Dolphins and porpoises are small whales that are similar to each other but that comprise several distinct groups. The popular bottlenose dolphin, *Tursiops*, has a smooth tail fin and conical teeth. Porpoises lack a distinct beak (nose) and have a notched tail fin and spadelike teeth.

Baleen whales are filter-feeders. Their mouths are equipped with long comblike plates, called baleen, which they use to filter small animals such as copepods, krill, and in some cases even fish, from the water. Swallowing is facilitated by their large tongue, which can be used to lick food off the baleen plates. Although most whales are pelagic feeders, the California gray whale feeds on small benthic organisms such as amphipods.

The walruses, seals, sea lions, and sea otters are carnivorous, feeding on animals such as clams, squid, crabs, sea urchins, and fish.

DECOMPOSERS

The decomposers are crucial in the sea, as they break down the excess organic matter produced by the rest of the organisms and return the materials to the sea in the form of inorganic nutrients, which are required for plant growth. If it were not for the decomposers, life in the sea would probably be greatly diminished, because dead organisms and feces would simply accumulate on the sea floor. Some of this material, even in the absence of decomposers, would be consumed by deposit-feeders, but still, much less of the nutrient substances would be recycled. Consequently the primary production of plants would decline, causing less food to be produced for the consumers, which would in turn greatly diminish in abundance. The decomposers are bacteria and fungi. These are generally considered to be neither plant nor animal.

Most bacteria (Fig. 11.41) are found attached to particulate matter, especially detritus. Organic materials tend to adhere to particles suspended in the water or lying in the sediment and provide nourishment for the bacteria associated with them.

Bacteria are found at all depths of the sea and in all kinds of environments, performing their essential role of converting detritus and dissolved organic matter into the inorganic nutrients used as raw materials in photosynthesis. All of the organic remains of marine organisms, such as crab shells and oil, are susceptible to decay by bacteria.

Bacteria, in addition to regenerating nutrients, have other functions in the sea. Many bacteria secrete organic compounds, such as vitamin B_{12}, which are essential for the growth of many diatoms and other phytoplankton. In addition, bacteria are important as a source of food for protozoans and sediment-feeding animals, such as many worms.

Some bacteria live in pouches on certain fish and produce a cold light or bioluminescence that may aid the fish in attracting prey. This phenomenon was discussed more fully in the previous chapter.

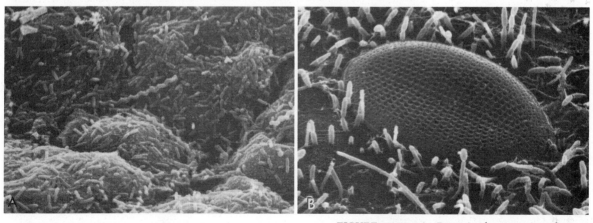

FIGURE 11.41 *A*, Bacteria decomposing the flesh of a young salmon with tail rot. (Scanning electron micrograph, 1500×.) *B*, A diatom (*Cocconeis*) and bacteria on turtle grass. (Scanning electron micrograph, 2750×.) (Courtesy of P. W. Johnson and J. McN. Sieburth, University of Rhode Island.)

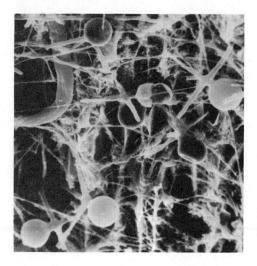

FIGURE 11.42 Fungi colonizing an oak lobster pot. (Scanning electron micrograph 425 ×.) (Courtesy of R. B. Brooks and J. McN. Sieburth, University of Rhode Island.)

Fungi play a role similar to that of bacteria. In comparison to terrestrial fungi, marine fungi are not nearly as important as decomposers. Many simple fungi, including yeasts and molds, decompose organic matter in the sea. They are either free-floating or attached to other marine organisms, especially in the shallow littoral zone, where they may be seen as cottony filaments attached to decaying algae and animals. Fungi also contribute to the breakdown of wood of terrestrial origin that enters the sea (Fig. 11.42).

SUMMARY

Life in the sea can be classified functionally as producers, consumers, and decomposers. Generally these are plants, animals, and bacteria, respectively. Based on structure, reproduction, and physiology, plants are subdivided into divisions and animals into phyla.

Most divisions and phyla include marine species. There is a tremendous diversity of species in the sea, each occupying a specialized niche in the ecosystem. The sea includes some of the most abundant organisms on earth. Copepods, for example, are probably the most abundant multicellular animals on earth, and the bristlemouth, *Cyclothone*, is very likely the most abundant vertebrate.

IMPORTANT TERMS

Chromatophore
Consumers
Decomposers
Deposit-feeders
Diatomaceous ooze
Division
Filter-feeders
Gill raker

Lophophore
Mantle
Notochord
Phylum
Predator
Producer
Scavenger
Secondary producer

1. Although the bristlemouth is one of the most abundant fish in the sea, few people have ever heard of it. Why is this so?
2. Distinguish between commensalism and parasitism and give an example of each relationship.
3. Outline the life cycles of a typical jellyfish and a crab. How are these life cycles related to dispersal and subsequent distribution of the jellyfish and crab?
4. Discuss the role of birds in the nutrient cycle of phosphorus.
5. Discuss the role of decomposers in the sea.
6. Why do some abyssal fish have disproportionately large mouths and expandable stomachs?
7. Why are diatoms sometimes called the grass of the sea?

Barnes, Robert D. 1973. *Invertebrate Zoology,* 3rd ed. Philadelphia, W. B. Saunders Co.

Caldwell, R. L. and Hugh Dingle. 1976 (January). Stomatopods. *Scientific American.*

Considine, James L., and John J. Winberry. 1978 (March). The Green Sea Turtle of the Cayman Islands. *Oceanus.*

Dawson, E. Yale. 1956. *How to Know the Seaweeds.* Dubuque, Iowa, William C. Brown.

Harbison, G. R., and L. P. Madin. 1979 (February). Diving—A New View of Plankton Biology. *Oceanus.*

Heezen, B. C., and C. D. Hollister. 1971. *The Face of the Deep.* New York, Oxford University Press.

Isaacs, J. D. 1969 (September). The Nature of Ocean Life. *Scientific American.*

Marshall, N. B., and Olga Marshall. 1971. *Ocean Life.* New York, Macmillan.

Murphy, Robert C. 1962 (September). The Ocean Life of the Antarctic. *Scientific American.*

Nicol, J. A. Colin. 1967. *The Biology of Marine Animals,* 2nd ed. London, Pitman

Ricketts, E. F., and J. Calvin. 1968. *Between Pacific Tides,* 4th ed. Revised by Joel W. Hedgpeth. Stanford, California, Stanford University Press.

Shaw, Evelyn. 1962 (June). The Schooling of Fishes. *Scientific American.*

Thorson, Gunnar. 1971. *Life in the Sea.* New York, McGraw-Hill.

Marine Production

12

Because of growing human population on a limited earth, we have long been aware of the need to increase the amount of food we get from the sea. Studies have been conducted to learn where marine food may be collected most easily. Efforts are being made to predict and control the changing populations of food organisms in the sea. Even mariculture (sea farming) is being attempted at this time in semienclosed embayments and marine ponds. Pollution, on the other hand, is undermining our attempts to utilize the sea at an optimal level. In addition, poisonous "red tides" pose a dangerous problem in the near-shore waters. In this chapter we will discuss some of the potentials and limitations of the sea as a source of food and recreation.

PRIMARY PRODUCTION

Plants use CO_2, sunlight, and inorganic nutrients in the presence of chlorophyll to produce organic matter by a process called *photosynthesis.* A simplified equation for photosynthesis is as follows: CO_2 + H_2O + light (in the presence of chlorophyll) $\rightarrow$ carbohydrate + O_2. This process results in the *primary production* of organic matter in the sea. By the process of photosynthesis, plants change the radiant energy of

Purse seining off the coast of Senegal. (FAO photo by J. Van Acker.)

sunlight into the chemical energy of sugars. Of all the radiant energy reaching the sea surface, perhaps 0.01 to 0.1 per cent is thus converted into chemical energy. In the sea the primary producers are mostly tiny (0.01 to one mm) drifting plants collectively termed *phytoplankton.* These include coccolithophores, diatoms, and dinoflagellates. On the other hand, some marine plants are quite large. The drifting *Sargassum* weed is a kelplike plant that grows in large masses in the otherwise-barren Sargasso Sea. The attached kelp *Macrocystis* may grow up to 60 meters long (200 feet) off the west coast of the United States.

Life in the sea depends on light from the sun as a source of energy. Most of the sunlight is absorbed by the water, organisms, and other particles in the water before a depth of about 200 meters is reached. Only near the surface is light of value to plants. Plants collect the light and convert the radiant energy into the stored energy in food (organic matter) by photosynthesis. This food in the form of plant life is available to herbivores, which are, in turn, food for carnivores.

For plants to live, they must have certain raw materials. These include carbon dioxide and water, required in the production of carbohydrates during photosynthesis. In addition, nutrients such as nitrate and phosphate are required if the plant is to produce essential proteins and nucleic acids. Additional elements essential for certain marine plants include iron, silicon, sulfate, and calcium. Carbon dioxide is present in plentiful supply as a dissolved gas from the air and as a waste product from the respiration* of bacteria, plants, and animals. Nutrients enter the sea as a result of the natural runoff from rivers and rainfall as well as from erosion and pollution. Nutrients may be recycled by bacterial decomposition of dead organisms and the waste products of living organisms. (Nutrient cycles are discussed in Chapter 4.) Nutrient-rich runoff causes increased production near shore. Since dead plants and animals usually sink toward the sea bottom, many nutrients are recycled in dark, deep water (Fig. 12.1), where they cannot be used by plants until they are brought back to the surface by upwelling and deep ocean circulation (see Chapter 8).

Biomass refers to the amount of organic matter in pounds or grams per given volume present at any one time; it may be compared to a *standing crop* in a field on land, or to a "forest" of seaweed. Productivity is the rate of increase or growth of the standing crop. Biomass and productivity are not necessarily related. For example, in a forest on land, the biomass is very large, but the productivity or growth rate is very low. Similarly, a large standing crop of *Sargassum* weed in the Sargasso Sea has a low productivity. On the other hand, high productivity in a polluted estuary may lead to a high biomass of unwanted organisms.

Productivity, the rate of increase or growth of the standing crop

---

*The respiration of both plants and animals may be summarized as follows: carbohydrate + $O_2 \rightarrow CO_2 + H_2O$ + energy (mostly heat).

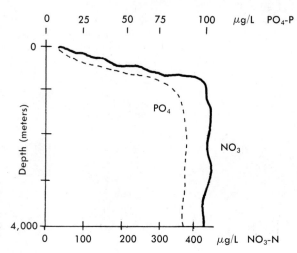

FIGURE 12.1 Distribution of two key nutrients, phosphate (PO_4) and nitrate (NO_3), in relation to depth. Note the depletion of nutrients near the surface and their abundance at greater depths. Note that the scales are not equal. (After Sverdrup et al., 1942.)

can be related to cattle grazing in a field of grass. If the crop of grass remains constant, the productivity of the grass balances the grazing rate of the cattle (and other herbivores). Similarly, many species of shrimplike copepods graze on a crop of diatoms in the sea. Copepods, the "cattle of the sea," are tremendously important as a link between phytoplankton and larger carnivores such as fish.

The primary production due to photosynthesis may be regarded as *gross production,* or the total amount of carbon assimilated by plants in a unit of time. This rate can be expressed in terms of mass, such as grams of carbon assimilated or grams of sugar produced per day beneath a square meter of sea surface, or in terms of energy, such as calories converted per day beneath a square meter of sea water. Actually, because the plants respire, they use up some of their assimilated energy in their own metabolism, releasing some energy as heat and carbon as CO_2. The amount of production thus available to the herbivores is known as *net production.* Net production (N) is the gross production (G) minus the respiration (R), or $N = G - R$. Generally N is about one half of G. An average rate of net production for the world's oceans is about 0.2 gm of sugar produced per square meter per day.

Productivity tends to be highest in and around nutrient-rich estuaries, or some enriched areas such as farmlands. Up to 25,000 kcal/square meter/yr of energy may be trapped by plants in these areas.* Open ocean waters are much like deserts with respect to productivity, at about 500 to 3000 kcal/square meter/yr (Fig. 12.2). The exploitation of shelf-fisheries is similar to open-range grazing of cattle, whereas estuarine production and mariculture are comparable to high-intensity agriculture on land.

*One kilocalorie (kcal) is the amount of heat energy required to raise the temperature of one kg (2.2 lb) of water from 14° C to 15° C.

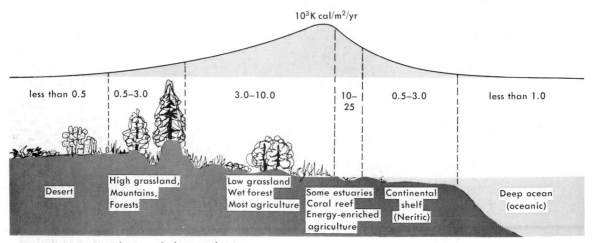

10^3K cal/m²/yr

| less than 0.5 | 0.5–3.0 | 3.0–10.0 | 10–25 | 0.5–3.0 | less than 1.0 |

Desert

High grassland,
Mountains,
Forests

Low grassland
Wet forest
Most agriculture

Some estuaries
Coral reef
Energy-enriched
agriculture

Continental
shelf
(Neritic)

Deep ocean
(oceanic)

FIGURE 12.2   Distribution of plant production over different types of areas. Values are in millions of calories produced in a square meter of area per year. (After Odum, 1971.)

In 1977, an unusual area of high productivity was found in deep-sea hot springs located in the Galapagos Rift along the northern boundary of the Nazca Plate (see Fig. 2.16). Several vents were found at about 3000 meters, out of which emerged water that was as much as 15° C warmer than the surrounding water. The water is rich in hydrogen sulfide, which is oxidized by chemosynthetic bacteria that apparently provide the basis for a richly productive food web including tube worms, clams, limpets, and crabs (Fig. 12.3). This region is especially remarkable because it supports a rich fauna without light, the energy source for most of the world's primary productivity.

Figure 12.4 shows the distribution of production over the open sea. We can see that regions of low productivity are in areas with warm surface water. Regions with high productivity include cooler areas such as polar and subpolar regions, where vertical mixing of nutrients occurs, and areas of upwelling. The high production that we find off the west coast of South America may be related to the South Equatorial Current, which flows westward from the land at the equator, and to upwelling along the coast, both of which processess may draw nutrient-rich water upward to replace water moving offshore at the surface.

BIOCHEMICAL OXYGEN DEMAND

The demand placed on the environment for oxygen by bacteria, animals, and plants is called *biochemical oxygen demand* (B.O.D.). The greater the amount of decaying wastes, the greater the B.O.D. and consequently the greater the amount of oxygen consumed.

FIGURE 12.3 *A*, Tube worms and fish; *B*, Mussels and clams on the sea floor at the Galapagos Rift. (Photos courtesy of Woods Hole Oceanographic Institution.)

As animals and bacteria feed and respire, they use up oxygen and release carbon dioxide. Carbon dioxide is taken by plants, which release oxygen. However, if the plants are unhealthy or receive insufficient light, they may consume more oxygen in respiration than they produce via photosynthesis. If there is an overabundance of dead organic matter (detritus) from plants and animals, including sewage and other biological wastes, animals and bacteria will require more oxygen than is produced by plants through photosynthesis. This is because the detritus is eaten and decomposed by detritus-feeding animals and bacteria, using up oxygen in the process. Consequently, the concentration of oxygen in the water will tend to decrease. If the oxygen level falls below a certain point, some species of fish and other animals will die, further increasing the amount of detritus and the demand for oxygen. Eventually the environment may become *anoxic,* or lacking in oxygen. Anaerobic bacteria thrive in an oxygen-free environment, respiring by a process similar to fermentation. Many of these bacteria produce hydrogen sulfide (H_2S) as a by-product, resulting in a "rotten-egg" odor characteristic of black, stagnant, organic-rich sediments.*

---

*One example of the sulfate reduction process is as follows:

$$CH_3COOH + SO_4^= \rightarrow 2CO_2 + H_2S + 2OH^-$$

| acetic acid | sulfate | carbon dioxide | hydrogen sulfide | hydroxyl |

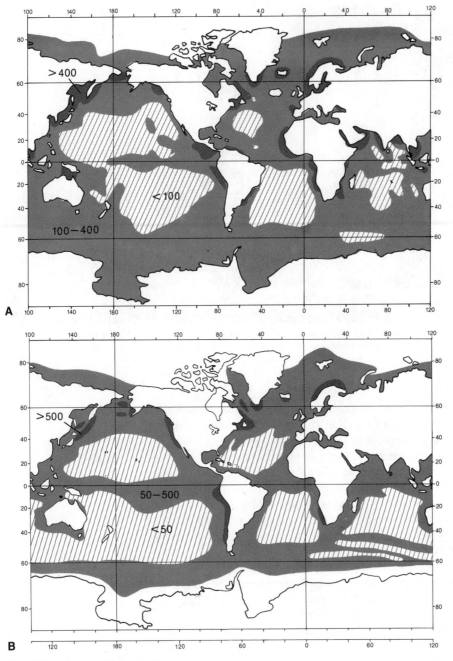

FIGURE 12.4 *A*, Phytoplankton production (mg carbon/m²/day); *B*, Concentration of zooplankton in the upper 100 meters of the oceans (mg/m³). (After FAO, 1971, and V. G. Bogorov, 1968.)

NUTRIENTS AND THEIR RELATION TO PRODUCTION AND DIVERSITY

In general, as inorganic nutrients such as nitrates and phosphates are added to a body of water, providing nourishment for the plants, the production of these plants increases, resulting in a higher biomass or standing crop. This may, if the plants are edible, supply food for zooplankton, nekton, or benthic animals. Sometimes the plant species that thrive on the added nutrients are, to most local fauna, inedible (such as some pond scums) or even harmful (such as some red-tide organisms).

The seasonal temperature structure of the ocean (see Chapter 5) plays a role in the distribution of nutrients. In shallower water in middle or high latitudes, during the winter months the water is about the same temperature at the surface and bottom. Uniform temperature results in little density stratification, and the water tends to be well mixed with regard to nutrients and salinity. During the warm summer, though, the surface water is warmed much more quickly than the deep water, and it becomes lighter. The water becomes stratified with regard to density, and little vertical mixing occurs. Nutrients near the surface are depleted, but nutrients are built up near the bottom because the "rain" of dead organisms from above is decomposed in the deeper water. Consequently, there is a reduction in primary production as the surface nutrients are depleted.

Warm, fresher water from an estuary tends to float over the salty ocean water. If the surface water in the estuary is polluted and nutrient-rich, it may trigger blooms of algae. If the nutrient-rich effluent from an ocean sewage outfall is warm and fresh (low salinity), it will tend to rise in the water column and enrich the surface water. If the effluent is colder than the surface water, it will tend to be mixed over a wider area by bottom currents and be more dilute when reaching surface waters. The enrichment of the surface water may be beneficial, increasing the amount of food available for fish. On the other hand, adverse conditions could cause a local red-tide bloom, a condition that may be harmful to people as well as to marine organisms. The red-tide problem is discussed later in this chapter.

Aquatic environments can be characterized as oligotrophic, eutrophic or hypertrophic, depending on the concentrations of nutrients and biomass and the diversity of species.

Oligotrophic (Greek *oligos* = few; Greek *trophe* = food) systems have low nutrient and biomass concentrations but may have many species of organisms. Owing to the rain of detritus into the deep, dark abyss, the surface waters in the oceanic areas contain low concentrations of nutrients. The nutrients are regenerated from the decomposing detritus, mostly in much deeper water. Consequently, production in bodies of water such as the open ocean and mountain lakes is very low.

Eutrophic (Greek *eus* = good) systems are rich in nutrients and biomass but still have many species. However, as the water becomes highly enriched with nutrients and polluted, there is a reduction in the number of species due to the death of less tolerant organisms. Streams or bays that are enriched by land runoff, especially from agricultural land, or by sewage pollution tend to be eutrophic. Of course, there is a wide range of eutrophism, depending on the level of nutrient addition. Bays enriched by runoff alone would tend to be "somewhat eutrophic." Those that are enriched by both runoff and sewage pollution tend to be more highly eutrophic, as, for example, Long Island Sound. This condition is generally indicated by a high biomass but a lower diversity of organisms than is found in the open ocean.

The next level of enrichment, called **hypertrophic** (Greek *hyper* = over, above), is characterized by high nutrient levels and biomass but low species diversity. Bays thus affected might be considered highly polluted. Extremely high nutrient levels, high B.O.D., or industrial pollutants may cause the loss of many species that are not tolerant of the high-stress environment implied by these factors. The few species that can survive are present in tremendous numbers because of the large amounts of available nutrients and limited competition. The survivors may have little economic importance or may be unfit for consumption due to contamination by poisons or bacteria. Newark Bay, New Jersey, may be considered an example of a hypertrophic bay. As few as three or four species may dominate the plankton; and the common killifish dominates the nekton of Newark Bay.

The distinction between eutrophic and hypertrophic water bodies is not easily delineated. It is based mainly on species diversity and nutrient concentration. The point of change also depends on many factors, such as temperature, salinity, tides, and the subjectivity of the observer. The change from a eutrophic to dystrophic condition may not be recognized until great destruction to the environment has occurred.

SUCCESSION

Succession is the change in species composition with time, whether on a substrate such as a rock, piling, or sediment; or suspended in the water, such as plankton or nekton. The change in communities that occurs on a new piling or an automobile body tossed into the water is an example of this process; another example is the seasonal succession of plankton species.

Benthic Succession

As an example of succession, let us look at marine fouling. Even the casual observer can notice that objects that are placed in the sea soon become covered with a great variety of sea life or "fouling" organisms. These organisms attach to rocks, lines, anchors, ship hulls, and buoys and then serve as a substrate on which other organisms grow. Fouling is usually not harmful and may be a source of food and hiding place for fish, shrimp, and other animals. Sunken ships and other introduced substrates soon become fouled and attract fish, benefiting sport fishing. On the other hand, some fouling is less desirable. A ship can be slowed in its movement by adherent organisms because the attached algae and animals increase the resistance to movement through the water. Currents pulling on a fouled buoy strain the mooring lines and may drag the buoy out of its assigned position. Burrowing animals may weaken ropes and pilings and cause them to break.

The stages of fouling that usually take place may be summarized as follows:

1. Preparation of the substrate. The first organisms to settle include bacteria. Many tend to adhere to virtually any solid substance placed in the water, producing a film suitable for certain tiny diatoms to become attached and grow in mats or chains. This prepares the substrate for the next stage.

2. Random settling. Other plants and animals settle and grow, including larvae of barnacles, mussels, hydroids, and young algae. Some of these survive and grow; others die when the tide goes out and exposes them to air. Some species may be crowded out by other species during the next stage of succession, which we may call a period of selection.

3. Selection. During this period most of the settlers die, are eaten, or are displaced by competition. Only the species best adapted to that particular environment survive. For example, of the settlers from Stage 2, only the mussels may survive. The hydroids and algae may die from exposure. The barnacles may be crowded out. At a higher point on the rock or piling, the mussels may be displaced by the barnacles. Hydroids and others, however, may arrive and survive at a later stage during a period of development.

4. Community development stage. At this stage the community is slowly developed. For example, new species enter the realm of the mussels and survive in the sheltered spaces or on the mussels themselves. Hydroids and sea squirts may flourish. Bryozoans, foraminiferans, and more diatoms may settle on the hydroids and grow, forming a "jungle" that collects detritus.

5. The climax community. A climax community (Fig. 12.5) is ordinarily named for the dominant species that is present. For instance, a fouled substrate in which mussels dominate would be called a mussel community. In such a community, the mussels make up most of the biomass and play a key role in determining the nature of the community. For example, the mussels may provide a substrate for hydroids,

FIGURE 12.5 A climax community. A slug-like nudibranch crawling over encrusting sponges, which are also inhabited by sea anemones. (Photo by Harold Wes Pratt.)

protection for amphipods, and food for sea stars. Individuals of smaller species such as amphipods might be present in greater numbers than the mussels, but the success of the mussels would determine the success of the overall community.

Coral reefs are examples of stable climax communities. Although plants may comprise most of the organic matter, the nature of the reef is determined by the coral animals and their stony skeletons that form the basis of the reef.

The fouling process is essentially the same sort of succession that can occur on the hard substrates of rocky coasts after the rocks are occasionally stripped bare by severe wave action.

The problem of marine fouling has been of concern for centuries to people who live near the sea. Efforts are made to reduce fouling of ropes, pilings, and ships' hulls with various chemicals, including creosote and special paints that are toxic to marine organisms. Part of this attempt has involved investigations of the succession process— which organisms settle first, which come later, and which organisms represent the climax community. The climax community is the one that causes most of the problems, as it is usually the most massive and creates the greatest resistance to currents and destruction of substrates.

On a piling (Fig. 12.6), there can be many communities, with the dominant species depending on the position of the piling and the resul-

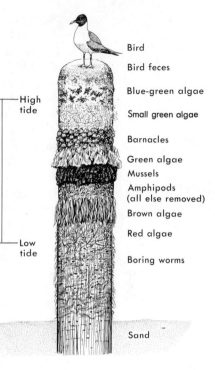

High
tide

Bird

Bird feces

Blue-green algae

Small green algae

Barnacles

Green algae
Mussels
Amphipods
(all else removed)
Brown algae

Red algae

Boring worms

Low
tide

Sand

FIGURE 12.6 Simplified view of typical zonation on a piling. The bird symbolizes the role of
birds as a source of nutrients for the plant life of
the sea.

tant effect of tides, temperature, salinity, light, and—of course—
conditions imposed by neighboring organisms. As can be seen in Figure 12.6, bird droppings may be a source of nutrients for the community of blue-green algae. The set of communities on a piling, in fact, is an
especially dramatic and condensed example of zonation, which was
discussed earlier (Chapter 10).

The climax community may develop within months, as in the case
of a rock periodically denuded by ice, or it may take years, as in the case
of a clam bed destroyed by dredging.

Pelagic Succession

Succession is not limited to growths on pilings and rocks. Succession of plankton communities occurs with seasonal temperature and
salinity changes, with one species providing nutrients or food for the
next species in succession. In this case, the substrate does not play a
great role except that it provides a home for some adults developing
from planktonic larvae, such as barnacles, crabs, and clams.

The species composition of the nekton (swimmers), benthic organisms (bottom-dwellers), zooplankton (drifting animals), and

phytoplankton (drifting plants) varies with the seasons. Species abundant in the spring may be scarce in summer, fall, or winter (Fig. 12.6*A* and *B*). The changes follow definite patterns that tend to be repeated annually, but extraneous factors such as pollutants or differences in termperature from year to year may alter the patterns of succession. In fact, it may be difficult to determine what "normal" succession really is in some cases.

A species increases in number when conditions are favorable for its survival and decreases when they are unfavorable. For example, in early spring the surface water in temperate latitudes is rich in nutrients, and the increasing temperature and light cause a sudden growth (bloom) of phytoplankton (Fig. 12.7*A* and 12.8). The phytoplankton soon decline in numbers, owing to nutrient depletion in the temperature-stratified surface water and increased grazing by zooplankton. In the fall, the thermocline breaks down and nutrients are mixed into the surface water by turbulence, and another smaller bloom of phytoplankton may occur, sometimes followed by another small burst of zooplankton herbivores. The fall bloom quickly dies out because of cooling of the water and decreasing light. Small blooms may also occur during the summer, when nutrients are brought into the surface water by upwelling.

The phytoplankton blooms really consist of a succession of smaller blooms of several plant species (Fig. 12.7*A*). Each species is adapted to different conditions of temperature, light, and chemicals secreted by the previous bloom. Thus, species that are abundant in the spring are frequently different from those abundant during the other seasons.

The zooplankton most prevalent during the summer in temperate water also exhibit a succession of dominant species (Fig. 12.7*B*). Temperature and salinity requirements and chemicals secreted by previous plant and animal bursts may control the seasonal succession of these species. All of these successions contribute to the ever-changing communities of interacting organisms.

The seasonal pattern of productivity varies greatly from one region to another. In subpolar regions, where nutrients are in plentiful supply, the availability of light limits the high productivity to the summer months. In tropical regions, seasons are not well defined. Where a strong thermocline restricts nutrient enrichment of tropical surface water, productivity is always low. On the other hand, in tropical regions subject to upwelling, productivity is high in all seasons.

FOOD WEBS

A food web is simply a pattern of feeding relationships among organisms; an ecologist would call it a diagram of *trophic relationships.* Each community has a characteristic food web, although communities

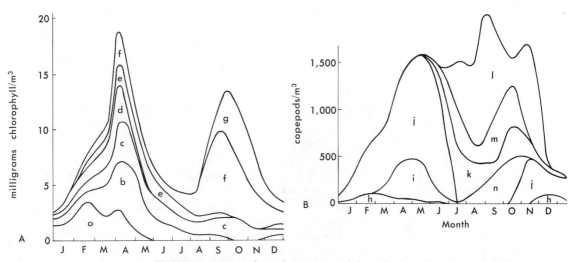

FIGURE 12.7 Hypothetical succession of phytoplankton (A) and zooplankton (B) species in temperate coastal waters. Top lines represent the total chlorophyll (in A) or numbers of copepods (in B) present in a cubic meter of water. The differently labeled areas represent the contribution of the individual labeled species present. For example, phytoplankton species "f" contributes approximately half the chlorophyll present in October.

that are fairly close to each other are actually linked in a much larger web, because all communities ultimately depend on the primary production of the local phytoplankton (and other plants), and because bacteria decompose detritus from all communities.

Plants collect the sun's energy and store it in the form of organic matter, such as sugar and fat. Seaweed and phytoplankton are eaten by animal herbivores such as some amphipods, copepods, clams, and some fish. The herbivores are, in turn, eaten by primary carnivores such as jellyfish, sea stars, and most fish. Primary carnivores are eaten by larger fish or invertebrates (the secondary carnivores). These, in turn, are eaten by tertiary carnivores, and so on. This sequence may be illustrated by a food chain such as the following:

trophic levels	*example*
plants	diatoms
↓	↓
herbivores	copepods
↓	↓
primary carnivores	anchovies
↓	↓
secondary carnivores	tuna
↓	↓
people	people

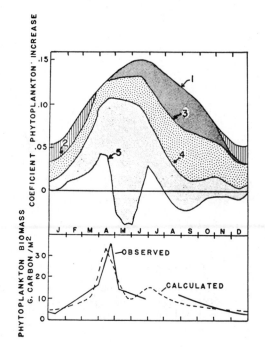

FIGURE 12.8 Effects of five limiting factors on the phytoplankton (upper) and the observed and calculated phytoplankton densities (lower) off the New England coast. The top line is the expected rate of increase in the phytoplankton population, in arbitrary units, considering factor 1, the available light and temperature. Each of the other factors takes away from the expected rate of increase when the previously considered factors have been applied. For example, the dotted area under line 3 indicates the limiting effect of factor 4. Factor 2 = turbulence, which carries cells below the photic zone; 3 = phosphate depletion; 4 = phytoplankton respiration; and 5 = zooplankton grazing, which produces the final coefficient of change in the phytoplankton, allowing for all of these factors. Only during the spring and late summer are conditions favorable for the rapid growth of the population. (After Riley, 1952; from Odum, 1971.)

A food chain shows the direction of movement of organic matt energy as it passes from one organism to another, but it does not i cate amounts of organic matter (biomass) or energy (calories) req at each level. A *pyramid of biomass* or a *pyramid of energy* illust these factors (Fig. 12.9).

Each level of the pyramid is called a trophic level (feeding lev we assume that each transfer of food energy is 10 per cent efficie takes about 10 kg of phytoplankton to produce one kg of copep since copepods may feed directly on phytoplankton. Similarly, it about 1000 kg of phytoplankton, which are transferred through additional trophic levels, to produce one kg of tuna. The copepod o tuna utilizes only 10 per cent of its food for building its body (pro tion). Efficiencies may vary considerably among species and wit species as well as at different times in the life cycle. Ranges of 10 per cent efficiency may be normal, but efficiencies as high as 30 per or more are possible, especially in areas of nutrient enrichment. Mc the energy consumed is lost as respiratory heat and undigested w that are decomposed by bacteria, fungi, and detritus feeders (sca gers). The energy budget of a typical salt-marsh fish is shown in F 12.10.

The efficiency of energy transfer from an organism to the orga that eats it *(trophic efficiency)* depends on the nature of the food,

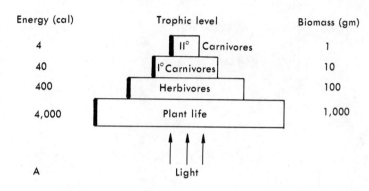

Energy (cal)	Trophic level	Biomass (gm)
4	II° Carnivores	1
40	I° Carnivores	10
400	Herbivores	100
4,000	Plant life	1,000

A

Light

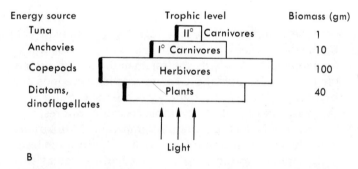

Energy source	Trophic level	Biomass (gm)
Tuna	II° Carnivores	1
Anchovies	I° Carnivores	10
Copepods	Herbivores	100
Diatoms, dinoflagellates	Plants	40

B

Light

FIGURE 12.9 *A,* Pyramid of energy and biomass for a marine eco-system, showing quantities required for growth. The quantities given are in approximately correct ratios to the other levels of the pyramid if we assume 10 per cent efficiency throughout.

B, Pyramid of biomass for productive coastal water, showing relative amounts present at one time. The apparent paradox in the greater amounts of herbivores than plant food is due to the fact that the plants are produced very rapidly and quickly replace those eaten.

C, Pictorial representation of a pyramid of biomass of the type shown in *A.*

C

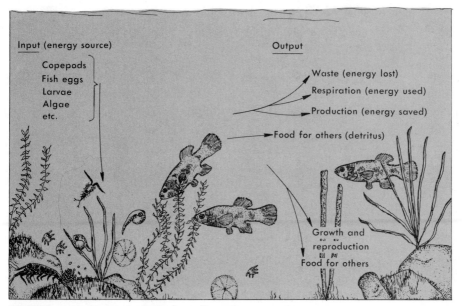

FIGURE 12.10 Energy budget of a salt-marsh fish. The total energy input equals the energy output.

food value, and its ease of capture. For example, a copepod may be able to eat a large diatom or a colony of diatoms but not the smallest of phytoplankton *(nannoplankton),* which are less than 0.1 mm in diameter. The larger the chains of diatoms, the easier it is for copepods and other larger herbivores to eat them. Less energy is used to capture food. If the phytoplankton are very small, for example coccolithophores, they may be eaten by protozoans such as foraminiferans and radiolarians, which are then eaten by carnivorous copepods at a higher trophic level. The larger phytoplankton are found mostly in nutrient-rich areas such as neritic or upwelling regions. The minute phytoplankton tend to predominate in the less nutrient-rich offshore oceanic regions.

Energy may pass through as many as six oceanic trophic levels before reaching people. In the rich upwelling areas, only two or three trophic levels are required to produce edible anchovies. Most neritic regions contain at least three or four trophic levels.

Because of the greater efficiency and shorter food chains found in upwelling areas, tremendous amounts of fish are produced in a small area (only about 0.1 per cent of the world's oceans). These relationships are summarized in Table 12.1. Ten per cent of the ocean (upwelling and other neritic areas) may provide 240 million metric tons (1 metric ton = 1000 kg) of fish, whereas 90 per cent of the oceans (the oceanic area) may provide only 1.6 million metric tons of fish per year. The 0.1 per cent of the ocean that lies in upwelling areas equals the fish production of the other 9.9 per cent of the ocean that is regarded as neritic. The high

Table 12.1 TOTAL ANNUAL POTENTIAL PRODUCTIVITY ESTIMATES FOR THE WORLD'S OCEANS, ASSUMING VARIOUS EFFICIENCIES OF ENERGY TRANSFER*

	OCEANIC AREAS	*NERITIC AREAS*	*UPWELLING AREAS*
% of Ocean Area	90	9.9	0.1
Trophic Efficiency (estimated %)	10		20
Mean Production (grams carbon/ square meter/yr)	50	100	300
Organic Carbon Produced per year (in millions of tons)			
phytoplankton	16,000 nannoplankton	3600 diatoms, dinoflagellates	100 colonial phytoplankton
herbivores	1600 protozoa, copepod larvae	540 copepods, clams	20 copepods, krill, clams, anchovies
I° carnivores	160 copepods	81 anchovies, benthic fish	4 anchovies, larger fish
II° carnivores	16 chaetognaths	12.2 larger fish	
III° carnivores	1.6 small fish		
IV° carnivores	0.16 larger fish		
Average number Marine Trophic Levels	6	4	2½
Total fish Production (in millions of tons *fresh* wt/yr)	1.6	120	120

*From data of Ryther, 1969.

productivity that is possible in upwelling areas is largely due to the nutrient-rich water, high trophic efficiency, and short food chains. These are also areas that are being heavily exploited for food fish because of the small amount of fishing effort needed to catch the abundant fish crops.

The food chains described in the preceding discussion are extremely simplified. Of course, a copepod or a clam eats many different kinds of phytoplankton. Any one species of phytoplankton may be eaten by a number of species of herbivores. Clams may be eaten by starfish, people, fish, or other carnivores. The food chain or pyramid of biomass does not show these trophic details, which are, however, illustrated by the ***food web.*** Figure 12.11 shows a greatly simplified marine food web.

All members of the food web eventually are consumed by other organisms or decompose upon death.* The complex organic compounds (fats, sugars, starches, proteins, and so forth) are converted by bacteria and other detritus-feeders into simple inorganic compounds, which

*Diatoms and other phytoplankton that reproduce asexually by division are sometimes considered "immortal." An individual cell may live a long time if it is not eaten or killed under adverse conditions.

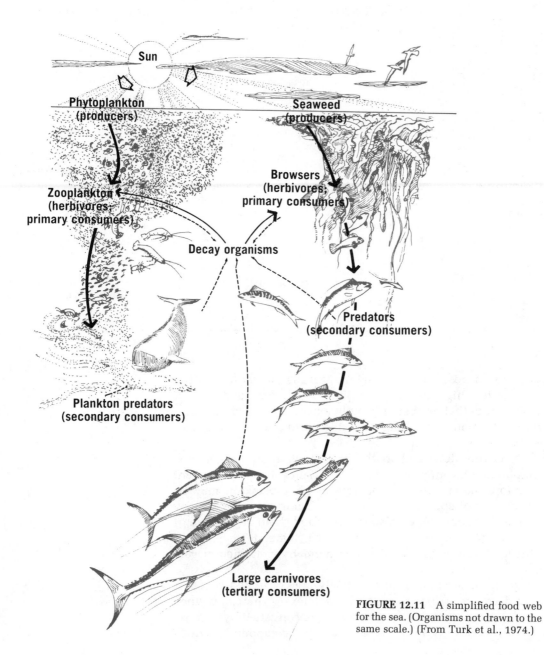

FIGURE 12.11 A simplified food web for the sea. (Organisms not drawn to the same scale.) (From Turk et al., 1974.)

dissolve in sea water and provide the nutrient raw materials for plant production. Nutrients are recycled, but energy that is lost as heat is not recycled. Therefore the propagation of life on earth is dependent on the continuing input of energy from the sun.

Figure 12.12 illustrates the flow of energy through communities in the photic and aphotic zones. Some of the export of energy as detritus from the photic community may become input of energy to the aphotic community as a "rain" of detritus.

As an ecosystem becomes polluted, some species die out and others replace them; or the remaining "pollution-tolerant" species flourish on the rich organic and inorganic nutrients frequently associated with pollution. Figure 12.13*A* shows a partial food web of a healthy estuarine salt marsh. Some of the changes that occur when the marsh is disturbed by human intervention are illustrated in Figure 12.13*B*, which shows the food web existing now in the Hackensack Meadowlands surrounding Secaucus, New Jersey. Note that many species have disappeared. However, some rugged species such as killifish have become even more abundant, since they no longer must compete with other fish, such as sticklebacks, silversides, and sheepshead minnows. Important food fish such as bluefish, striped bass, and flounders; and shellfish such as oysters and mussels are lost because they cannot tolerate the highly polluted conditions.

The change from healthy to disturbed marsh is sometimes indicated by the rapid shift in growth from salt-marsh hay *(Spartina)* to plume grass *(Phragmites)* as deposits of sediment and detritus accumulate on the marsh. Plume grass is not as desirable a food for marsh animals as is salt-marsh hay, although it is important as a food and habitat for muskrats, rabbits, mosquitoes, and killifish, and as a nesting ground for shore birds. The Hackensack Meadowlands of New Jersey are a good example of a disturbed salt marsh where plume grass is a dominant flowering plant and killifish abound.

Disturbed marshes such as the Hackensack Meadowlands may be beyond repair, but they should be protected from needless further destruction because of their present importance as a source of food and habitat for the organisms that live in them and because they help to protect the surrounding waters by absorbing excess nutrients carried by polluted water.

RED TIDES AND OTHER BLOOMS

For thousands of years, people have observed that the surface of the ocean occasionally turns reddish near the shore or in bays and estuaries. Frequently, catastrophic fish kills are associated with these red waters. The beaches become littered with the dead and dying bodies of fish and the stench drives tourists away from resort beaches. Red-tide

Text continued on page 391.

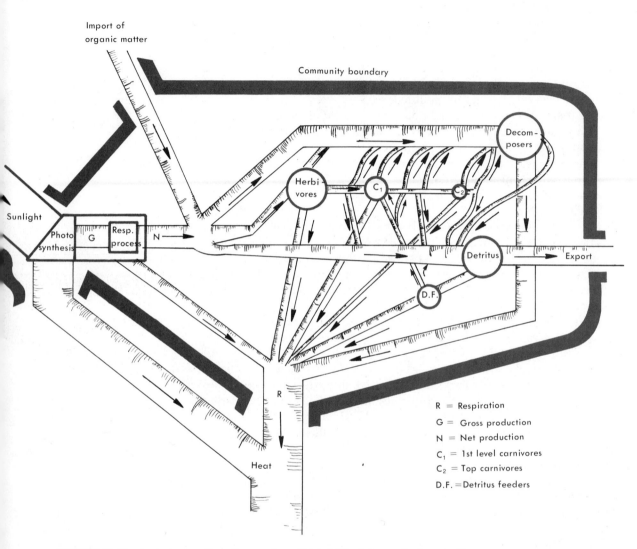

FIGURE 12.12 *A*, An energy-flow diagram for a lighted (photic) zone in the sea. The arrows indicate the direction of energy flow. (Modified after Odum, 1956.)

Illustration continued on the opposite page.

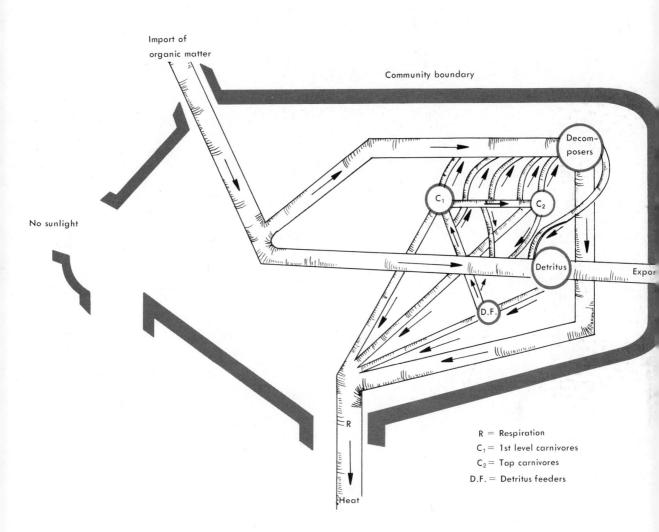

Import of
organic matter

Community boundary

No sunlight

Decom-
posers

C_1

C_2

Detritus

Expor

D.F.

R

Heat

R = Respiration
C_1 = 1st level carnivores
C_2 = Top carnivores
D.F. = Detritus feeders

FIGURE 12.12   B, An energy-flow diagram for the deep sea (aphotic zone). (Modified
after Odum, 1956.)

THE ECOLOGY OF AN UNPOLLUTED MARSH/ESTUARY

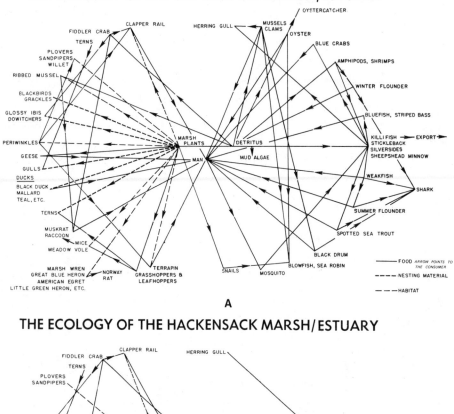

A

THE ECOLOGY OF THE HACKENSACK MARSH/ESTUARY

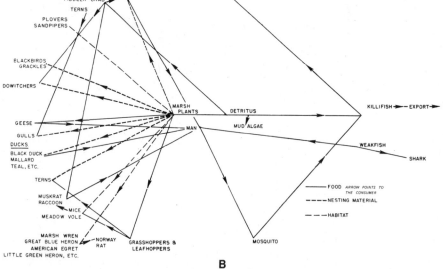

B

FIGURE 12.13 *A,* The ecology of an unpolluted salt marsh estuary. *B,* The ecology of a disturbed salt marsh estuary. (Food web diagrams by E. E. MacNamara, courtesy of the Hackensack Meadowlands Development Commission.)

organisms may also contaminate filter-feeding shellfish. Clearly this is a critical problem for the fish and for people. Only recently have we begun to find the cause and possible remedies for this crisis.

When surface waters are enriched with nutrients, tremendous growths *(blooms)* of phytoplankton may occur. Usually these blooms are harmless or even beneficial to production of small crustaceans (especially copepods) and fish. However, the bloom occasionally consists of certain species of dinoflagellates that produce toxic substances and which sometimes but not always cause an area of the sea to turn red. Red-tide organisms such as the dinoflagellate *Gonyaulax* are eaten by shellfish, which may concentrate the toxic substances, causing "paralytic shellfish poisoning" in people who eat the affected shellfish. The poison is heat-stable and is not destroyed by cooking. Red tides may occur naturally, but the frequency as well as the severity may be increased due to human-caused pollution. Agricultrual runoff, nutrient-rich river flow, or sewage effluents may contribute. Factors such as temperature, light, salinity, turbidity, and nutrient content are also involved, so it is not always possible to assign a specific cause to a bloom.

One example of periodically recurring red tides takes place off the west coast of Florida. They have been attributed to increased amounts of phosphate and iron added to the sea by unusually heavy rainfall runoff. Tannic acid also enters the sea with the rainfall runoff and increases the solubility of iron in sea water. The extra iron stimulates the growth of *Gymnodinium breve,* a notorious toxic red-tide dinoflagellate.

From many other examples, scientists have pieced together a general model of how most red tides probably originate. Among the most favorable conditions for red tides are the following:

1. Upwelling, tidal mixing, and runoff, causing enrichment of nutrients in the surface water. Runoff from agricultural and phosphate-rich land may be especially important, as for example in the Florida red tides.

2. Diluted sea water (lowered salinity), which appears to aid the growth of red-tide organisms. This condition is particularly apt to occur near the mouths of rivers.

3. Vitamin B_{12} from soil, marshes, blue-green algae, and bacteria, which may aid the growth of dinoflagellates, as many of them cannot produce this vitamin on their own.

4. Iron and tannic acid from industrial and swamp runoff.

5. Presence of small quantities of red-tide organisms that can survive when nutrient concentrations are low.

6. Other conditions favorable for growth of bloom organisms, including optimum light, temperature, and organic and inorganic nutrients. Possible ways to control red tides include the following:

1. Introduction of vitamin B_{12}–destroying bacteria.

2. Encouragement of natural predators that eat phytoplankton.

3. Copper poisons to control production (copper is toxic to many aquatic plants, but unfortunately also to many animals).

The above controls may have unwanted side effects and may also be rather expensive. A fourth method of control and probably the most scientifically sound one is to limit concentrations of nutrients entering the sea. This might involve the use of certain plants to extract the nutrients from sewage effluents before they enter the sea, or to stringently control the needless use of large quantities of fertilizers.

We should note that not all red waters are toxic. *Noctiluca,* a dinoflagellate that may turn the sea the color of tomato soup, is not toxic. The Red Sea and Gulf of California at times get their red color from certain species of "blue-green algae" that actually have a reddish color.

FISHERIES

We have long depended on the sea as a source of food. The sea provides about 17 per cent of the meat protein consumed by people. Terrestrial and freshwater animals provide the rest. About 75 per cent of the fish caught and processed worldwide are used for human food. The rest is used for other purposes, such as pet food, livestock food supplement, and fertilizer. Table 12.2 lists the most abundantly caught groups of fish.

These resources have been regarded as inexhaustible until at the beginning of the twentieth century it was found that some species of the fish were being depleted by "over-fishing." Over-fishing is generally credited to the development of efficient trawl nets (Fig. 12.14) in the nineteenth century. Toward the middle of the twentieth century, commercial fishermen began to develop "less efficient" nets that would catch only the larger fish and allow the young to remain free to grow and reproduce. However, recent advances in the technology of making food and fertilizer from any size or kind of fish has put tremendous economic pressure on fishermen to sweep up everything in sight. Today, even the factory has gone to sea with its fleet of trawlers (Fig. 12.15).

Table 12.2 MOST ABUNDANTLY CAUGHT FISH FOR 1977, COMPILED FROM FAO FISHERIES STATISTICS

FISH GROUP	*METRIC TONS CAUGHT*
Anchovies, herring, sardines, etc.	12,962,071
Cod, hake, haddock, etc.	10,694,688
Jacks, mullets, sauries, etc.	8,683,055
Bass, redfish, etc.	5,139,170
Mackerel, etc.	3,556,580
Tuna, bonito, etc.	2,334,109
Flounder, sole, etc.	1,083,770

FIGURE 12.14 A modern stern-trawler towing an otter trawl over the sea floor.

FIGURE 12.15 The Soviet factory ship *Trudovaya Slava*, one of the mother ships of a large foreign fleet trawling off the Atlantic Coast of the United States in 1969. (Official U.S. Coast Guard photo.)

Fortunately, fish populations apparently can "bounce back" from depletion if the condition is not too severe. During the First and Second World Wars, for instance, the North Sea fishing grounds were closed because of the mines placed in the waters. Much greater catches of fish were made just after the wars compared with just before (Fig. 12.16). Apparently the fish stocks were renewed during the forced moratorium on fishing.

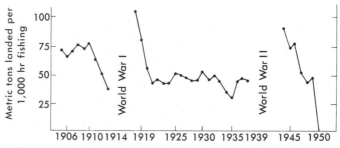

FIGURE 12.16 Catches of haddock by Scottish vessels in the North Sea, showing the effects of the moratorium on fishing during the world wars. (After Russell-Hunter, 1970, and Graham, 1956.)

Declines in fish populations may be due to factors other than fishing. When the *mortality* rate increases for any reason, more fish die, and the fish stock declines. Any change in the environment—for instance, the nutrients available, the presence of predators, or increased pollution—can kill fish and reduce the numbers available for human consumption.

Over-fishing, especially when it coincides with natural catastrophies, can have devastating effects on local fisheries. As a result of the upwelling of nutrients from deeper water, the anchovy fishery off the west coast of South America is one of the world's most productive regions. Fish catches there increased dramatically from 1955 to 1970 (Fig. 12.17*A*) in spite of the detrimental effects of the 1957 and 1965 El Niños, which brought warm, nutrient-poor water closer to shore and caused an interruption in the upwelling process. The decreased productivity resulted in temporary declines in the fish population. The maximum sustainable yield for this area was estimated to be about 9.6 million tons per year. Over-fishing between 1966 and 1972 eventually resulted in much-reduced catches of anchovies. Then the 1972 El Niño caused even further reduction of the fish stock.

Factors that tend to increase the biomass of fish include *reproduction* and *growth.* Over-fishing occurs if fishing and mortality take place at a rate greater than the natural reproduction and growth rates. On the other hand, if the fishing rate is less than the reproduction and growth rates, we can say that the fishing grounds are "under-fished." A *sustained yield* is maintained when fish catch and mortality equal reproduction and growth. These relationships are illustrated in Figure 12.17*B* and by the following equation developed in 1931 by E. S. Russell, an expert on fisheries management:

$$S = S_0 + (A + G) - (M + C)$$

Where

S = biomass of fish at the end of the year
S_0 = biomass of fish at the beginning of the year
A = amount reproduced for the year
G = growth for the year
M = mortality (natural and pollution-caused) for the year
C = fish catch for the year

During a period of sustained yield, therefore, $(M + C) = (A + G)$. We might infer from this relationship that a fish or whale with a slow growth rate (and generally a long life) and low reproductive rate would be more apt to be over-fished than one that grows quickly and produces many offspring. Whales, which grow slowly and produce only one offspring every other year, have in fact fared badly at the hands of whalers.

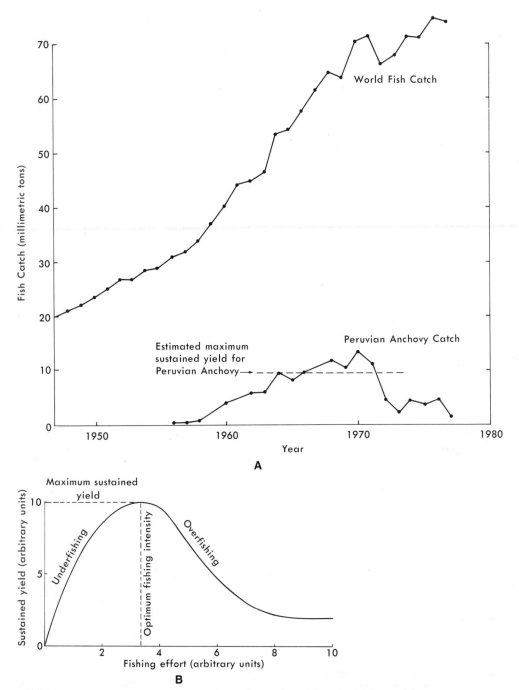

FIGURE 12.17 *A*, Peruvian anchovy catch in relation to world catch of fish 1955–1977. (Based on FAO Fisheries statistics). *B*, The relationship between sustained yield and fishing intensity. Notice that as fishing effort is increased, the yield of fish is increased until a maximum is reached, beyond which fish catches decline. Beyond the optimum fishing intensity, increased effort results in smaller catches, owing to depletion of the fish population. (After Russell-Hunter, 1970.)

Intensive whaling has resulted in the near-extinction of several species, including the northern right whale. In the Antarctic, 8000 blue whales were caught in the summer of 1948; 1684 in the summer of 1958, and only 20 in the summer of 1965. Although an international commission was set up in 1946 to control Antarctic whaling, it apparently did not benefit the blue whale! If sustained-yield whaling had been practiced in the beginning, an estimated annual catch of 6000 blue whales could have been maintained. Meanwhile, protective legislation has enabled the population of California grey whales to climb from a few hundred in 1938 to more than 10,000 in 1967, permitting a certain amount of fishery exploitation.

At one time, fishermen threw away fish that they regarded as "trash fish," such as sea robins and sharks. These were usually discarded dead and were recycled into the ecosystem by scavengers. However, the effort expended in catching them was wasted. Many other fish were too small to bother with, except in the case of delicacies such as sardines and anchovies. Today most of the anchovy catch, as well as formerly unwanted fish, form the basis of a great industry in fish meal, which is fed to poultry, swine, and pets. Further treatment of the fish meal to extract the protein is being done on a small scale. This product, known as *fish protein concentrate (FPC)*, may go a long way toward improving the nutrition level of the hungry people of the world, if maximum sustained yield can be achieved. FPC is nutritious, almost tasteless, and simple to store without refrigeration. It can be easily added as a nutritious "extender" to other foods such as bread and soup. Further exploitation of the sea's resources without over-fishing will probably involve more technological advances like protein extraction. If this protein can be extracted from the lower trophic levels, the protein yield of the sea could be greatly increased.

Although anchovies, cod, pollack, mackerel, herring, and hake are the most frequently caught fish (about 40 per cent of the total catch of marine life), new potential sources of food include shrimplike krill *(Euphausia superba)*, a staple food of baleen whales; lantern fish (myctophids); and the pelagic "red crab" *(Pleuroncodes)*, an important food of tuna in the Pacific.

Another by-product of fish that cannot be ignored is edible oil. We all know of iron-rich cod-liver oil. But how many realize that the sea is a major source of the hydrogenated oil used in the production of some margarine? Anchovies, herring, and whales are important sources of this oil.

Despite the decline in the catch of certain "over-fished" species, the total world catch of marine organisms (plant and animal) increased from about 20 million metric tons in 1938 to about 40 million metric tons in 1960 and to over 70 million metric tons in 1978. This trend paralleled the development of large factory ships and more sophisticated fishing techniques, such as the use of special nets and electronic lures.

It has been estimated that the harvest of food from the sea might be increased 2½ times, or to about 150 million metric tons per year. This is possible considering the great potential of FPC and the further use of marine algae, but would require vastly increased utilization of neglected fisheries and a return to maximum sustained yield for over-fished regions. About 30 per cent of the world's protein requirement for a population of 6 billion in the year 2000 would be satisfied by 150 million tons of seafood. Of course, this assumes moderate success in population control. Can we or will we be sensible enough to control our future?

MARICULTURE

Whereas fisheries biologists are most concerned with sustained yield and fishing rates, mariculturists are interested in increasing the reproductive rate and especially the growth rate of marine food resources. As the limits of fishing potentials become more pronounced and areas suitable for the natural growth of shellfish become more limited, we are turning toward "sea farming," or *mariculture,* more than ever before. Productivity is greater under controlled conditions, and the problems of maintaining a fleet of large fishing vessels are eliminated.

There are a number of mariculture operations that are currently profitable. Many of these are relatively small operations. Most involve the growing of marine organisms such as oysters, clams, mussels, and seaweed on racks or ropes suspended from floats or placed on the sediment, usually in a nutrient-rich estuary. A smaller number of operations are based on land in buildings, using purified water from the nearby ocean or bay. The most efficient of these use *polyculture,* or the growing of a variety of trophically interrelated organisms. Phytoplankton diatoms are cultured and fed to larval and adult clams, some fish, and copepods that are also food for small fish. The walls of the tanks are kept clean by grazing gastropods such as snails and sea hares. All of the organisms are marketable to restaurants, schools, laboratories, or individuals. The clams are sold as food and the other organisms are usually sold as laboratory organisms for study, experimentation, or testing, especially for assaying the toxicity of industrial wastes.

Much larger-scale mariculture operations may use tanks, pens, and impoundments to contain oysters, shrimp, lobsters, trout, and salmon. These animals are fed special diets and grow at above-normal rates. Several large corporations have invested heavily in large-scale mariculture. Among them are Union Carbide (salmon), Carrier Corporation (oysters), Ralston Purina (shrimp), and Sanders Associates (lobsters). When fish are grown in restrictive impoundments surrounded by a submerged mesh fence, the operation becomes "fish farming" and may be the most economical way to increase the productivity of fish.

In Java, about 60 per cent of the fish protein comes from the milk fish *(Chanos chanos)*. The fish are collected at sea as juveniles and placed in shallow brackish ponds. Organic fertilizers such as sewage are added. The fertilizer enables a mat of algae to grow on the bottom of the pond, providing food for the fish, which are grown to maturity in captivity.

Oysters usually grow in nature on the bottom, on a hard substrate such as a shell or rock in shallow water, especially in estuaries rich in phytoplankton food. Production may be limited by pollution, including industrial and domestic sewage, and by the effects of dredging, which may increase sedimentation on the oyster beds themselves. In 1967, 29,000 tons of eastern oysters were harvested in the United States, whereas the production in 1880 was 76,000 tons. Increased pollution, dredging, and covering of large areas of suitable oyster habitats with dredge spoils or other fill caused much of the decline over the years. Pollution abatement may increase production of oysters. The growth of oysters is also limited by the area that is populated and by natural siltation, which may bury the oysters or may clog their filtering apparatus.

Oysters are frequently grown either on racks suspended from rafts or attached to ropes suspended from rafts or floats (Fig. 12.18). The

FIGURE 12.18 Oyster culture farm in Ago Bay, Japan. The oysters are placed in wire cages which are suspended in the calm ocean waters from rafts and carefully tended and protected for three to five years. (Photo courtesy of Japan Information Service.)

racks keep the oysters out of the sediment but in the tidal waters, which are rich in phytoplankton food. Whereas bottom-grown oysters are limited by the *area* covered, raft-grown oysters are limited by the *volume* of water, since more than one oyster may be grown on a vertically hung rope. Many more oysters may be grown under a given area of rafts than on the bottom, assuming a plentiful source of food. By growing oysters on ropes or racks, it is possible to increase production from less than half a metric ton per acre for bottom-grown oysters to as much as 30 metric tons per acre. They are also protected from predation by sea stars.

In the past few years, experimental culture of various crustaceans, including shrimp and lobsters, indicates that (1) plenty of food and (2) water that is warmer than normal for the animals contribute to very rapid growth of these organisms (four times the normal growth rate for lobster). This suggests that commercial cultivation of these crustaceans in tanks or impoundments, possibly utilizing the now wasted heated water from power plants, may soon be possible.

Some marine plants are also of economic importance. Most of the ice cream that we eat is stabilized with a substance called algin, extracted from the giant seaweed (kelp), *Macrocystis*. Algin is used as a smoothing agent and helps to keep the ice cream from becoming watery. Algin is also used in salad dressings and candy bars and in the production of synthetic fibers and plastics. *Macrocystis* is harvested off the coast of California by huge barges that cut the tops of the kelp with blades, much as a lawn mower trims a lawn. This seaweed (Fig. 12.19) grows up to 60 meters long in water from eight to 30 meters deep, off Southern California. The new shoots and reproductive organs grow from the base, which is attached to rocks on the sea bed. The flexible stem with leaflike fronds is held upright by gas-filled floats which the plant produces. In order to maintain the "forest" in a healthy condition, only the portion within four feet of the surface is harvested. Harvesting reduces the shade on the young shoots, probably increasing their rate of production by exposing them to more sunlight.

Another plant, the red alga *Porphyra,* is grown by the Japanese on sticks placed in the sea. It is periodically harvested, dried, and eaten in dishes such as seaweed soup.

Mariculture may even derive benefit from human pollution. From highly populated and industrial shore cities, tremendous amounts of nutrients are dumped into our estuaries and near-shore ocean as sewage. If uncontrolled, these nutrients and organic matter can be dangerous. Proper treatment, however, can make these nutrients available to grow phytoplankton in tanks as a food for clams and oysters. The attached algae that grow on the sides of the tanks may be grazed upon by snails, which in turn may someday be eaten by people, or at least ground up as animal feed.

FIGURE 12.19 The giant kelp *Macrocystis pyrifera*, a major source of algin.

OTHER USES OF MARINE ORGANISMS

Marine organisms are tremendously important in biomedical research. The giant nerves of the squid are ideal for the study of the travel of nerve impulses. The basic mechanisms of reproduction, fertilization, and development have been discovered by using echinoderms such as sea urchins and sea stars.

Many chemicals of pharmacologic interest have been extracted from marine animals. The simple sponge produces a chemical that appears to be effective in the treatment of some virus infections and of leukemia in mammals. Poisons found in sea stars and sea cucumbers include digitalis, an important drug used in treating heart disorders. Other chemicals from echinoderms have slowed the growth of tumors in laboratory animals. The potential of marine organisms as useful sources of medical information and drugs has only begun to be realized.

Marine organisms will continue to be a vital source of food as well as a source of drugs. Mariculture may ease the pressure on the sea fisheries, but even mariculture is a finite source of food. Productivity is limited by the energy reaching the sea from the sun. Therefore, the production of fish and other economically important marine organisms is also limited.

SUMMARY

The survival of all organisms depends on an adequate source of energy. Photosynthetic plants derive their energy from sunlight and convert it into chemical energy in organic compounds. Animals get their energy from the food they eat. Nearly all of the energy in marine food sources was originally captured as light by plants near the sea surface.

Productivity of the sea varies from place to place, tending to be greatest where light and inorganic nutrients such as nitrate and phosphate are abundant. These areas include relatively shallow coastal regions, where vertical mixing renews the nutrient supply, and regions subject to the upwelling of deeper nutrient-rich water toward the surface. Great fisheries have been established in such areas, as off the west coast of South America. The nutrient supply off that region, however, is sometimes lessened by the phenomenon known as El Niño. During El Niño years, the upwelling temporarily weakens and results in decreased food for fish and consequently great fish mortality (see also Chapter 8). In recent years El Niño and over-fishing have combined to bring about a near collapse of that fishery. The total worldwide fish catch, however, has continued to rise in spite of such local problems.

The abundance of species of marine organisms is always changing. Increases and decreases in abundance are normal occurrences. The changes may follow definite seasonal trends or may be shorter or longer in duration. One species declines and another takes its place in the ecosystem, as part of a succession of species. Succession occurs as a clean surface becomes fouled by changing communities of marine life, resulting in the final stable climax community. Succession of plankton species also takes place in temperate and colder regions. Each species has its own environmental requirements which may be met during part of the year. Furthermore, one species sometimes releases chemical by-products that inhibit or promote the growth of other species.

The feeding (trophic) relationships of marine organisms are complex and interrelated. No form of life in the sea is completely isolated from any other. They are all tied to one another through a vast food web. Plants are food for a variety of herbivores which are, in turn, eaten by a variety of carnivores. When plants and animals die, their dead bodies are food for scavengers, which are eaten in turn by carnivores. Waste material and other organic remains are decomposed by bacteria, returning nutrients to the sea for the production of more plants.

One future use of the sea as a source of food will require careful management in cooperation with natural processes. Careful control of contamination by domestic and industrial wastes must be achieved. Regulation of fish catches in relation to the productivity of the fish population will help to avoid over-fishing of popular fish species. Effective use of the potential productivity of near-shore environments through mariculture offers further opportunities for increased food yield from the sea.

IMPORTANT TERMS

Anoxic
Biochemical oxygen demand
Biomass
Climax community
Eutrophic
Fish protein concentrate
Food web
Gross production
Hypertrophic
Nannoplankton
Net production
Oligotrophic
Photosynthesis
Phytoplankton
Polyculture
Primary production
Pyramid of biomass
Pyramid of energy
Standing crop
Succession
Sustained yield
Trophic efficiency
Trophic relationships

STUDY QUESTIONS

1. What is the difference between a food chain and a food web? Give an example of each.
2. Discuss the comment, "Polluted salt marshes are worthless except for providing space for housing developments and shopping centers."
3. Why is the efficiency of energy transfer so high in upwelling waters?
4. Discuss the stages in succession from a clean wooden piling to one that is well fouled.
5. Make a list of organisms that are lost when a marsh becomes polluted (use Fig. 12.14 as a source). Compare this list with the organisms that remain in the disturbed marsh. What percentage of each list consists of birds, of fish, of mollusks (clams, snails, etc.)? Discuss possible reasons for the difference in the composition of these lists.
6. Discuss the causes of and possible remedies for the red-tide problem.
7. Why did the North Sea catch of haddock increase immediately after World Wars I and II compared with prewar statistics? Why, in subsequent postwar years, did the fish catch quickly return to lower levels?
8. Why are more anchovies caught each year than any other fish when there are others, such as *Cyclothone*, that are more abundant?
9. Discuss the polyculture method of mariculture.

SUGGESTED READINGS

Dale, Barrie, and Clarice M. Yentsch. 1978 (March). Red Tide and Paralytic Shellfish Poisoning. *Oceanus.*
Galapagos Biology Expedition Participants 1979 (February). Galapagos '79: Initial Findings of a Deep-Sea Biological Quest. *Oceanus.*
Gulland, J. A. (ed.). 1971. *The Fish Resources of the Ocean.* West by Fleet, Surrey, England, Fishing News (Books) Ltd.

Holt, S. J. 1969 (September). Food Resources of the Oceans. *Scientific American*.

Hunter, S. H., and John McLaughlin. 1958 (August). Poisonous Tides. *Scientific American*.

Idyll, C. P. 1973 (June). The Anchovy Crisis. *Scientific American*.

Karl, D. M., C. O. Wirsen, and H. W. Jannasch. 1980. Deep-Sea Primary Production at the Galápagos Hydrothermal Vents. *Science*, 207(4437):1345--1347.

Parsons, T. R. and M. Takahashi. 1973. *Biological Oceanographic Processes*. New York, Pergamon Press.

Pinchot, Gifford B. 1970 (December). Marine Farming. *Scientific American*.

Raymont, J. E. G. 1963. *Plankton and Productivity in the Oceans*. New York, Macmillan.

Russell-Hunter, W. D. 1970. *Aquatic Productivity*. New York. Macmillan.

Ryther, John. 1969. Photosynthesis and Fish Production in the Sea. *Science*, 178:72--76.

Ryther, J. H. 1979. Fuels from marine biomass. *Oceanus*, 22(4):48–58.

Maps and Navigation

LATITUDE AND LONGITUDE

The position of a point on the earth's surface may be defined by its latitude and longitude. *Latitude* is the angular distance of a point on the earth's surface northward or southward from the equator. The latitude of a point on the equator is 0° and that of a point at the north or south pole is 90° north (N) or south (S), respectively. A parallel (of latitude) is a line connecting points of equal latitude. Parallels are circles drawn on the earth parallel to the equatorial plane of the earth (Fig. A.1).

The length of a degree of latitude remains nearly constant everywhere on the earth. For most navigational purposes, the length of a degree of latitude is assumed to be equal to 60 nautical miles; that is, one minute of latitude equals one nautical mile (6076 ft or 1852 meters).

The *longitude* of a point on the earth's surface is the angular distance of that point east or west from the *prime meridian,* which is a half circle on the earth extending from the north pole to the south pole and passing through Greenwich, near London, England. The prime meridian is the 0° longitude line. Longitudes increase eastward and westward of the prime meridian, and are referred to as east (E) or west (W) longitudes, respectively (Fig. A.1). The greatest longitude is 180° and is found along the meridian halfway around the earth from the prime meridian. The 180° meridian coincides approximately with the International Date Line in the Pacific Ocean. All longitude lines converge at the poles.

The length of a degree of longitude at the equator is 60 nautical miles, the same as the length of a degree of latitude. Since all longitude lines converge at the poles, the length of a degree of longitude decreases toward the poles, and at the poles themselves it is zero.

405

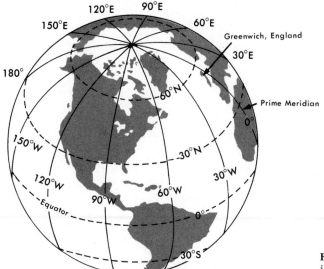

FIGURE A.1 Parallels (dashed curves) and meridians (solid curves) of the earth.

GREAT CIRCLES AND RHUMB LINES

A *great circle* is a line of intersection of the earth's surface and any plane passing through the center of the earth. The equator and the meridians, if extended completely around the earth, are special examples of great circles. Great circles are important in navigation because the arc of a great circle connecting two points represents the shortest route between these two points. A *rhumb line* is a line on the earth that indicates a constant direction. A rhumb line makes the same oblique angle with all meridians. Since the meridians and parallels, including the equator, always indicate constant directions, they can be thought of as special cases of rhumb lines (Fig. A.2).

MAP PROJECTIONS

A map is a very important oceanographical tool, not only for displaying oceanographical data, but also for use as a navigational aid. Many different kinds of map projections are currently in use and belong to the following major categories: *cylindrical projection, conic projection,* and *azimuthal projection* (Fig. A.3).

A cylindrical projection is obtained by placing a cylinder tangent to the earth along the equator or any great circle. If the cylinder touches the equator, as shown in Fig. A.3*A*, the meridians will appear as vertical lines on the cylinder and the parallels of latitude will be peripendicular to the meridians.

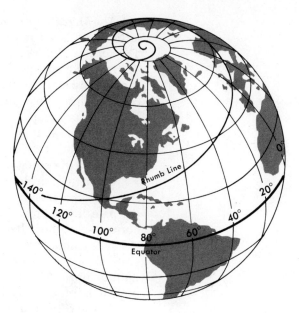

FIGURE A.2 A rhumb line on the earth traces a line with constant direction. Note that the latitudes and longitudes are all rhumb lines. (After Bowditch, 1977)

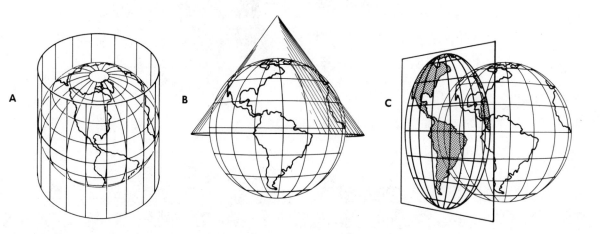

FIGURE A.3 Principles of map projection: A, cylindrical projection; B, conic projection; C, azimuthal projection.

The *Mercator projection* (Fig. A.4) is a special type of cylindrical projection and is the most commonly used one for navigational purposes. The spacings between parallels progressively increase toward the poles. The main advantage of using the Mercator map is that directions are true; that is, any straight line drawn on this map corresponds to a constant compass direction. Consequently, all rhumb lines appear as straight lines on a Mercator map. As seen earlier, all meridians and parallels are also rhumb lines, as they also indicate true directions. All great circles on the earth except the equator and the meridians appear as curved lines concave to the equator.

Since the meridians are all parallel to each other on Mercator projections, and parallels are spaced farther apart toward the poles, areas near the poles are highly exaggerated in size, although their shapes remain correct. Another disadvantage of the Mercator projection is that distances cannot be measured using a single scale for the whole map. The scale expands toward the poles proportional to the expansion of the latitude lines. As we have seen earlier, the length of a degree of latitude anywhere on earth is about 60 nautical miles.

In the conic projection, a cone is placed on the globe and points on

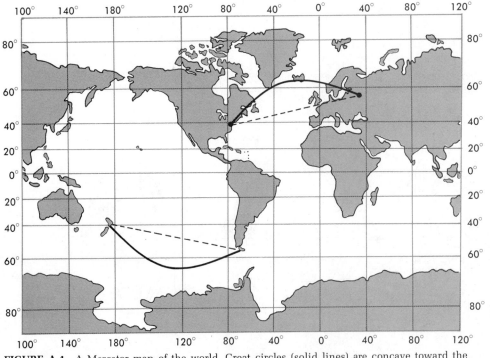

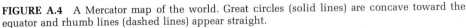

FIGURE A.4 A Mercator map of the world. Great circles (solid lines) are concave toward the equator and rhumb lines (dashed lines) appear straight.

the earth are projected on the cone. If the axes of the earth and the cone coincide, a simple conic projection results (Fig. A.5). On this projection, the meridians are straight lines converging toward the pole and parallels are concentric circles. If a series of cones are used, the resulting projection is called a *polyconic projection* (Fig. A.6). The familiar topographic maps published by the U.S. Geological Survey are polyconic maps.

The azimuthal projection is obtained by projecting points on the earth to a plane tangent to the earth. A special kind of azimuthal projection is the *gnomonic projection* (also called great-circle charts), in which points on the earth are projected geometrically from the center of the earth. Fig. A.7 illustrates a gnomonic map obtained by placing the plane tangent to the pole. In the azimuthal projection, all great circles appear as straight lines and bearings from the point of tangency always indicate true directions. Great-circle charts, especially of the polar regions, are commonly used in marine navigation.

NAVIGATION

Navigation on the sea involves the determination of the position of a ship, that is, its latitude and longitude. In *celestial navigation,* latitude and longitude are determined by measuring the relative positions of certain stars. For example, in the Northern Hemisphere, the latitude of

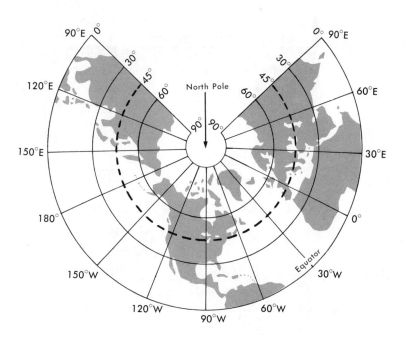

FIGURE A.5 A simple conic map of the Northern Hemisphere. The cone was tangential to the globe at 45° latitude (dashed line). (After Bowditch, 1977)

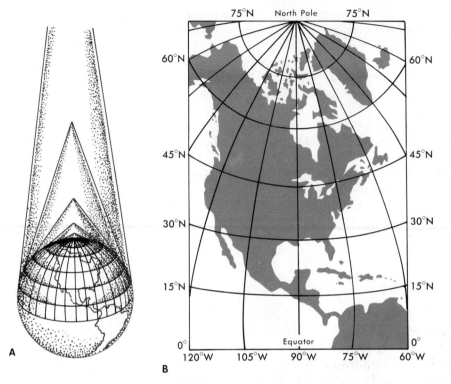

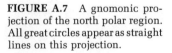

FIGURE A.6 A polyconic projection (*A*) and the resulting map (*B*).

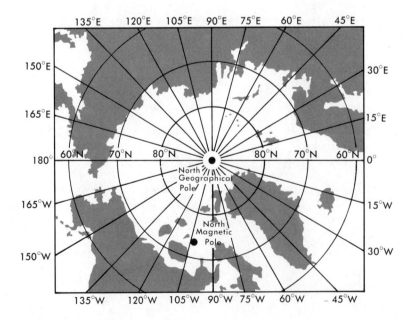

FIGURE A.7 A gnomonic projection of the north polar region. All great circles appear as straight lines on this projection.

a place is the same as the altitude of Polaris, the North Star. Altitude is the angular elevation of a star measured above the horizon. Polaris is almost directly overhead at the north pole. Since stars are not generally visible during the day, the latitude of a place can be determined by measuring the altitude of the sun at noon. Reference to tables is then required to account for seasonal changes in the relative orientation of the earth to the sun.

The longitude of a place can be determined if we know accurately the local time and Greenwich time (the time in Greenwich, England). Greenwich time can be determined from an accurate clock (called a *chronometer*) onboard the ship or from radio broadcasts. Local time can be obtained from the altitude of the sun, usually measured at noon. The difference between local time and Greenwich time can be used to calculate the longitude. The earth rotates from west to east 15° every hour. For example, when the sun is directly overhead in New York City (noon local time), the Greenwich time is 4:56 P.M. (From this we can determine that New York City is at 76° W longitude.)

A compass is an essential tool for navigation. It provides direct information as to the direction of the earth's magnetic poles. Since the locations of the poles are well known and are near the earth's geographic poles (Fig. A.7), simple corrections may be applied to determine the direction of true north.

DEAD RECKONING

Dead reckoning is the technique of determining the position of a ship at sea relative to a previously known position. It can be done if the direction and speed of the ship are known accurately. The position determined in this manner is only approximate as the speed and direction are affected by wind and ocean currents. It is, however, a basic, useful technique, especially when celestial navigation is impeded by cloud cover.

ELECTRONIC NAVIGATION

There are many electronic navigational techniques in operation today that are far superior to the ones mentioned above. One of these is LORAN (*LO*ng *RA*nge *N*avigation). Synchronized radio signals are transmitted from two shore-based stations. A ship's position can be determined by knowing the time difference in the reception of the signals from these two stations. The range of LORAN is of the order of 1000 km. Many areas of the world ocean are now covered by LORAN stations.

The familiar RADAR (*RA*dio *D*etection *A*nd *R*anging) usually works on the basis of reflection of a radio signal from a target. It is an important tool for determining the position of a ship, especially for safe navigation during reduced visibility. The range of RADAR, however, is limited to line of sight.

SATELLITE NAVIGATION

For navigation over the open sea, the most accurate method is satellite navigation. Special satellites orbit the earth and are continuously monitored by land-based tracking stations. The satellites emit signals at predetermined frequencies that give their precise locations as corrected by signals prepared by the tracking stations. Specialized satellite navigation equipment onboard the ship receives the satellite signals and the navigator determines the ship's position, taking into account information received regarding the satellite's location and changes in the radio frequency due to the *Doppler effect.* The Doppler effect is the phenomenon by which the radio waves, as well as light and sound waves, increase in frequency as the transmitter approaches the receiver and decrease as it moves away. Satellite navigation is accurate to within about 400 meters.

Refraction of Waves

To a diver in the water, the sun appears to be closer to the vertical than it really is. This is because the light waves are *refracted,* or bent, as they travel from a medium in which they have a higher speed (air) into one in which they have a slower speed (water). In fact, all waves, whether they are light waves, water waves, or sound waves, undergo refraction under certain conditions. The principle governing the refraction of waves (Fig. B.1) is known as *Snell's Law* and is stated as follows:

$$\frac{\text{Sine } i}{C_a} = \frac{\text{Sine } r}{C_b}$$

where
    i is the angle of incidence, which equals the angle of reflection (R),
    r is the angle of refraction,
    C_a is the speed of the wave in the first medium, and
    C_b is the speed of the wave in the second medium.

Thus waves tend to be refracted toward the perpendicular when they slow down and are refracted away from the perpendicular when they speed up. However, no refraction occurs as the waves enter the second medium perpendicular to the interface; that is, when the angle of incidence (i) is zero, the angle of refraction (r) is zero. Note that only a portion of the wave energy enters the lower medium. Part of it is reflected by the surface. The "media" that we are discussing here need not be as distinctly different as air and water; they may be waters or rocks of slightly different densities. Light waves decrease in speed, whereas sound waves increase in speed, as they enter media of greater densities. In the case of water waves, "the media" refers to regions of different water depths. These waters slow down as they enter shallower water. Water waves, however, are refracted *only* when they "feel the bottom"; that is, when the water depth is less than half the wavelength.

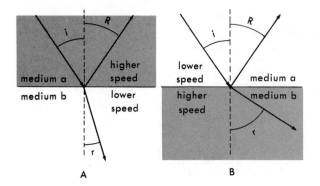

FIGURE B.1 *A*, The refraction of waves as they enter a medium in which they have a lower wave speed. *B*, Refraction as they enter a medium in which they have a higher wave speed. The angle of incidence = i; r is the angle of refraction; and R is the angle of reflection and equals i.

Figure B.2 illustrates the refracted paths of some seismic (sound) waves through different rock layers of increasing density with depth. In general, increasing rock density results in an increasing seismic wave speed. Note that each seismic ray, after refraction, penetrates to a maximum depth, at which it may travel along the interface with the velocity that it would have in the lower layer. As it travels along the interface it sends seismic waves back toward the surface, as shown in the figure.

Not all media are arranged in discrete layers. In most cases, density changes continuously with depth. Figure B.3 illustrates the paths of some seismic waves as they travel through rocks in which the density (and, consequently, the velocity) increases gradually with depth. The resulting wave paths are curved, as shown in the figure. Sound waves in the ocean behave in a similar way.

The refraction of light in the sea is shown in Figure B.4. As light penetrates deeper into the water its speed decreases gradually, owing to the increasing density with depth, and hence the light rays are curved downward. Water waves are refracted in a manner similar to that of light waves. As they enter shallower water near the shore, they slow down and are refracted as shown in the figure. After refraction, water waves tend to arrive perpendicular to the shore.

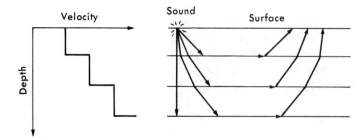

FIGURE B.2 The paths of some sound waves through layers (rock or water) of increasing density and consequently of increasing wave speeds.

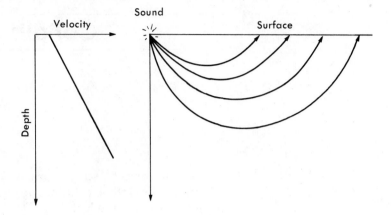

FIGURE B.3   The paths of some sound waves through rock (or water) in which density increases gradually with depth.

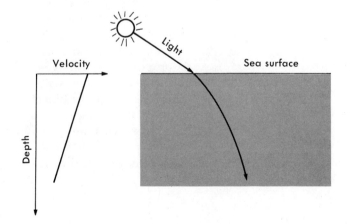

FIGURE B.4   Path of a light ray in water in which density increases gradually with depth and the velocity of the light decreases with depth. As water waves enter shallower water (lower speed) near the shore, they are refracted in a similar manner. After refraction these waves are more nearly perpendicular to the shore than before refraction.

Appendix C

Coriolis Force

Moving objects, such as air, water, airplanes, and satellites, are affected by the earth's rotation. The result is an apparent deflection in the paths of the objects, which is due to the fact that the moving objects are observed from a frame of reference fixed to a rotating earth. For example, a missile fired initially in a straight direction will appear to an observer on the earth to be deflected to the right of its direction of travel in the Northern Hemisphere and to the left of its direction of travel in the Southern Hemisphere (Fig. C.1). However, to an observer in space the missile would appear to be moving in a straight path over a rotating earth. Of course, there would be no deflection at all if the earth were not rotating. Water particles associated with ocean currents also tend to be deflected to the right in the Northern Hemisphere and to the left in the Southern Hemisphere on the rotating earth, assuming no counteracting forces exist.

This phenomenon (also known as the Coriolis Effect) results in a force called the *Coriolis force,* named after Gaspard Gustav de Coriolis, the French mathematician who described it in 1835. The Coriolis force is defined by the following equation:

$$\text{Coriolis force per unit mass} = 2\Omega V \sin \phi$$

where Ω is the angular velocity of the earth (360°, or 2π radians, per 24 hours), V is the velocity of the object relative to the earth, and ϕ is the latitude. The Coriolis force is zero for an object moving directly over the equator ($\phi = 0$) and increases with increasing latitude. Coriolis force is directed perpendicular to the direction of the object's movement and is proportional to its velocity. However, the amount of deflection becomes less as the velocity of the object increases (at a given latitude). This is because an object moving rapidly (such as a missile) over a given distance allows less rotation of the earth than would an object moving much more slowly (such as an ocean current) over the same distance. Let us now take a closer look at Coriolis deflection.

An object fixed or resting on the earth's surface will be subjected

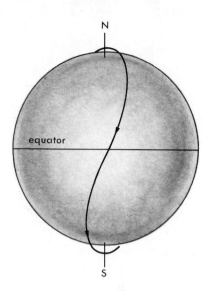

FIGURE C.1 The path of a missile as it travels around the earth. Note that it is deflected toward the right in the Northern Hemisphere and toward the left in the Southern Hemisphere.

to a gravitational force toward the earth's center and to a centrifugal force due to the earth's rotation directed away from the earth's axis of rotation, as shown in Figure C.2. The centrifugal force increases from zero at the poles to a maximum at the equator. It is the horizontal component of the centrifugal force (Fig. C.3) that has resulted in the flattening of the earth at the poles (the polar radius of the earth is about 22 km shorter than the equatorial radius) and a bulging at the equator. This probably occurred very early in the earth's history, before even the oceans were formed. Every particle of the earth, including water particles, is in equilibrium, provided it is not in motion.

Consider a water particle moving from west to east in the Northern Hemisphere. It will be moving eastward at a speed equal to that of a fixed point on the earth at that latitude (Fig. C.4) plus the particle's own speed with respect to the earth. Thus, the eastward-moving water particle will be subjected to an increased centrifugal force (and hence a greater horizontal component of this force) than that of a fixed point at the same latitude. On the other hand, a particle moving due west will experience a decreased centrifugal force as compared with a particle at rest.

Figure C.5 shows the differences in centrifugal force between particles that are moving eastward or westward and those that are at rest at the same latitude in the Northern Hemisphere. Their horizontal components, directed 90° toward the right of the particle motion, represent the Coriolis force.

The Coriolis deflection of objects moving northward or southward over the earth's surface can be explained in the following manner: Consider an object moving southward in the Northern Hemisphere. In addition to its southward speed, it also maintains an eastward motion

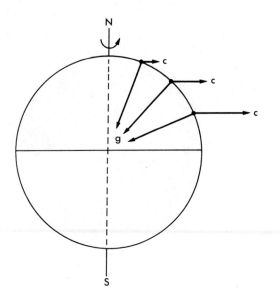

FIGURE C.2 The centrifugal (c) and gravitational (g) forces acting on particles at rest on the earth's surface. The centrifugal forces shown are highly exaggerated, as they are actually much smaller in magnitude than the gravitational force.

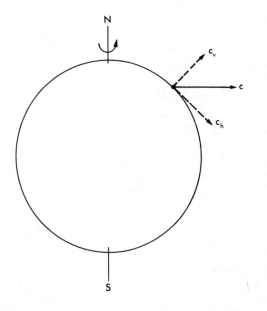

FIGURE C.3 The horizontal (c_h) and the vertical (c_v) components of the centrifugal force (c) acting on a particle at the earth's surface.

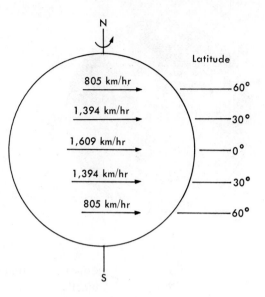

FIGURE C.4   The linear (tangential) velocities of fixed particles at different latitudes as the earth rotates on its axis.

FIGURE C.5   *A*, The difference (d) in centrifugal force of a particle that is moving eastward and one that is resting at the same latitude. *B*, Difference in centrifugal force of a westward-moving particle and a resting particle. d_h and d_v represent the horizontal and vertical components of these differences, respectively. d_h represents the Coriolis force; d_v simply diminishes or augments the effect of gravity on the particle.

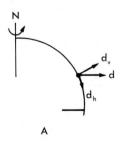

A

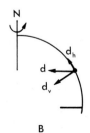

B

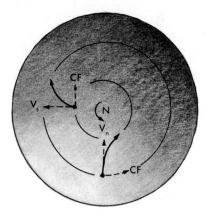

FIGURE C.6 Polar view of the Coriolis forces (CF) acting upon particles initially moving southward (V_s) and northward (V_n) in the Northern Hemisphere. The heavy arrows indicate the paths of the particles' movements.

equivalent to that of a fixed particle on the earth at the starting point. The object will be deflected to the right of its initial path, since the earth's surface under its path has a greater eastward speed than does the moving object (Fig. C.6). On the other hand, if the object moves northward, its eastward component of velocity is retained as it travels toward a point where the eastward motion of the earth's surface is less. The object tends to drift eastward, or to the right.

It is important to note that Coriolis force acts on all moving objects, regardless of the direction in which they are moving, and that it tends to deflect them to the right of their direction of motion in the Northern Hemisphere and to the left in the Southern Hemisphere.

Conversion Factors
and Constants

DISTANCE

1 kilometer (km) = 1000 meters = 0.6214 statute mile = 0.5396 nautical mile

1 statute mile (mi) = 5280 feet = 0.8684 nautical mile = 1609 meters

1 nautical mile (naut mi) = 6076 feet = 1.152 mi = 1852 meters

1 meter (m) = 100 centimeters = 3.281 feet = 0.5468 fathom

1 foot (ft) = 0.3048 m

1 fathom (fm) = 6 ft = 1.829 m

1 centimeter (cm) = 10 millimeters = 0.3937 inch

1 millimeter (mm) = 1000 microns (μ) or micrometers (μm) = 0.03937 inch

1 inch (in) = 2.540 cm

SPEED

1 km/hr = 27.78 cm/sec = 0.6214 mi/hr = 0.5396 knots

1 m/sec = 3.6 km/hr = 2.237 mi/hr

1 naut mi/hr (knot) = 1.151 mi/hr = 1.852 km/hr = 51.44 cm/sec

1 mi/hr = 0.8684 knot = 1.6093 km/hr

1 ft/sec = 0.6818 mi/hr = 0.5921 knot = 1.097 km/hr = 30.48 cm/sec

VOLUME

1 cubic (cu) km = 0.2399 cu mi

1 cu m = 10^6 cu cm (cc) = 3.531 cu ft = 1.308 cu yards = 264.2 gallons = 1000 liters

1 cc = 0.06102 cu in

1 cu ft = 0.0283 cu m

1 cu in = 16.39 cc

1 liter (l) = 1000 cc = 1.05 quarts = 0.26 gallon

1 gallon = 3.785 l

MASS

1 kilogram (kg) = 1000 grams (gm) = 2.205 pounds (lb) = 35.27 ounces (oz)
1 metric ton = 1000 kilograms = 1.102 short tons
1 short ton = 0.9 metric ton
1 pound = 453.5 grams

DENSITY

1 gm/cc = 1000 km/cu m = 62.3 lb/cu ft

TEMPERATURE

° Centigrade or celsius (°C) = (°F − 32) $\frac{5}{9}$
° Fahrenheit (°F) = ($\frac{9}{5}$ °C) + 32
° Kelvin (°K) = °C + 273.18

CONSTANTS

Volume of the oceans = 1.4×10^9 km³ (1.4 billion cubic kilometers)
Total mass of sea water = 1.4×10^{24} gm
Average depth of the oceans = 3800 meters
Average depth of the Pacific Ocean = 3900 meters
Average depth of the Indian Ocean = 3800 meters
Average depth of the Atlantic Ocean = 3600 meters
Average depth of the Arctic Ocean = 1100 meters
Total area of the oceans = 362×10^6 km²
Average elevation of land = 840 meters
Equatorial radius of the earth = 6378 km
Polar radius of the earth = 6356 km
Average density of the earth = 5.5 gm/cc
Average density of the moon = 3.3 gm/cc
Average density of the sun = 1.4 gm/cc
Mass of the earth = 5.976×10^{27} gm
Mass of the moon = 7.347×10^{25} gm
Mass of the sun = 1.971×10^{33} gm
Gravitational acceleration of the earth = 980 cm/sec² = 9.8 meters/sec²
Mean distance to moon from earth = 384,393 km (238,860 mi)
Mean distance to sun from earth = 149,450,000 km (92,900,000 mi)

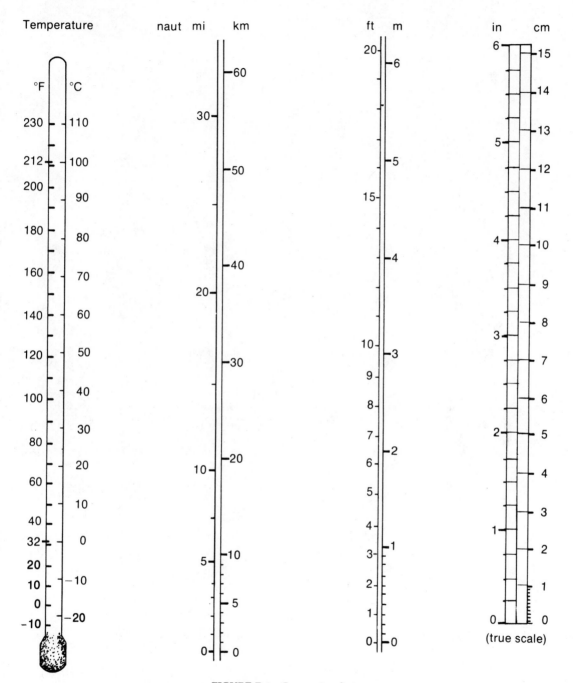

FIGURE D.1 Conversion factors.

Glossary

A

abyssal hills. Small extinct volcanoes (usually less than a kilometer in height) distributed over large areas of the ocean floor.

abyssal plain. Almost featureless or flat region of the ocean floor, produced by the burial of abyssal hills by sediments.

abyssopelagic. Pertaining to the deep oceanic pelagic environment between 3,000 and 6,000 meters of depth.

aerobic. Requiring free oxygen for organic growth.

alga (pl., **algae**). The simplest plants, having no stems, roots, or leaves. They may be single-celled (such as diatoms) or quite large (such as the seaweeds). The term *algae* is most commonly applied to the seaweeds.

amphidromic point. A point at which there is no tidal fluctuation but around which tides (cotidal lines) rotate.

Annelida. Phylum comprising segmented worms, including clamworms and earthworms.

anoxic. Lacking in free oxygen. Organisms not requiring free oxygen are also termed *anaerobic*.

aphotic. Devoid of light.

asthenosphere. See *low velocity layer*.

atoll. A ringlike island made up of coral reefs.

azoic zone. A zone in the sea originally postulated to be lifeless. We now know that life exists at all depths.

B

barrier island. An offshore island formed parallel to the shore.

barrier spit. Similar to *barrier island* except that one end is attached to the mainland.

basalt. A fine-grained, dark-colored, igneous rock, commonly produced by volcanic eruptions. The oceanic crust and the lower continental crust are believed to be basaltic in nature.

bathyal. Pertaining to the environments of the continental slope. Organisms living at this depth are called *bathyal* organisms.

bathypelagic. Pertaining to the oceanic pelagic environment between 1,000 and 3,000 meters.

bathythermograph. A device that measures and records the change in temperature with depth.

beach. The region of unconsolidated materials extending from the low tide line to the upper-most limit of wave action, usually represented by cliffs, sand dunes, or permanent vegetation.

beach drifting. The process of lateral transport of beach material, caused by obliquely approaching waves.

benthic. Pertaining to the environments of the sea floor.

berm. A nearly horizontal surface located above the high tide line and produced by the deposition of beach materials by wave action.

biochemical oxygen demand (BOD). A drain on the oxygen supply of an environment produced by chemical oxidation and biological respiration

bioluminescence. The production of a cool light by certain plants, animals, and bacteria.

biomass. The quantity (weight) of living organisms present in a given volume or area.

bloom. A sudden burst or growth of phyto-plankton, leading to massive populations of the bloom organism and sometimes resulting in discolored or toxic water (red-tides).

brackish. Pertaining to water, generally estuarine in which the salinity ranges from about 0.50 to 17 parts per thousand ($^o/oo$) by weight.

breakwater. An offshore artificial structure built parallel to the shore to protect beaches and harbors from large waves.

C

calcareous. Composed of calcium carbonate ($CaCO_3$).

centrifugal force. The force directed away from the center of curvature of a body moving in a curved path.

chlorinity. The chloride content of sea water expressed in parts per thousand ($^0/_{00}$) by weight. It is now defined as the weight in grams of silver required to precipitate the chloride, bromide, fluoride, and iodide in 0.3285233 kg of sea water.

chromatophore. Pigmented patches, on the surface of an animal, that can be expanded or contracted, changing the animal's color.

ciliate. A single-celled animal that moves by beating many tiny hairlike threads called cilia.

climax community. The temporally stable assemblage of interrelated organisms sharing a common habitat.

coccolith. Calcareous ($CaCO_3$) buttonlike plates covering the single-celled plants known as coccolithophores.

continental rise. A broad sedimentary wedge that may be present at the foot of the continental slope.

continuous seismic profiling (sparker profiling). Method of obtaining a continuous profile record of geological structures below the sea floor by the use of a strong energy (sound) source.

core. The central region of the earth, extending from a depth of 2,900 to 6,370 km (the center of the earth). The outer core (2,900 to 5,100 km) resembles a liquid, and the inner core (below 5,100 km) resembles a solid.

core method. The technique of tracing the movements of deep ocean currents by the use of temperature-salinity curves.

Coriolis effect. The deflection of moving particles near the earth's surface due to the earth's rotation.

cotidal line. A line connecting points of simultaneous high tide in the ocean.

Cromwell current. The equatorial undercurrent in the Pacific Ocean.

crust. The thin cover of the earth above the mantle. In oceanic regions, the crust has an average thickness of about 10 km, and in continental areas it has an average thickness of 35 km.

D

decomposer. An organism that causes the breakdown of organic matter and the production of inorganic nutrients.

deep scattering layer (DSL). A subsurface stratified population of animals that scatter sound. They may cause the appearance of a "false bottom" on echograms.

detritus. Dead organic matter.

diatom. A single-celled plant with two overlapping, siliceous (SiO_2), porous shells. Diatoms are sometimes called the "grass of the sea" because of their importance as food for marine herbivores.

diatomaceous earth. A sedimentary rock, largely composed of diatom shells.

dinoflagellate. A single-celled, flagellated plant. Some are bioluminescent; others may cause toxic red-tides.

disphotic. Pertaining to regions where some light exists but not enough for plant growth.

diurnal. Daily, as in *diurnal* tides, which comprise one high tide and one low tide each day.

dust cloud. A diffuse cloud of dust and gas, composed mostly of hydrogen. According to the *dust cloud hypothesis,* all celestial bodies were formed from dust clouds.

E

echinoderm. A spiny-skinned, radially symmetrical animal. This group includes sea stars, sea urchins, and sea cucumbers.

echogram. A continuous record of the ocean floor profile produced on board a moving ship by the principle of echo sounding.

Ekman transport. The net transport of a surface layer of water 90° to the right (Northern Hemisphere) or left (Southern Hemisphere) of the wind direction. This is an expression of the Coriolis effect.

El Niño. A periodic decline in productivity and resulting death of marine life, due to an inundation of warm nutrient-poor water along the west coast of South America.

epi-. Prefix meaning: *on, upper,* or *attached to.* From the Greek *epi* = on.

epifauna. Animals that live upon the sea floor.

epipelagic. Pertaining to the well-lighted surface waters of the sea.

equatorial countercurrent. Current moving eastward near the equator between the North and South Equatorial Currents in all oceans.

equatorial undercurrents. Eastward-flowing subsurface current found beneath the equator in all oceans.

estuary. A bay or river mouth where sea water and fresh water meet and mix and where sea water is measurably diluted.

euphotic. Pertaining to the well-lighted regions where plants may live and grow.

eutrophic. Pertaining to nutrient-enriched conditions, with large biomass and moderately high diversity of species.

F

feldspar. A mineral composed of silicates of potassium, sodium, or calcium.

fetch. The distance over which the wind blows in the wave-generating area.

fiord (fjord). A narrow deep arm of the sea; they are usually formed by glacial erosion.

flagellate. A single-celled organism that moves by beating a whiplike thread, or flagellum.

food web. The feeding (trophic) relationships among members of a community.

foraminiferan. A single-celled animal with pseudopodia and a calcareous ($CaCO_3$), porous shell.

G

granite. A coarse-grained, light-colored igneous rock, formed within the earth by the cooling of molten materials. The upper continental crust is believed to be granitic in nature.

groin. A short artificial structure built perpendicular to the shoreline in an effort to control beach erosion by trapping sediments carried by longshore currents. Deposition occurs on one side of the groin but beach erosion continues on the other side.

guyot. A flat-topped seamount. The flat surface is believed to have been produced by wave erosion.

gyre. A large, closed system of surface current circulation.

H

hadal. Pertaining to the environment of the trenches.

hadopelagic. The pelagic environments of the trenches (generally at depths greater than 6,000 meters).

halocline. A sharp change in salinity with depth.

heat capacity. The ratio of the heat in calories per gram absorbed (or released) by a substance to the corresponding temperature (°C) rise (or fall). For example, one calorie is required to raise the temperature of one gram of liquid water by 1°C.

hot spot. A rising plume of heat from deep within the mantle, believed to be responsible for chains of volcanic islands that are not near plate margins.

hypertrophic. Extreme productivity due to high levels of nutrient enrichment, associated with pollution.

I

infauna. Animals that live within the sediment.

in situ. A Latin term meaning *in place*; in the original or natural position.

internal wave. A subsurface wave that occurs at the boundary between discrete layers of light and dense water.

ion. An atom or group of atoms bearing an electrical charge, either positive (cation) or negative (anion).

iso-. Prefix meaning *alike, equal* or *same*. From the Greek *isos* = equal.

isohaline. A line or plane of equal salinity.

isotherm. A line or plane of equal temperature.

J

jetty. A long artificial structure built perpendicular to the coast, usually to protect the mouths of rivers and harbors.

L

larva. The young, generally planktonic stages in an animal's life cycle.

lime. Calcium carbonate ($CaCO_3$); for example, that which is found in the shell of a clam.

lithosphere. The entire region of the earth located above the low velocity layer.

littoral. Intertidal; the benthic zone between the high and low tide levels.

littoral transport. The movement of suspended material parallel to the shore by longshore currents.

longshore current. A current flowing parallel to the shore, formed when waves strike the coast obliquely.

lophophore. A ciliated looped, spiraled, or ring-like structure, used in feeding and respiration in the lophophorate animals.

low velocity layer. A region in the upper mantle of the earth (between about 100 and 300 kilometers below the earth's surface) where seismic wave speeds are lower than in regions above and below it.

luciferase. An enzyme that promotes the oxidation of luciferin in the process of bioluminescence.

luciferin. A chemical substance that may be oxidized to produce a cool blue-white light during bioluminescence.

M

manganese nodules. Rounded objects on the sea floor formed by the precipitation of minerals rich in manganese, copper, cobalt and nickel, and other metals. They are formed around sharp or rough objects such as rock particles and shark teeth.

mantle. 1. The region of the earth between the crust and the core. The mantle extends downward to a depth of about 2,900 kilometers. 2. The fleshy sheetlike structure that surrounds the internal organs of a mollusk; it also secretes any shell that these organisms may have.

mariculture. Sea farming, or the cultivation of marine organisms in the ocean, estuaries, or saltwater tanks, frequently with artificial nutrient enrichment.

mean sea level. The average height of the sea surface for all stages of tides over a period of 18.6 years.

meroplankton. Plankton that become nektonic or benthic at a certain stage in their life cycle.

mesopelagic. Pertaining to the "twilight zone" of the sea, between 200 and 1,000 meters.

Moho (Mohorovičić discontinuity). The boundary between the crust and the mantle.

mollusk. A soft-bodied animal with gills and a mantle, which in most mollusks, secretes a shell. This group includes clams, snails, and squid.

N

nannoplankton. Minute plankton that are less than 0.1 mm (100 microns) in size and which thus can pass through the finest plankton nets.

neap tide. The tide that occurs when the tide-producing forces of the sun and the moon are at right angles to each other. Neap tides occur every two weeks (about a week after full and new moons), and they have the smallest tidal ranges.

nekton. The swimmers, such as fish and squid, that can swim against ocean currents.

neritic realm. The water near shore, or over the continental shelf.

niche. The set of functional relationships of an organism to the environment that it occupies.

nodal line. A line (across a basin) along which no tidal fluctuations occur.

notochord. A flexible strengthening rod located along the back in the members of the Phylum Chordata, present at least in the larval or embryonic stages.

nutrient. A chemical, such as nitrate or phosphate, required for the growth of plants.

O

oceanic realm. The waters of the open ocean, beyond the continental shelf.

oligotrophic. Pertaining to nutrient-poor conditions, with low biomass and a high diversity of species.

ooze. Fine-grained sediments composed of at least 30 per cent skeletal remains of organisms.

opal. Hydrated silicon dioxide, similar to glass. Opal is a major constituent of the siliceous shells or skeleta of some marine organisms (e.g., diatoms).

osmosis. Diffusion across a selectively permeable membrane.

osmotic pressure. The pressure exerted on a membrane by osmosis.

P

P-Wave. See *Primary wave.*

paleomagnetism. The ancient magnetism of rocks that was produced when they were initially formed.

paleontology. The study of ancient life, through the use of fossil evidence.

Pangaea. A "supercontinent" that fragmented and drifted to form the present continents.

pelagic. Pertaining to the water between the surface and the sea floor. The pelagic environment is distinct and separate from the sea floor itself. Pelagic sediments are formed primarily in the pelagic environment.

phosphorite. Deposits of phosphate rich minerals precipitated on some continental shelves and slopes.

photophore. A light-producing organ.

photosynthesis. The utilization of light energy in the synthesis of organic matter by a plant.

plankton. Drifting organisms that cannot swim against ocean currents, including phytoplankton (plant plankton) and zooplankton (animal plankton).

plate. A broken piece of the lithosphere. The lithosphere of the earth is believed to consist of about seven major plates and several smaller ones.

primary producer. A plant that produces organic matter by photosynthesis.

primary wave (P-wave). A seismic wave in which the particles vibrate parallel to the direction of wave advancement. See also *Secondary wave.*

productivity. The rate of production of organic matter by the organisms in an area.

pseudopodia. The threadlike or fingerlike projections of the semifluid body of certain single-celled animals in the Phylum Protozoa. They are used in locomotion and in the entrapment of food.

pycnocline. A sharp change in density with depth.

Q

quartz. A crystalline mineral consisting of silicon dioxide (SiO_2).

R

radiolarian. A single-celled animal with pseudopodia and a siliceous (SiO_2) skeleton.

residence time. The length of time a substance remains in the ocean before removal.

respiration. The consumption of oxygen and the release of energy from stored organic matter by a plant or animal.

rift valley. A valley that sometimes characterizes the axis of a mid-ocean ridge.

S

S-wave. See *Secondary wave.*

salinity. The dissolved salt content of water, usually expressed in parts per thousand ($^o/_{oo}$) by weight.

Sargasso Sea. A region of warm, sluggish, nutrient-poor water in the west central Atlantic, north of the equator.

sea. 1. Ocean or a subdivision of an ocean, such as a semi-enclosed body of water connected to an ocean. 2. Waves in the generating (storm) area.

seamount. Undersea volcanic peak having a height of over one kilometer.

secondary producer. Herbivore, or plant-eater.

secondary wave (S-wave). A seismic wave in which the particles vibrate perpendicular to the direction of wave advancement. They can propagate through solids only. See also *Primary wave.*

seismic sea wave. See *Tsunami.*

seismic waves. Waves (vibrations) produced by earthquakes, explosions, or other energy sources and transmitted through the earth; a type of sound wave.

semidiurnal. Occurring twice daily, as in semi-diurnal tides, which comprise two high and two low tides each day.

shadow zone. A region into which little sound energy penetrates when the sound source is at or near the surface. It is produced when a sound velocity maximum occurs near the surface (usually at a depth of about 100 meters).

significant wave height. The average wave height of the highest 1/3 of the waves resulting from a particular storm.

siliceous. Composed of silicon dioxide (SiO_2).

sill. A shallow ridge across the mouth of a basin or a deep estuary (such as a fiord).

siphon. A fleshy tube, found in some mollusks through which water flows. It is frequently used to carry water and food into (incurrent) and out of (excurrent) the animal.

SOFAR. See *Sound channel.*

SONAR (*SO*und *N*avigation *A*nd *R*anging). The method of locating submerged objects by the technique of sound reflection.

sound channel. A region in the ocean where sound velocity is at a minimum (usually at a depth of about 1,000 meters). Sound originating in this region can be propagated thousands of kilometers without appreciable loss of energy. Also known as the SOFAR (SOund Fixing and Ranging) channel.

splash zone. The sea spray zone above high tide levels.

spring tide. The tide that occurs when the earth, sun, and moon are on a straight line. It occurs every two weeks (during full and new moons) and has the greatest tidal ranges.

standing crop. The total amount (biomass) of organisms in a volume of water at one time.

storm surge (storm wave, storm tide). An abnormal rise in water level caused by a hurricane or other severe storm.

subduction. The process whereby older sea floor descends and is consumed into the earth. This usually occurs near trenches.

sublittoral.   Pertaining to the environment of the continental shelf below low tide.

submarine canyon.   A V-shaped, rocky canyon located across the continental shelf and slope.

surf.   The zone of breakers from the shoreline to the outermost breakers, or a collective term for breakers.

surface tension.   The attraction of molecules near the surface of a liquid toward the interior of the liquid, tending to result in decreased surface area and a measurable tension at the surface.

swell.   Ocean waves outside the generating area, generally having longer periods and flatter crests than those within the generating area (sea).

swim bladder.   A gas-filled sac found in certain fish. It is used in the regulation of the fish's buoyancy. Owing to the difference in density between water and the gas, swim bladders reflect sound well and may cause deep scattering layers.

T

thermocline.   A sharp change in temperature with depth.

terrigenous sediment.   Land-derived sediment.

tidal flow.   1. The total volume of sea water flowing into or out of an estuary with the incoming or outgoing tide. 2. The difference between the mean high water volume and the mean low water volume of an estuary, also known as the *tidal prism.*

transform fault.   A horizontal offset of the axis of a mid-ocean ridge.

trophic.   Pertaining to feeding or food.

T-S curve.   A curve connecting points representing temperature and salinity values on a T-S diagram.

T-S diagram.   A grid with temperature, salinity and density scales, upon which are plotted temperature and salinity data.

tsunami (seismic sea wave).   A surface ocean wave produced by an abrupt vertical displacement of the sea floor caused by an earthquake, submarine landslide, or volcanic explosion.

turbidity.   Reduced water clarity owing to the presence of suspended material.

turbidity current.   A dense fast-flowing body of sediment-laden water.

U

upwelling.   The upward movement of water, near the coast, caused by longshore wind and Ekman transport.

V

viscosity (internal friction).   The tendency of a liquid to resist flowing.

W

water mass.   Water having a range of temperature and salinity values, represented by a straight line on a T-S diagram.

water type.   Water with specific temperature and salinity values, represented by a point on a T-S diagram.

X

XBT.   Expendable bathythermograph, a device that electronically measures and records temperature versus depth. The probe is used only once and is then replaced.

Bibliography

BAKER, B. B., Jr., W. R. DEEBEL and R. D. GEISENDERFER (eds.). 1966.
Glossary of Oceanographic Terms, Sp 35. Washington, D.C., U.S. Naval Oceanographic Office.
BARAZANGI, M. and J. DORMAN. 1969.
World Seismicity Maps Compiled from ESSA, Coast and Geodetic Survey, Epicenter Data 1961–1967. *Seismological Society of America Bulletin,* **59**:369–380.
BARDACH, J. E., J. H. RYTHER and W. O. McLARNEY. 1972.
Aquaculture. New York, Wiley.
BARNES, R.D. 1974.
Invertebrate Zoology, 3rd ed. Philadelphia, W. B. Saunders Co.
BASCOM, W. 1960. (August).
Beaches, *Scientific American.*
BASCOM, W. 1964.
Waves and Beaches. Garden City, N.Y., Doubleday.
BIGELOW, H. B. and W. C. SCHROEDER. 1953.
Fishes of the Gulf of Maine, Fisheries Bulletin 74. Washington, D.C., U.S. Fish and Wildlife Service.
BLAXTER, J. H. S. 1970.
Light: Animals; Fishes. In: *Marine Ecology,* Vol. I, Part 1. Otto Kinne (ed.). New York, Wiley-Interscience.
BOWDITCH, N. 1977.
American Practical Navigator. Washington, D.C., Government Printing Offiice.

CALMAN, W. T. 1909.
Crustacea. In: *Treatise on Zoology,* Vol. 8. E. R. Lankester (ed.). London, A & C Black.
CARMODY, D. J., J. B. PEARSE and W. E. YASSO. 1973.
Trace Metals in Sediments of New York Bight. *Marine Pollution Bulletin.* **4**:132–135.
CARR, A. 1955 (May).
The Navigation of the Green Turtle. *Scientific American.*
CARRIKER, M. R. 1969.
Excavation of Boreholes by the Gastropod, *Urosalpinx. American Zoologist.* **9**:917–933.
CARTWRIGHT, D. E. 1969.
Deep Sea Tides, *Science Journal.* **5**:60–67.

CLARKE, G. L. and E. J. DENTON. 1962.
Light and Animal Life. In: *The Sea,* Vol. 1. M. N. Hill (ed.). New York, Wiley-Interscience.
COUNCIL ON ENVIRONMENTAL QUALITY. 1970.
Ocean Dumping–A National Policy. Washington, D.C., Government Printing Office.
CULKIN, F. 1965.
The Major Constituents of Seawater. In: *Chemical Oceanography,* Vol. 1. J. P. Riley and G. Skirrow (eds.). New York, Academic Press.

DEFANT, A. 1961.
Physical Oceanography. New York, Pergamon Press.
DEWEY, J. F. and J. M. BIRD. 1970.
Mountain Belts and the New Global Tectonics. *Journal of Geophysical Research.* **75**:2625–2645.
DIETRICH, G. 1963.
General Oceanography. New York, Wiley.
DIETZ, R. S., 1961.
Continent and Ocean Basin Evolution by Spreading of the Sea Floor. *Nature.* **190**:854–857.
DOTY, M. S. 1957.
Rocky Intertidal Surfaces. In: *Treatise on Marine Ecology and Paleoecology,* Vol. 1, Ecology. J. W. Hedgpeth (ed.). New York, Geological Society of America.

EUKEN, A. 1948.
Zür Struktur des Flüssigen Wassers. *Angew. Chemie A.* **60**:166.

FAO, 1972.
Atlas of the Living Resources of the Sea. Rome: United Nations, Food and Agricultural Organization.
FAO. 1978.
Yearbook of Fisheries Statistics–1977. Rome: United Nations, Food and Agricultural Organization.
FITZGERALD, R. A., D. C. GORDON and R. E. CRANSTON. 1974.
Total Mercury in Sea Water. *Deep-Sea Research.* **21**:139–144.
FLEMING, N. C. 1969.
Archaeological Evidence for Eustatic Change of Sea Level and Earth Movements in the Western Mediterranean During the Last 2000 Years. Special Paper 109. Boulder, Colorado, Geological Society of America.
FLEMING, R. H. 1957.
Features of the Oceans. In: *Treatise on Marine Ecology and Paleoecology,* Vol. 1, Ecology. J. W. Hedgpeth (ed.). New York, Geological Society of America.

GARLAND, G. D. 1979.
Introduction to Geophysics, 2nd ed. Philadelphia, W. B. Saunders Co.

GIESBRECHT, W. 1892.
Systematik und Faunistik der Pelagischen Copepoden. *Fauna und Flora Golfes Neapel.* **19**:1–831.
GOLDBERG, E. D. 1965.
Minor Constituents in Seawater. In: *Chemical Oceanography,* Vol. 2. J. P. Riley and G. Skirrow (eds.). New York, Academic Press.
GRAHAM, M. 1956.
Sea Fisheries. London, Edward Arnold, Ltd.
GREEN, J. 1961.
A Biology of Crustacea. Chicago, Quadrangle Books.
GROEN, P. 1969.
The Waters of the Sea. London, Van Nostrand Reinhold.

HARDY, A. C. 1956.
The Open Sea. London, Collins.
HAYES, D. E. and A. C. PIMM. 1972.
Bathymetric, Magnetics, and Seismic Reflection Data. *Initial Reports of the Deep Sea Drilling Project.* **14**:341–376.
HEDGPETH, J. W. 1953.
An Introduction to the Zoogeography of the Northwestern Gulf of Mexico, with Reference to the Invertebrate Fauna. *Publications of the Institute of Marine Science of the University of Texas.* **3**:109–224.
HEDGPETH, J. W. (ed.). 1957.
Treatise on Marine Ecology and Paleoecology, Vol. 1, Ecology. New York, Geological Society of America.
HEEZEN, B. C. and M. EWING. 1952.
Turbidity Currents and Submarine Slumps, and the 1929 Grand Banks Earthquake. *American Journal of Science.* **250**:849–873.
HEEZEN, B. C. and C. D. HOLLISTER, 1971.
The Face of the Deep. New York, Oxford University Press.
HEIRTZLER, J. R., G. O. DICKSON, E. M. HERRON, W. C. PITMAN, III, and X. LE PICHON. 1968.
Marine Magnetic Anomalies, Geomagnetic Field Reversals, and Motions of the Ocean Floor and Continents. *Journal of Geophysical Research.* **73**:2119–2136.
HEIRTZLER, J. R., X. LE PICHON and J. G. BARON. 1966.
Magnetic Anomalies over the Reykjanes Ridge. *Deep Sea Research.* **13**:427–443.
HESS, H. H. 1962.
History of Ocean Basins. In: *Petrologic Studies.* A. E. J. Engel et al. (eds.). New York, Geological Society of America.
HYMAN, L. H. 1940–1959.
Invertebrates, Vols. 1–5. New York, McGraw-Hill.

ISACKS, B., J. OLIVER and L. R. SYKES. 1968.
Seismology and the New Global Tectonics. *Journal of Geophysical Research.* **73**:5855–5899.

JONES, M. M. et al. 1976.
Chemistry, Man and Society. Philadelphia: W. B. Saunders Co.

KESLER, S. E. 1976.
Our Finite Mineral Resources. New York, McGraw-Hill.
KING, C. A. M. 1960.
Beaches and Coasts. London, Edward Arnold, Ltd.
KINNE, O. (ed.). 1970–1972.
Marine Ecology, Vol. 1, in 3 parts. New York, Wiley-Interscience.

LARSON, R. L. and W. C. PITMAN III. 1972.
Worldwide Correlation of Mesozoic Magnetic Anomalies and Its
    Implications. *Bull. Geol. Soc. Am.* **83**:3645–3661.
LOWRIE, A. and E. ESCOWITZ. 1969.
Sea-floor Spreading and Continental Drift. *Journal of Geophysical Research* **73**:3661–3697.
LOWRIE, A. and E. ESCOWITZ. 1969.
Global Ocean Floor Analysis and Research Data Series, Kane-9.
    Washington, D.C. U.S. Naval Oceanographic Office.

MACGINITIE, G. E. and N. MACGINITIE. 1968.
Natural History of Marine Animals, 2nd ed. New York. McGraw-Hill.
MARSHALL, N. B. and O. MARSHALL. 1971.
Ocean's Life. New York, Macmillan.
MORGAN, W. J. 1968.
Rises, Trenches, Great Faults, and Crustal Blocks. *Journal of Geophysical Research.* **73**:1959–1982.

NATIONAL ACADEMY OF SCIENCES. 1972.
Understanding the Mid-Atlantic Ridge. Washington, D.C., National Academy of Sciences.
NEUMANN, G. 1968.
Ocean Currents. Amsterdam, Elsevier.
NEUMANN, G. and W. J. Pierson, Jr. 1966.
Principles of Physical Oceanography. Englewood Cliffs, N.J., Prentice-Hall.
NEWELL, G. E. and R. C. NEWELL. 1966.
Marine Plankton; A Practical Guide. London, Hutchinson.

ODUM, E. P. 1971.
Fundamentals of Ecology, 3rd ed. Philadelphia, W. B. Saunders Co.
ODUM, H. T. 1956.
Efficiencies, Size of Organisms, and Community Structures.
Ecology, **37**:592–597.
ODUM, H. T. and H. P. ODUM. 1955.
Trophic Structure and Productivity of a Windward Coral Reef
    Community on Eniwetok Atoll. *Ecological Monographs.*
    25:291–320.

OTHMER, D. and O. ROELS. 1973.
Power, Freshwater and Food from Cold Deep Seawater. *Science.* **182**:121–125.

PEARSE, A. S., H. J. HUMM and G. W. WHARTON. 1942.
Ecology of Sand Beaches at Beaufort, North Carolina. *Ecological Monographs.* **12**:136–190.
PICKARD, G. L. 1963.
Descriptive Physical Oceanography. New York, Pergamon Press.
PIERSON, W. S., G. NEUMANN and R. W. JONES. 1955.
Practical Methods of Observing and Forecasting Ocean Waves by Means of Wave Spectra and Statistics. H.O. Publication 603. Washington, D.C., U.S. Naval Oceanographic Office.
PRITCHARD, D. W. 1952.
The Physical Structure, Circulation, and Mixing in a Coastal Plain Estuary. *Chesapeake Bay Institute, The Johns Hopkins University, Technical Report 3.*

RAFF, A. D. and R. G. MASON. 1961.
Magnetic Survey of the West Coast of North America, 40°N.–52°N. Latitude. *Bulletin of the Geological Society of America.* **72**:1267–1270.
RAYMONT, J. E. G. 1963.
Plankton and Productivity in the Oceans. New York, Macmillan.
RILEY, G. A. 1952.
Biological Oceanography. In: *Survey of Biological Progress.* **2**:79–104. New York, Academic Press.
RILEY, J. P. and G. SKIRROW (eds.). 1965.
Chemical Oceanography. New York, Academic Press.
RUSSEL, F. S. 1927.
The Vertical Distribution of Marine Macroplankton. V. The Distribution of Animals Caught in the Ring-trawl in the Daytime in the Plymouth Area. *Journal of the Marine Biological Association of the United Kingdom.* **14**:557–608.
RUSSEL-HUNTER, W. D. 1970.
Aquatic Productivity. New York, Macmillan.
RYTHER, J. 1969.
Photosynthesis and Fish Production in the Sea. *Science.* **178**:72–76.

SEGERSTRÅLE, S. G. 1957.
Baltic Sea. In: *Treatise on Marine Ecology and Paleoecology,* Vol. 1, J. W. Hedgpeth (ed.) New York, Geological Society of America.
SHEPARD, F. P., G. A. MacDONALD and D. C. COX. 1950.
The Tsunami of April 1, 1946. *Bulletin of the Scripps Institution of Oceanography.* **5**:391–455.
SHEPARD, F. P. 1973.
Submarine Geology, 3rd ed. New York, Harper and Row.
SMITH, F. WALTON. 1973.
The Seas in Motion. New York, Thomas Y. Crowell Co.

SMITH, R. L. 1966.
Ecology and Field Biology. New York, Harper and Row.
STOMMEL, H., E. D. STROUP, J. L. REID and B. A. WARREN. 1973.
Transpacific Hydrographic Sections at Latitudes 43°S. and 28°S.: the SCORPIO Expedition-I. *Deep Sea Research.* **20**:1–7.
SVERDRUP, H. O., M. W. JOHNSON and R. H. FLEMING. 1942.
The Oceans. Englewood Cliffs, N.J., Prentice-Hall.
SWALLOW, J. C. 1955.
A Neutral-buoyancy Float for Measuring Deep Currents. *Deep Sea Research.* **3**:74–81.

TAIT, R. S. and R. V. DE SANTO. 1972.
Elements of Marine Ecology. New York, Springer-Verlag.
TÉTRY, A. 1959.
Classe des Sipunculiens. In: *Traité de Zoologie,* Vol. 5. P. Grassé (ed.). Paris, Maison et cie.
THORSON, G. 1971.
Life in the Sea. New York, McGraw-Hill.
TREWARTHA, G. T. 1954.
An Introduction to Climate, 3rd ed. New York, McGraw-Hill.
TURK, A., J. TURK, J. T. WITTES and R. WITTES. 1974.
Environmental Science. Philadelphia, W. B. Saunders Co.

U.S. ARMY COASTAL ENGINEERING RESEARCH CENTER. 1966.
Shore Protection, Planning and Design, TR-4, 3rd ed. Washington, D.C., Government Printing Office.
U.S. ARMY COASTAL ENGINEERING RESEARCH CENTER. 1973.
Shore Protection Manual. Washington, D.C., Government Printing Office.
U.S. NAVAL OCEANOGRAPHIC OFFICE. 1944.
Breakers and Surf; Principles in Forecasting, H.O. Publication 234, Washington, D.C., Government Printing Office.
U.S. NAVAL OCEANOGRAPHIC OFFICE. 1944.
Wind Waves at Sea, Breakers and Surf, H.O. Publication 602. Washington, D.C., Government Printing Office.
U.S. NAVAL OCEANOGRAPHIC OFFICE. 1950.
Sea and Swell Observations, H.O. Publication 606-e. Washington, D.C., Government Printing Office.
U.S. NAVAL OCEANOGRAPHIC OFFICE. 1965.
Oceanographic Atlas of the North Atlantic Ocean, H.O. Publication 700. Washington, D.C., Government Printing Office.
U.S. NAVAL OCEANOGRAPHIC OFFICE. 1966.
Handbook of Oceanographic Tables, S.P. 68. Washington, D.C., Government Printing Office.
U.S. NAVAL OCEANOGRAPHIC OFFICE. 1968.
Instruction Manual for Oceanographic Observations, 3rd ed., H.O. Publication 607. Washington, D.C., Government Printing Office.

VETTER, R. C. (ed.). 1973.
Oceanography. New York, Basic Books.

VILLEE, C. A., W. F. WALKER and R. D. BARNES. 1973.
General Zoology, 4th ed. Philadelphia, W. B. Saunders Co.

VINE, F. J. 1966.
Spreading of the Ocean Floor: New Evidence. *Science.* *154*:1405–1415.

VINE, F. J. 1968.
Magnetic Anomalies Associated with Mid Ocean Ridges. In: *The History of the Earth's Crust, A Symposium.* R. A. Phinney (ed.). Princeton, Princeton University Press.

VINE, F. J. 1969.
Sea-floor Spreading—New Evidence. *Journal of Geological Education.* *17*:6–16.

VINE, F. J. and D. H. MATTHEWS. 1963.
Magnetic Anomalies Over Oceanic Ridges. *Nature.* *199*:947–949.

VON ARX, W. S. 1962.
Introduction to Physical Oceanography. Reading, Mass., Addison-Wesley.

WARREN, B. A. 1966.
Oceanic Circulation. In: *Encyclopedia of Oceanography.* R. W. Fairbridge (ed.). New York, Reinhold.

WILLIAMS, J., J. J. HIGGINSON and J. D. ROHRBOUGH. 1968.
Oceanic Surface Currents. In: *Sea and Air, the Naval Environment.* Annapolis, Naval Institute Press.

WILSON, B. W. and A. TØRUM. 1968.
The Tsunami of the Alaskan Earthquake, 1964: Engineering Evaluation, Technical Memorandum No. 25. Washington, D.C., U.S. Coastal Engineering Research Center.

WATTENBERG, H. 1933.
Über die Titrations-alkalinität und der Kalziumkarbonatgehalt des Meerwassers. *Deutsche Atlantische Exped. Meteor 1925–1927. Wiss. Erg. 8*:122–231.

WEGENER, A. 1966.
Origin of Continents and Oceans. New York, Dover Publications. (A translation of the original German edition of 1929.)

WÜST, G. 1936.
Die Stratosphere des Atlantischen Ozeans. *Deutsche Atlantische Exped. Meteor 1925–1927.* *6*:109–251.

WÜST, G. 1954.
Gesetzmässige Wechselbziehungen Zwischen Ozean und Atmosphäre in der Zonalen Verteilung von Oberflächenzalzgehalt, Verdunstung und Niederschlag. *Archiv Für Meteorologie Geophysik und Bioklimatologie 7A*:305–328.

WYRTKI, KLAUS. 1979.
Sea Level Variations: Monitoring the Breath of the Pacific. *Trans. Am. Geophysical Union.* *60(3)*:25–27.

YONGE, C. M. 1949.
The Seashores. London, Collins.

Index

Page numbers in *italics* indicate illustrations; those followed by (t) indicate tables.

A

Absolute temperature, 140
Abyssal hills, 70, *72*, 425
Abyssal plain, 70, *72*, 425
Abyssal zone, 295, *296*
Abyssopelagic zone, 291, 425
Acorn worms, 341, 342, *343*
Aerobic, 425
Africa, break-up of, 59
    rift valley, 46, *48*, 429
Agar, 323
Agassiz, Louis, 11
Agulhas Current, 225
Air-sea interaction, 133
Alaskan Earthquake of 1964, *190*
Alaskan gyre, 227
Algae, 320–323, 425
    blue-green, 320, *320*, 321
    brown, 321, 322, *322*
    green, 321, *321*
    red, 322, 323, *323*
Algin, 321, 322
Alps, 59
American Practical Navigator, 14(t)
Amoco Cadiz, 278
Amphidromic point, 209, 425
Amphipods, 351, *352*
Anaerobic, 373
Anchoa, 359, 360
Anchovy, 5, *316*, 359, *360*
Angler fish, 304, *304*, 359–361, *360*, *362*
Annelida, 337, *338*, *339*, 425
Anoxic, 124, 425
Antarctic Bottom Current, 239
Antarctic Circumpolar Current, 239
Antarctic Convergence, 242
Antarctic Intermediate Current, 239
Anthophyta, 327
Aphelion, 204
Aphotic zone, 290, 425

Apogee, 204
Appalachians, 59
Arenicola, 337, 339
Argo Merchant, 278
Arrow worms, 337, *340*
Arthropoda, 346–352
Asteroidea, 353(t)
Asthenosphere, 42, *58*. See also *Low velocity layer.*
Atlantic Ocean, widening of, 51
Atmospheric circulation, 222–224, *222*
Atomic structure of elements, 109(t)
Atoll, 72–74, *73*, *75*, 425
Aves, 362, *363*
Azimuthal projection, 409
Azoic zone, 10, 425

B

Back-radiation, 140
Backshore, 253
Bacon, Francis, 44
Bacteria, 318, 319, 365, *365*
Baltic Sea, 273
Barnacles, 292, *350*, 351
Barrier islands, 257–260, *258*, *259*, 425
Barrier reef, 74
Barrier spits, 259, *259*, 425
Basalt, 425
Basins, 273
Basket star, 353(t)
Bathyal zone, 295, 425
Bathypelagic zone, 291, 425
Bathythermograph, 149, 150, 425
Bay of Fundy, *205*, 207, *211*
Bdelloura, 336, *336*
Beach, 250, 425
    cusps, 253, *253*
    drifting, 255, *255*, 425
    face, *252*, 253
    materials, 250

S